# The Anoles of Honduras

Systematics, Distribution, and Conservation

JAMES R. MCCRANIE AND GUNTHER KÖHLER

Bulletin of the Museum of Comparative Zoology

Special Publications Series, No. 1

# The Anoles of Honduras

Systematics, Distribution, and Conservation

JAMES R. MCCRANIE AND GUNTHER KÖHLER

Harvard University Museum of Comparative Zoology

2015

Distributed by Harvard University Press

Cambridge, Massachusetts and London, England

Printed in the United States of America

ISBN 978-0-674-50441-7

# THE ANOLES (REPTILIA: SQUAMATA: DACTYLOIDAE: *ANOLIS*: *NOROPS*) OF HONDURAS. SYSTEMATICS, DISTRIBUTION, AND CONSERVATION

JAMES R. MCCRANIE[1] AND GUNTHER KÖHLER[2]

## CONTENTS

ABSTRACT. In this review, we provide thorough descriptions of the 39 named species of the family Dactyloidae (commonly referred to as anoles) we recognize as occurring in Honduras. We follow two recent phylogenetic analyses suggesting eight genera of anoles are recognizable, two of which occur in Honduras (*Anolis* and *Norops*). Each species account contains a synonymy, statements on its geographic distribution, description, diagnosis/similar species statements, a list of illustrations, a remarks section, natural history comments, an etymology section, and a list of specimens examined; for a few species, a list of other records is included. The synonymy for each species includes the original description of the species with the original proposed combination, the type specimen(s), and direct quotes of the species' type locality. Each synonymy also includes the first use of the currently used scientific name and all known references to Honduran specimens and/or localities. Distribution maps with the known Honduran localities plotted are included for each species (with one exception), and other records are included on a few of those maps. Color photographs of an adult of each species are included as well as color photographs of the male dewlap of all species. Following the species accounts are two dichotomous identification keys (English and Spanish versions) to help the reader identify any specimen in hand. Following the identification keys is a section on species group relationships of each species known from the country. A distribution section that contains the distribution of each anole species in Honduras by department, forest formation, elevation, physiographic region, and ecophysiographic area; broad patterns of geographic distribution; historical units; and a discussion of Honduras as a distributional endpoint. A section on conservation discusses vulnerability gauges, IUCN Red List categories, and a section detailing each species' occurrence in Honduran protected areas.

Key words: Anoles, Dactyloidae, *Anolis*, *Norops*, Honduras, Taxonomy, Systematics, Distribution, Conservation

[1] 10770 SW 164th Street, Miami, Florida 33157-2933. Author for correspondence (jmccrani@bellsouth.net).

[2] Forschungsinstitut und Naturmuseum Senckenberg, Sektion Herpetologie, Senckenberganlage 25, D-60325 Frankfurt am Main, Germany.

Some years ago—I am unable to date the event more accurately—word seems to have been passed among the herpetological fraternity that I knew something about anoles. How the rumor was started, I am unable to say, as at that time I had no enemies who would have stooped so low in retaliation for some imaginary ill-treatment at my hands. Notwithstanding, the rumor grew rapidly, with the result that since that time a great majority of the anoles collected in northern Central America and in much of Mexico have passed through my hands. From these I have learned something of variation in the genus, but the more specimens I have examined the more convinced I have become that the rumor as to my knowledge is without substantial foundation. I take this opportunity, therefore, to present the few data that I have collected over the years on Guatemalan anoles, which should definitely silence it (Laurence Cooper Stuart, 1955: 1).

The trouble with these anoles … is that no one sticks to the job long enough. If you will make a real review of the Guatemalan ones, I have the feeling that the rest of the Central American ones will fall into line without so much effort (Karl Patterson Schmidt *in* Stuart, 1955: 1).

## INTRODUCTION

Identifying anoles, especially preserved specimens, can be very challenging, as so well expressed by Stuart's quote. Few herpetologists up to Stuart's time were willing to take the time to tackle systematic studies of anoles (as expressed in Schmidt's quote) occurring in their geographical area of expertise. Thus, our knowledge of the systematics of anoles on the mainland has lagged far behind those of other speciose lizard groups (i.e., *Sceloporus*). Additionally, fieldwork and examining living anole specimens is essential to gaining good knowledge of this lizard group. Recording color notes in life and photographing living specimens (especially that of the male dewlap) to go along with color notes is extremely valuable to improving one's knowledge of some of the most important characters of any given species. One cannot overstress the importance of fieldwork and that of recording exact locality data for anole specimens. Some anole species are difficult to distinguish from similar species on the basis of preserved specimens alone (i.e., *Norops limifrons* from *N. zeus*, *N. rodriguezii* from *N. yoroensis*). McCranie has had extensive fieldwork in Honduras for about 38 years and during that time has collected and photographed all anole species recognized in this work. Köhler also has undertaken fieldwork from Mexico to Panama over the last 18 years in an effort to study anole systematics. The fieldwork studies by us are used as a background to identify the anole species and their distributions within Honduras. Finally, tissues of 13 species collected by the first author in Honduras were used in the recent phylogenetic study of Nicholson et al. (2012), and tissues of 20 other species collected in Honduras by McCranie await study.

Anoles have a wide geographical distribution, occurring in the southeastern United States, throughout the Bahamas, the Greater and Lesser Antilles, and from northern Mexico into South America as far as southeastern Peru, south-central Paraguay, and southeastern Brazil (and many islands close to the mainland). Anoles also have a wide elevational distribution occurring from sea level to about 3,750 m. It is common to find several species occurring syntopically. In those cases, ecological segregation between species is usually evident. Niche partitioning in anoles has been documented with respect to perch height, type of vegetation, body size, and timing of foraging activities, particularly in the West Indies (reviewed in Losos, 2009). On the other hand, niche partitioning of anole species occurring on the mainland has been relatively poorly studied.

Despite their wide occurrence, and the wishful thinking of Stuart and Schmidt as quoted above, the systematics of anole species remains poorly known from much of their mainland range, including much of Central America. In recent years, Savage (2002) studied the anole species occurring in Costa Rica and Köhler et al. (2005b) studied those occurring in El Salvador. Additionally, Köhler and Acevedo (2004) studied those species occurring at low and moderate elevations on the Pacific versant of Guatemala.

Meyer and Wilson (1973) provided a checklist of the lizard species known to occur in Honduras that included 15 species of anoles and 646 total anole specimens listed in their locality records. The 40 years since the Meyer and Wilson (1973) checklist have seen much additional fieldwork in Honduras, including numerous cloud forest and mid-elevation broadleaf forest localities. Meyer and Wilson almost entirely limited their fieldwork to low elevation localities with access by roads. Subsequent fieldwork, especially that in isolated cloud forest localities, has discovered several new species. Additional fieldwork at low- and moderate-elevation localities has documented several range extensions of species into Honduras from nearby countries. More recently, detailed studies of several wide-ranging "species" have demonstrated those species to represent complexes composed of several cryptic taxa (i.e., Köhler, 2010; Köhler and Vesely, 2010), a few of which occur in Honduras. This report presents a systematic review of the 39 named anole species known to occur in Honduras, including the Islas de la Bahía, Islas del Cisne, and the small islands in the Golfo de Fonseca.

## MATERIALS AND METHODS

For this study we examined 5,150 specimens (plus 34 skeletons, four cleared and stained [C&S] specimens, and four eggs) of Honduran anoles, which represent the vast majority of those in museums worldwide. A list of the specimens examined is provided in the respective species accounts. The numbers of specimens given for each species by Meyer and Wilson (1973) are included in brackets in our Specimens Examined section following the number we examined for this revision. We have tried to examine all known Honduran specimens housed in U.S. museums, and those of the Senckenberg Museum in Germany and the Natural History Museum in England. We have also examined select specimens in several other museums worldwide. Much variation exists in the number of specimens available for each species, with numbers varying from only two for *Norops carpenteri* to 981 specimens for *N. tropidonotus*. Abbreviations for museum collections follow those of Leviton et al. (1985) and Leviton and Gibbs (1988). Terminology for hemipenial morphology follows that of Myers et al. (1993) and Savage (1997). Characters used in the species descriptions generally follow those used by Smith (1967), Williams et al. (1995), and especially the recent detailed methods of Köhler (2014). Snout-vent length (in mm) was measured using calipers for most species, except for larger species in which a ruler was used. All other measurements were made using calipers and were rounded to the nearest 0.1 mm. Head length was measured from the tip of the snout to the anterior margin of the ear opening. Tail height and width were measured at the point reached by the heel of the extended hind leg. Dorsal and ventral scales were counted along the midline of the body. Methods for counting male dewlap scales follow Fitch and Hillis (1984). Subdigital lamellae were counted on phalanges ii–iv of the fourth toe of each hind limb. Abbreviations used in the species descriptions are: HL (head length); SHL (shank [tibia] length); SVL (snout-vent length); TAL (tail length); TH (tail height); TW (tail width), and TOL (total length). The capitalized colors and color codes (the latter in parentheses) are those of Smithe (1975–1981), except in a few recent instances in which Köhler (2012) was used. The latter is noted when used.

Species synonymies contain the original description of each species, including the museum number(s) for the type specimen(s), the type locality as originally stated, the first use of the name combination recognized herein, and all known references to Honduran specimens or localities. The majority of Honduran localities listed herein are included in the Gazetteer in McCranie

(2011), and the newer ones will be published in McCranie (in preparation).

The dichotomous keys are meant to be significantly detailed to allow the user to identify any anole in hand to species. However, in some cases it might be necessary to consult the photographs, diagnosis/similar species sections, and/or the descriptions of the species in question. Figures are included with the keys to help the user.

General geographical statements of the overall known distributions of each genus and species are included. The elevational statements of Stuart (1963) are slightly modified and used in those statements (low elevations = sea level to 600 m; moderate elevations = 601–1,500 m; and intermediate elevations = 1,501–2,700 m). The natural history comments contain information on what is known about Honduran species in the field, and for the nonendemic species, general habitat information from outside the country is also included based on the literature. Notes from the literature are also included on diet and reproduction when available. Forest formations used are slightly modified from those used by Holdridge (1967). All information in the Natural History Comments is based on Honduran specimens unless otherwise noted.

Distribution maps are included for each species. Closed symbols on all maps (except that of *Norops nelsoni*) represent specimens examined, whereas an open symbol represents a reliable record listed in the other records section. A single symbol may represent more than one nearby locality. Shading to illustrate the general distribution is not used on those maps because of the complex topography of Honduras.

Color photographs in life are included for all recognized species. Dewlap photographs are also included for males of all species. The specimen photographed in each case is noted in the figure legend, as is the person who took the photograph.

All literature cited in this work was examined to verify the correct spelling of scientific names, the correct citation of type localities and type specimens, pagination, and titles of the literature.

This publication includes results of fieldwork and examination of the known literature through 31 June 2013.

## Justification for the Use of the Generic Name *Norops*

Modern studies of the systematics of the anole lizards had its beginnings with Etheridge (1959), who placed all anole species known to him into two groups based on condition of the caudal vertebrae. The beta group "is characterized by the presence of a pair of long, bifurcate, forward-directed processes" on the autotomic caudal vertebrae (Etheridge, 1959: 132), with the alpha section lacking those processes. The Etheridge study was confined to mostly skeletal characteristics; thus, Etheridge (1959) did not study scale characteristics.

Williams (1976a) studied the West Indian anoles and placed those lizards into three genera (*Chamaeleolis* Cocteau *in* Sagra, 1838; *Chamaelinorops* Schmidt, 1919; and *Anolis* Daudin, 1802). *Anolis* was further divided into two groups, the alpha and beta sections, following the work of Etheridge (1959). Williams (1976a) also relied almost completely on skeletal features; no mention was made of scale data. Williams (1976b) summarized his concepts of the species groups of South American anoles, again relying heavily on the alpha and beta caudal vertebrae division of Etheridge (1959). Scale data were again almost completely ignored because, in his opinion, "There is unfortunately no external characters by which the two sections [alpha-beta] can be separated" (Williams, 1976b: 263). Williams et al. (1995) developed a protocol they titled "A computer approach to the comparison and identification of species in different taxonomic groups." Williams et al. (1995) did use some scale data, but as the title suggests, those data were not thought to be informative above the species level. As a result, few scientists currently working on upper level anole systematics use scale data in their work.

Poe (2004) used a few of the Williams et al. (1995) scale characters in his phylogenetic analysis of the anoles, but none that involved counting of more than ca. 10 scales.

Guyer and Savage (1987) attempted a cladistic analysis (again, external morphology was almost completely ignored) of *Anolis* (sensu lato) and recognized five clades as genera (*Anolis*; *Ctenonotus* Fitzinger, 1843; *Dactyloa* Wagler, 1830; *Norops* Wagler, 1830; and *Semiurus* Fitzinger, 1843 [Savage and Guyer, 1991 noted that *Semiurus* is a junior synonym of *Xiphosurus* Fitzinger, 1826]). With that proposal, those authors suggested that eight anole genera be recognized (those just listed plus *Chamaeleolis*; *Chamaelinorops*; and *Phenacosaurus* Barbour, 1920; but see below). However, the Guyer and Savage efforts were met with strong opposition. Williams (1989) bluntly criticized the databases of the Guyer and Savage (1987) analysis and Cannatella and de Queiroz (1989) criticized the Guyer and Savage analytical procedures plus their osteological categorizations. Guyer and Savage (1992) argued point by point to the critiques by Cannatella and de Queiroz (1989) and Williams (1989), added more data (no scale data, however) to their analysis, and generated a new phylogenetic study that recovered "essentially the same set of relationships" as they recovered in their 1987 publication (Crother, 1999: 292). Curiously, there were no published responses to the new Guyer and Savage (1992) efforts and data sets. Crother (1999: 291) pointed out "the 1992 work by Guyer and Savage is almost never cited" even by workers citing and rejecting the Guyer and Savage (1987) study. Even Poe (2013), in his rejection of the Nicholson et al. (2012) proposed generic scheme, referred to the Guyer and Savage (1987; dated 1986) publication several times, but did not cite Guyer and Savage (1992). Based on these types of responses, Crother (1999: 291) asked if the Guyer and Savage "studies and subsequent classification [have] been given a fair scientific treatment?" As a result, few workers on anoles have followed the Guyer and Savage taxonomy. Addition-

ally, both of us have submitted manuscripts using *Norops*, one of the genera recognized by Guyer and Savage for most of the beta anoles of Etheridge (1959), but the name was changed to *Anolis* by the editors of those scientific journals.

Other authors (see Glor et al., 2001; Nicholson, 2002; Poe, 2004; Nicholson et al., 2005, 2012; Alföldi et al., 2011) have performed phylogenetic analyses that demonstrated *Norops*, the beta anoles of Etheridge (1959), to form a clade. However, some of these workers question the monophyly of at least some of the other proposed genera.

Recently, Nicholson et al. (2012) produced a new phylogenetic analysis of the anoles, again almost completely ignoring scale data. Those authors also recovered eight clades (although not the exact same eight clades as recovered by Guyer and Savage, 1987, 1992) they recognized as genera (*Anolis*; *Audantia* Cochran, 1934; *Chamaelinorops*; *Ctenonotus*; *Dactyloa*; *Deiroptyx* Fitzinger, 1843; *Norops*; and *Xiphosurus*). Nicholson et al. (2012) combined previously published morphological and molecular data, along with new molecular data from four anole species. Pyron et al. (2013: 15) provided the next phylogenetic analysis, which was based entirely on molecular data, with their results supporting "the monophyly of all the genera recognized" by Nicholson et al. (2012). Losos (2013, 2014) noted two exceptions, one of which was the result of Nicholson et al. (2012) having listed one species (*christophei* Williams, 1960, of Haiti) in two genera in their tables (*Chamaelinorops* and *Xiphosurus*). The most recent phylogenetic analysis of the anoles is that of Gamble et al. (2014; again, without any scale data). Losos (2014) reported that three of the 216 species used by Gamble et al. (2014; in addition to the error made by Nicholson with *christophei* noted above) fell in different clades than they did in the Nicholson et al. (2012) study. As a result, those four species make three of the Nicholson et al. genera nonmonophyletic. The two genera of Nicholson et al. (2012) occurring on the

Latin American mainland (*Dactyloa* and *Norops*; *Anolis allisoni* Barbour, 1928, and *Ctenonotus cristatellus* [Duméril and Bibron, 1837] have been recently introduced by man to at least one Central American mainland locality each) remain monophyletic in the Pyron et al. (2013) and Gamble et al. (2014) studies.

Poe (2013) responded to the Nicholson et al. (2012) study and concluded that "most" of their proposed genera are not monophyletic. The subsequent studies of Pyron et al. (2013) and Gamble et al. (2014; also see Losos, 2013, 2014) have shown that "most" of the Nicholson et al. genera are, in fact, monophyletic using only molecular data. Thus, the "most" statement used several times by Poe (2013) is not supported by the literature. Poe's (2013) concern about not enough taxa and not enough characters used in the Nicholson et al. (2012) study is certainly correct. Simply put, more species and more characters need to be added to the databases of future phylogenetic analyses of the anoles. Despite those shortcomings, we recognize two genera (*Anolis* and *Norops*) for the anoles occurring in Honduras. In doing so, we realize that opponents of recognizing more than a single genus of anoles, with about 400 species, will continue to argue that there is no compelling need to divide a monophyletic *Anolis* into eight genera. However, we feel that this conservative approach is misguided. As far as is known, the anterolaterally directed transverse process (see Etheridge, 1967, fig. 3; Guyer and Savage, 1987, fig. 1) on the autotomic caudal vertebrae of *Norops* is "*unique in structure among all lizards*" (emphasis ours; Guyer and Savage, 1992: 105). We argue that *Norops* is so distinct that it warrants its own genus (other similar examples probably also exist, i.e., *Chamaelinorops barbouri* Schmidt, 1919, for which Schmidt proposed the new genus *Chamaelinorops*, in part because of unique osteological characters in that form). If the consequence of our action renders *Anolis* nonmonophyletic, then that stresses the point that additional work does need to be done. Our argument to recognize *Norops* as a valid genus is supported by all phylogenetic analyses done to date that were based entirely on, or nearly completely on, molecular data. The unique tail structure found in *Norops* also helps to support a monophyletic *Norops*. Those two independent types of evidence strongly influenced our decision to elevate *Norops* to a full genus despite the concerns expressed by Poe (2013) and others.

One of our concerns with the Nicholson et al. (2012), and all other phylogenetic analyses of the anoles conducted to date, is that scale data, including difficult and time-consuming scale counts, have been presumed to be uninformative above the species level and, thus, ignored. Ignoring a type of data that is presumed, but not proven, to be uninformative is not a sound scientific approach. Unfortunately, only two of the eight genera (clades) recovered by Nicholson et al. (2012) are represented in Honduras (*Anolis* and *Norops*), with *Anolis* having only one species in the country. Thus, we are unable to apply scale morphology to those Nicholson et al. clades in this study. Adalsteinsson et al. (2009), Hedges et al. (2009, 2014), Harvey et al. (2012), and Hedges and Conn (2012) have recently incorporated time-consuming scale counts, in most cases in association with molecular data, in their generic divisions of several snake (Leptotyphlopidae, Typhlopidae, and Dipsadidae) and lizard (Scincidae and Teiidae) taxa, so why are we ignoring these types of data in anole phylogenetic studies? We have given above several examples of scale data having been largely ignored in upper level anole systematic studies because we believe we have been ignoring potentially informative data. One of the clades, along with *Norops*, that renders the remaining "*Anolis*" paraphyletic is *Dactyloa*, the only other anole clade occurring naturally on the Latin American mainland. Individuals of that clade look so different from the remaining anoles that it seems likely that a suite of scale and other morphological characters can be found to distinguish *Dactyloa* from the remaining anoles, in addition to the molecular studies that have already accomplished that (Castañeda and de Queiroz, 2011 [also see Castañeda and de Queiroz, 2013, who added osteological and some basic scale data to their molecular analysis]; Nicholson et al., 2012; Pyron et al., 2013; Gamble et al., 2014). We are in no way suggesting that everyone is compelled to follow our suggested recognition of *Norops* as a genus. We are asking, however, that workers involved in future taxonomic studies of anoles add scale counts and configurations and skeletal data sets to their molecular data bases before performing additional phylogenetic studies. We are also hoping that with time (the "revision shock" of Hedges, 2013), the anole community will accept *Norops* and other anole clades as valid genera in the Linnaean taxonomic scheme. Also, any worker who disagrees with our suggested Linnaean taxonomy could apply the Phylo-Code to our conclusions (see de Queiroz and Cantino, 2001, for an introduction to that nomenclatural system).

After the above was finished and sent to copyediting, Nicholson et al. (2014) published a response to Poe's (2013) critique of their 2012 revision of anole classification. Those authors concluded that Poe made several errors and misrepresentations in his critique. Nicholson et al. (2014) discussed point-by-point each of Poe's critiques and stated that they had explained their positions in their 2012 publication regarding the few "unstable" species that Poe used to call the Nicholson et al. (2012) work flawed. Nicholson et al. (2014: 109) also stated that Poe (2013) "missed the opportunity to present an alternative comprehensive taxonomy to replace the one against which he argues so strenuously." We completely agree with that statement and with the Nicholson et al. (2014: 109) statement that commentary regarding published taxonomic revisions should be "constructive, objective, and scientifically accurate." Nicholson et al. (2014) concluded that Poe's (2013) report was lacking in those points. Time will tell if the scientific community adopts or rejects the Nicholson et al. (2012, 2014) reclassification of the anoles.

## Ecomorphs, Ecomodes, or Just Niche Partitioning

Niche partitioning in anoles, especially on Caribbean Islands, has been documented with respect to perch height, vegetation types, body size, and timing of foraging activities. The wide variety of behavioral traits exhibited by anole species has been demonstrated to have a strong correlation with morphology among West Indian species regardless of phylogenetic relationships. Thus, West Indian species having the same or similar ecological roles have been characterized as "ecomorphs" (reviewed in Losos, 2009). On the other hand, Schaad and Poe (2010; also see Irschick et al., 1997) were able to assign only 15 of the 123 mainland species they analyzed to a West Indian ecomorph. Nicholson et al. (2012) introduced the term "ecomode" to replace ecomorph for the mainland members of the Dactyloidae; however, those authors did not adequately define their intended meaning of that term. Based largely on habitat statements in the literature, Nicholson et al. (2012) placed a large number of mainland species of anoles into one of their eight ecomodes (those authors also recognized a polymodal category for species they thought regularly occurred in more than one of their eight categories). Returning to the ecomorph category used so successfully for West Indian anoles, Losos (2012) noted that those ecomorph assignments are objective because the existing quantitative data are testable statistically. On the other hand, the Nicholson et al. (2012) ecomode assignments are subjective decisions based on literature statements varying from short, limited summaries to well-studied species. Most anole species also use a variety of habitats. Thus, where does one draw the line to distinguish one ecomode from another (i.e., trunk anole versus trunk-ground anole)?

Although we applaud the Nicholson et al. (2012) efforts to classify mainland anole species into habitat categories, we think the data for the vast majority of the mainland species are inadequate at present. Also, based on our own fieldwork, many anole species show such extensive variation in habitat preference and "ecomode" variation that it is not possible to place many species into one of the eight ecomodes recognized by Nicholson et al. (2012). Therefore, we do not try to force each anole species occurring in Honduras into one of the Nicholson et al. (2012) ecomode categories; rather, we just give statements about where we have observed individuals of each species relative to their available habitats. Statements of habitat use from the literature are also given for most nonendemic species.

## Environment of Honduras

McCranie (2011) recently discussed the general description, physiography, climate, and forest formations of Honduras. Interested readers are referred to that book for information on those subjects (also see distribution of anoles in Honduras herein for reproductions of several pertinent maps taken from McCranie [2011]).

## A BRIEF HISTORY OF THE STUDY OF THE ANOLE LIZARDS IN HONDURAS

For much of the 20th century, anoles and their close relatives were placed in the family Iguanidae Gray (1827: 56). However, Frost and Etheridge (1989) performed a phylogenetic analysis of iguanian lizards based on morphology and divided the former Iguanidae into several families. The phylogeny recovered by Frost and Etheridge (1989) suggested that the "anoloid iguanid lizards" of Etheridge and Williams (1985: 1) formed a monophyletic lineage that they placed in the family Polychridae (= Polychrotidae) Fitzinger (1843, type genus *Polychrus* Cuvier, 1816). Frost et al. (2001) performed a new phylogenetic analysis of the polychrotid lizards based on both morphological and molecular data that demonstrated the Polychrotidae recovered in the Frost and Etheridge (1989) study was not monophyletic. Thus, Frost et al. (2001) restricted the family Polychrotidae to the

genera *Polychrus* and *Anolis*. Frost et al. (2001) did not follow the suggested partitioning of *Anolis* into eight genera by Guyer and Savage (1987). Subsequently, Townsend et al. (2011) performed a phylogenetic analysis of iguanian lizards based on molecular data and recovered a polyphyletic Polychrotidae. To rectify the polyphyletic Polychrotidae, Townsend et al. (2011) restricted the family Polychrotidae to the genus *Polychrus* and placed *Anolis* (sensu lato) in the family Dactyloidae Fitzinger (1843; type genus *Dactyloa* Wagler, 1830, family name originally spelled Dactyloae by Fitzinger, 1843). Within the Dactyloidae, two of the eight genera, *Anolis* and *Norops*, occur in Honduras. The type species of the former genus is *A. carolinensis* Voigt, 1832, a species native to the southeastern United States, and that of the latter genus is *N. auratus* Bonnaterre, 1789, a species that occurs in South America and lower Central America.

The year 1834 saw the first descriptions of any anole species that occurs in Honduras. Wiegmann (1834) described *Norops biporcatus* and *N. laeviventris* (both as *Dactyloa*), with the former having its type locality in Guatemala and the latter in Mexico. Fifteen of the anole species occurring in Honduras were described during the 1800s (Table 1), none of which have their type localities in Honduras. The first four anole species with type localities occurring in Honduras were described during the early 1900s. Barbour (1914) described *N. nelsoni* (as *Anolis*) from the Islas del Cisne, Gracias a Dios; Barbour (1928) described *A. allisoni* from Isla de Roatán, Islas de la Bahía; Dunn and Emlen (1932) described *N. smithus* (as *Anolis*) from what is now part of Parque Nacional La Tigra, Francisco Morazán; and Schmidt (1936) described *N. loveridgei* (as *Anolis*) from Portillo Grande, Yoro. Three other species of anoles occurring in Honduras were described during the period from the late 1930s to the early 1970s. These are *N. wellbornae* Ahl (1939; as *Anolis ustus wellbornae*, type locality in El Salvador), *N. heteropholidotus* Mertens (1952a; as *Anolis*, type locality in El

TABLE 1.    A LISTING OF THE ORIGINAL DESCRIPTIONS OF THE HONDURAN SPECIES OF *ANOLIS* AND *NOROPS*, AUTHORS, AND ABBREVIATED TYPE LOCALITIES.

| Original Name | Author(s) | Abbreviated Type Locality |
|---|---|---|
| *Anolis allisoni* | Barbour, 1928 | Isla de Roatán, Honduras |
| *Norops amplisquamosus* | McCranie, Wilson, and Williams, 1992 | El Cusuco, Honduras |
| *Anolis beckeri* | Boulenger, 1882 | Yucatán, Mexico |
| *Norops bicaorum* | Köhler, 1996b | Isla de Utila, Honduras |
| *Dactyloa biporcata* | Wiegmann, 1834 | Santa Rosa de Pansos, Guatemala |
| *Anolis (Draconura) capito* | Peters, 1863 | "Costa Rica" |
| *Anolis carpenteri* | Echelle, Echelle, and Fitch, 1971 | near Turrialba, Costa Rica |
| *Anolis crassulus* | Cope, 1864 | Coban, Guatemala |
| *Anolis cupreus* | Hallowell, 1861 | "Nicaragua" |
| *Norops cusuco* | McCranie, Köhler, and Wilson, 2000 | El Cusuco, Honduras |
| *Anolis heteropholidotus* | Mertens, 1952a | Santa Ana, El Salvador |
| *Anolis johnmeyeri* | Wilson and McCranie, 1982 | WSW of Buenos Aires, Honduras |
| *Norops kreutzi* | McCranie, Köhler, and Wilson, 2000 | near La Fortuna, Honduras |
| *Dactyloa laeviventris* | Wiegmann, 1834 | "Mexico" |
| *Anolis (Gastrotropis) lemurinus* | Cope, 1861 | Veragua = Panama |
| *Anolis (Dracontura) capito* | Cope, 1862 | Veragua = Panama |
| *Anolis loveridgei* | K. Schmidt, 1936 | Portillo Grande, Honduras |
| *Anolis morazani* | J. Townsend and Wilson, 2009 | Cataguana, Honduras |
| *Norops muralla* | Köhler, McCranie, and Wilson, 1999 | Cerro de Enmedio, Honduras |
| *Anolis nelsoni* | Barbour, 1914 | Swan Islands, Honduras |
| *Norops ocelloscapularis* | Köhler, McCranie, and Wilson, 2001 | near Quebrada Grande, Honduras |
| *Anolis oxylophus* | Cope, 1875 | eastern Costa Rica |
| *Anolis petersii* | Bocourt, 1873, *in* A. H. A. Duméril, Bocourt, and Mocquard, 1870–1909 | Vera Paz, Guatemala |
| *Norops pijolense* | McCranie, Wilson, and Williams, 1993 | Pico Pijol, Honduras |
| *Norops purpurgularis* | McCranie, Cruz, and Holm, 1993 | near La Fortuna, Honduras |
| *Anolis quaggulus* | Cope, 1885 | San Juan River, Nicaragua |
| *Norops roatanensis* | Köhler and McCranie, 2001 | Isla de Roatán, Honduras |
| *Anolis rodriguezii* | Bocourt, 1873 | "Pansos Guatemala" |
| *Norops rubribarbaris* | Köhler, McCranie, and Wilson, 1999 | San Luís de los Planes, Honduras |
| *Anolis sagrei* | Cocteau, *in* A. M. C. Duméril and Bibron, 1837 | "Cuba" |
| *Anolis sminthus* | Dunn and Emlen, 1932 | San Juancito Mountains, Honduras |
| *Anolis tropidonotus* | W. Peters, 1863 | Huanusco, Mexico |
| *Anolis uniformis* | Cope, 1885 | "Guatemala" and "Yucatan" |
| *Anolis unilobatus* | Köhler and Vesely, 2010 | Awasbila, Honduras |
| *Norops utilensis* | Köhler, 1996a | Isla de Utila, Honduras |
| *Norops wampuensis* | McCranie and Köhler, 2001 | Ríos Aner-Wampú, Honduras |
| *Anolis ustus wellbornae* | Ahl, 1939 | "El Salvador" |
| *Norops yoroensis* | McCranie, Nicholson, and Köhler, 2002 | near La Fortuna, Honduras |
| *Norops zeus* | Köhler and McCranie, 2001 | Liberia, Honduras |

Salvador), and *N. carpenteri* Echelle, Echelle, and Fitch (1971; as *Anolis*, type locality in Costa Rica).

During 1979, McCranie made his first of what was to become numerous trips to isolated cloud forest and many broadleaf forest localities throughout Honduras. That first cloud forest trip, to El Cusuco, Cortés, resulted in the discovery of three anole species new to science (*Norops amplisquamosus* McCranie, Wilson, and Williams, 1992; *N. cusuco* McCranie, Köhler, and Wilson, 2000; and *N. johnmeyeri* Wilson and McCranie, 1982). Köhler began fieldwork in Honduras during 1993. The combined fieldwork of McCranie and Köhler, along with examination of Honduran anoles in museum collections, re-

TABLE 2. THE ANOLE SPECIES LISTED FROM HONDURAS BY DUNN AND EMLEN, 1932, AND THEIR CURRENT TAXONOMY (IN SAME ORDER AS USED BY DUNN AND EMLEN).

| Dunn and Emlen, 1932 | Current Taxonomy |
|---|---|
| *Anolis tropidonotus* | *Norops tropidonotus* |
| *Anolis sminthus* | *Norops sminthus* |
| *Anolis palpebrosus* | *Norops lemurinus* |
| *Anolis sallaei* | *Norops unilobatus, N. wellbornae* |
| *Anolis copei* | *Norops biporcatus, N. capito* |
| *Anolis limifrons* | *Norops zeus* |

sulted in the descriptions of 13 additional species of anoles from Honduras (Table 1). The most recent anole species with a name change from Honduras is *N. unilobatus* Köhler and Vesely (2010; formerly considered *N. sericeus* Hallowell, 1857), a species that occurs across northern and central Honduras. However, the most recently described Honduran endemic anole is *N. morazani* (Townsend and Wilson, 2009), a species known only from northern Francisco Morazán.

It now seems that fieldwork has been sufficient enough that it is not likely that populations of new species remain to be discovered in Honduras. However, future studies of hemipenial morphology and molecular data from Honduran species seem likely to discover cryptic species among the 39 recognized species. Also, a

few species are poorly defined, and such additional studies might result in their placement in the synonymy of a closely related species (i.e., *N. wampuensis* with *N. tropidonotus*).

Aside from the species descriptions outlined above and listed in Table 1, only two summations of anole collections from Honduras have appeared in the literature. Dunn and Emlen (1932) in their review of reptiles and amphibians from Honduras (principally from the ANSP and MCZ collections), listed six anole species (all placed in the genus *Anolis*) from Honduras, one of which they described as a new species (*Norops sminthus*). The exact number of specimens they examined was not given, but Table 2 gives the Dunn and Emlen (1932) species' identifications and the taxonomy we use. Meyer and Wilson (1973) provided a list of the anole species from various museum collections known to them. Those authors listed 646 specimens that they placed in 15 species. Table 3 gives the Meyer and Wilson taxonomy and the current taxonomy. Because Meyer and Wilson (1973) listed actual museum numbers, we are able to include the numbers of specimens (in parentheses) listed by those authors. We were also able to reidentify some of their included material.

TABLE 3. THE ANOLE SPECIES LISTED FROM HONDURAS BY MEYER AND WILSON, 1973, AND THEIR CURRENT TAXONOMY (NUMBERS IN PARENTHESES ARE THOSE IDENTIFIED TO THAT SPECIES BY MEYER AND WILSON).

| Meyer and Wilson, 1973 | Current Taxonomy |
|---|---|
| *Anolis allisoni* | *Anolis allisoni* (53), *Norops roatanensis* (1), *Basiliscus vittatus* (1) |
| *Anolis biporcatus* | *Norops biporcatus* (6) |
| *Anolis capito* | *Norops capito* (3) |
| *Anolis crassulus* | *Norops crassulus* (9), *N. tropidonotus* (1) |
| *Anolis cupreus* | *Norops cupreus* (28) |
| *Anolis humilis* | *Norops quaggulus* (1) |
| *Anolis laeviventris* | *Norops laeviventris* (3) |
| *Anolis lemurinus* | *Norops bicaorum* (12), *N. biporcatus* (1), *N. lemurinus* (135), *N. roatanensis* (25) |
| *Anolis limifrons* | *Norops limifrons* (2), *N. rodriguezii* (15), *N. zeus* (10) |
| *Anolis loveridgei* | *Norops loveridgei* (6) |
| *Anolis pentaprion* | *Norops beckeri* (5) |
| *Anolis sagrei* | *Norops roatanensis* (2), *N. sagrei* (12) |
| *Anolis sericeus* | *Norops unilobatus* (101), *N. wellbornae* (16) |
| *Anolis sminthus* | *Norops heteropholidotus* (4), *N. sminthus* (20) |
| *Anolis tropidonotus* | *Norops lemurinus* (1), *N. tropidonotus* (213) |

## SYSTEMATIC SPECIES ACCOUNTS

### Family Dactyloidae Fitzinger, 1843

Anoles are a family of lizards characterized by having an extensible dewlap in males of almost all species (and in females of some species), widened lamellae on the under-surfaces of the digits (in almost all species, including all Honduran species) with the compressed terminal phalanx rising above and slightly proximal to the end of the digit (Savage, 2002), a completely or nearly completely divided mental scale (personal observation), and a reproductive mode of the female depositing one egg at a time while at the same time containing a developing egg in the other ovary (Savage, 2002; Losos, 2009). In addition, the male dewlap is often brightly colored and is used in courtship and territorial display (termed a "classic example of a complex signaling system" by Nicholson et al., 2007: 1). This group of New World lizards contains about 400 named species placed in eight genera (Nicholson et al., 2012). Of these eight genera (*Anolis* Daudin, 1802; *Audantia* Cochran, 1934; *Chamaelinorops* Schmidt, 1919; *Ctenonotus* Fitzinger, 1843; *Dactyloa* Wagler, 1830; *Deiroptyx* Fitzinger, 1843; *Norops* Wagler, 1830; and *Xiphosurus* Fitzinger, 1826), only *Anolis* (1 species) and *Norops* (38 species) occur in Honduras. Members of this family have a wide geographical distribution including the southeastern USA, much of Mexico southward into South America, the Greater and Lesser Antilles, and numerous other Caribbean and eastern Pacific islands.

All anoles have circular pupils and are diurnal (or in a few cases largely crepuscular), with many known to shift their behavior by alternating between basking in the sun to shifting to shaded or covered areas. However, some species apparently do not bask, instead spending their active time in shaded areas (Savage, 2002). Most anoles are primarily generalist feeders that consume a wide variety of insects and other invertebrates. Carnivory, frugivory and molluscivory have also been reported, but mostly in larger species and larger individuals of moderate-sized species (Losos, 2009). Herrel et al. (2004) suggested that frugivory in anoles is mediated by large body size. Anoles of many species usually sight their prey from elevated perches and then quickly grab the prey before typically returning to their perch site to finish their meal (Savage, 2002). However, some species pursue their prey in leaf litter and apparently do not frequently climb other objects (Savage, 2002).

### Genus *Anolis* Daudin

*Anolis* Daudin, 1802: 50 (type species: *Anolis carolinensis* Voigt, 1832: 71, by fiat of the International Commission on Zoological Nomenclature [Anonymous, 1986: 125]).

*Geographic Distribution and Content*. This genus occurs from extreme northeastern Mexico eastward to the southeastern United States and southward through Florida, USA, and on the Islands of the Bahamas and much of the Greater Antilles. The genus also occurs on the Islas de la Bahía, Honduras, and islands off the coast of Quintana Roo, Mexico, and Belize. It has also been introduced on the Honduran mainland at La Ceiba, Atlántida, and on Isla de Utila, Islas de la Bahía, Hawaii, and Guam, with other more recent introductions being occasionally reported. Forty-three living species are recognized (Nicholson et al., 2012), one of which occurs in Honduras.

*Remarks*. Nicholson et al. (2012) recognized five species groups in this genus. *Anolis allisoni*, the single species of *Anolis* occurring in Honduras, was placed in the *A. carolinensis* species group. *Anolis carolinensis* is the type species of *Anolis*.

*Etymology*. The name *Anolis* is from the French l'anole, which is derived from *anoli* (or *anolis*) or *anaoli* (or *anoali*); aboriginal West Indian words meaning "lizard" (see Nicholson et al., 2012, for more information on the origin of *Anolis*).

### *Anolis allisoni* Barbour
*Anolis allisoni* Barbour, 1928: 58 (holotype, MCZ R-26725; type locality: "Coxen Hole,

Ruatan, Bay Islands, Honduras"), 1930: 114; Flower, 1928: 49; Barbour and Loveridge, 1929b: 216; Schmidt, 1941: 493; Peters, 1952: 25; Cochran, 1961: 85; Ruibal and Williams, 1961: 184; Duellman and Berg, 1962: 196; Ruibal, 1964: 486; Smith et al., 1964: 39; Meyer, 1969: 213; Williams, 1969: 357; Peters and Donoso-Barros, 1970: 47; Meyer and Wilson, 1973: 15 (in part); Wilson and Hahn, 1973: 109; Schwartz and Thomas, 1975: 66; Gundy and Wurst, 1976: 116; Mac-Lean et al., 1977: 4; Hudson, 1981: 377; Kluge, 1984: 6; O'Shea, 1986: 67; Schoener, 1988: 22; Schwartz and Henderson, 1988: 71, 1991: 210; Wilson and Cruz Díaz, 1993: 16; Köhler, 1994: 4, 1995: 102, 1998b: 377, 382, 2000: 65, 2003: 95, 2008: 99; Lee, 1996: 226; Cruz Díaz, 1998: 29, *in* Bermingham et al., 1998; Monzel, 1998: 157, 2001: 27; Rodríguez Schettino, 1999: 221; Stafford and Meyer, 1999: 129; Grismer et al., 2001: 135; Lundberg, 2001: 26, 2002a: 7, 2002b: 7; Wilson et al., 2001: 134; Flores-Villela and Canseco-Márquez, 2004: 124; Henderson and Powell, 2004: 301; Glor et al., 2005: 2424; McCranie et al., 2005: 100; Wilson and Townsend, 2006: 105; Diener, 2008: 10; McCranie and Gutsche, 2009: 112; Martinez and Clayson, 2013: 624; McCranie and Valdés Orellana, 2014: 44.

*Geographic Distribution.* Anolis allisoni occurs on the Islas de la Bahía, Honduras, and cays and islands off the coast of Belize and southern Quintana Roo, Mexico. It also occurs at low elevations throughout much of Cuba (exclusive of the western and eastern ends). In Honduras, this species is found on Islas de Barbareta, Guanaja, Morat, Roatán, Utila, and the Cayos Cochinos in the Islas de la Bahía. It has also been introduced into La Ceiba, Atlántida, on the north coast of Honduras. Since the species was only recently found on Utila, it is likely that the one specimen known from that island is a recent introduction (the animal photographed by Martinez and Clayson, 2013, likely represents the same animal that was collected).

*Description.* The following is based on 14 males (SMF 75978–79, 78317–18, 80853; USNM 563084–89, 563093, 563097–98) and 10 females (KU 101379, 220120; SMF 78319–20, 79205–06, 80854; USNM 563091, 563094, 565425). *Anolis allisoni* is a moderately large anole (SVL 82 mm in largest Honduran male [SMF 75978] and female [KU 220120] measured for this study; maximum reported SVL 91 mm [Schoener, 1988]); dorsal head scales keeled (mostly multicarinate, some unicarinate, some smooth in large males) in internasal, prefrontal, frontal, and parietal areas; deep frontal depression present; shallow to deep (especially in large males) parietal depression present; 3–7 (4.7 ± 1.0) postrostrals; anterior nasal divided, lower section contacting first supralabial; usually 2 (occasionally 3) scales between circumnasal and rostral; 3–5 (4.1 ± 0.4) internasals; canthal ridge sharply defined, ridged in large males; scales comprising supraorbital semicircles keeled, ridged in large males, largest scale in semicircles larger than largest supraocular scale; supraorbital semicircles well defined; supraorbital semicircles separated by 1 scale row at narrowest point; 1–3 (1.7 ± 0.5) scales separating supraorbital semicircles and interparietal at narrowest point; interparietal well defined, distinctly enlarged relative to adjacent scales, surrounded by scales of moderate size, longer than wide, about same size as ear opening; 2–3 rows of about 3–8 (total number) enlarged, faintly multicarinate supraocular scales; enlarged supraoculars completely separated from supraorbital semicircles by 1 row of small scales; 1 large elongate superciliary; 3–5 (3.6 ± 0.7) enlarged canthals; 4–8 (5.1 ± 0.9) scales between second canthals; 5–7 (5.3 ± 0.6) scales between posterior canthals; loreal region slightly to distinctly concave, 15–27 (20.9 ± 3.0) mostly smooth or rugose (some weakly keeled) loreal scales in a maximum of 4–5 (4.1 ± 0.3) horizontal rows; 7–9 (7.8 ± 0.7) supralabials and 6–10

(8.2 ± 0.9) infralabials to level below center of eye; suboculars keeled, in broad contact with supralabials; ear opening longitudinally elongated; scales anterior to ear opening granular, slightly larger than those posterior to ear opening; 5–7 (6.0 ± 0.5) postmentals, outer pair several times larger than medial ones; keeled granular scales present on chin and throat; male dewlap relatively small, extending to level of axilla; male dewlap with keeled gorgetal-sternal scales, not arranged in rows, about 57–75 total scales (*n* = 8); 2–4 (modal number) anterior marginal pairs in adult male dewlap; female dewlap absent; a low nuchal crest and a low dorsal ridge present; middorsal scale rows not enlarged, dorsal scales mostly weakly keeled, juxtaposed to subimbricate, those lateral to middorsal series grading into granular lateral scales; no enlarged scales among laterals; 56–73 (65.9 ± 4.7) dorsal scales along vertebral midline between levels of axilla and groin in males, 57–76 (65.6 ± 6.6) in females; 46–64 (53.4 ± 6.1) dorsal scales along vertebral midline contained in 1 head length in males, 42–64 (51.8 ± 7.0) in females; midventral scales on midsection slightly larger than largest dorsal scales; midventral body scales weakly keeled with rounded posterior margins, subimbricate; 46–66 (56.5 ± 6.5) ventral scales along midventral line between levels of axilla and groin in males, 43–57 (51.9 ± 4.1) in females; 39–56 (47.1 ± 5.9) ventral scales contained in 1 head length in males, 36–52 (42.4 ± 4.9) in females; 96–117 (101.8 ± 5.6) scales around midbody in males, 94–110 (100.1 ± 4.6) in females; tubelike axillary pocket absent; precloacal scales keeled; pair of slightly enlarged postcloacal scales in males; tail varying from nearly rounded to slightly compressed, TH/TW 0.86–1.33 in 21; basal subcaudal scales keeled; lateral caudal scales keeled, homogeneous, although indistinct division in segments discernable; dorsal medial caudal scale row slightly enlarged, keeled, not forming crest; most scales on anterior surface of antebrachium weakly keeled, unicarinate; 36–49 (42.4 ± 3.3) subdigital

lamellae on Phalanges II–IV of Toe IV of hind limbs; 5–10 (7.1 ± 1.3) subdigital scales on Phalanx I of Toe IV of hind limbs; SVL 53.0–82.0 (70.2 ± 8.8) mm in males, 48.0–82.0 (65.8 ± 10.1) mm in females; TAL/SVL 1.66–2.22 in 12 males, 2.07–2.21 in seven females; HL/SVL 0.30–0.37 in males, 0.28–0.34 in females; SHL/SVL 0.20–0.24 in males, 0.19–0.24 in females; SHL/HL 0.58–0.75 in males, 0.62–0.76 in females; longest toe of adpressed hind limb reaching between shoulder region and ear.

Color in life of an adult male (USNM 563084): dorsal surfaces of body, tail, and limbs Lime Green (159); dorsal surface of head turquoise blue; belly pale green; subcaudal surface Lime Green; darker green cross-reticulations also present on dorsum of limbs; eyelids pale brown; iris brown with copper rim; dewlap Poppy Red (108A) with turquoise blue scales. Color in life of another adult male (KU 220121) was described by Wilson and Cruz Díaz (1993: 16): "dorsum of body lime green with a pale rust patina on the nuchal area; dorsum of the head lime green with turquoise blue cast; dark gray postorbital blotch present; pale (white) lip stripe begins below eye and continues to just behind ear opening; skin around eye rust brown; limbs lime green with gray crossbars; tail lime green; venter pale olive green; dewlap magenta with pale green scales."

Color in alcohol: dorsal surfaces of head and body brown, with or without patches of gray or faded turquoise blue; lateral surface of head brown, except suboculars white to bluish white, pale pigment continuing posteriorly as a postlabial stripe extending to anterior edge of ear opening; dorsal surfaces of limbs same as for body, except darker brown crossbars present in some; dorsal surface of tail same as for body, except turquoise blue pigment can form indistinct crossbands; ventral surfaces of head, body, and tail pale brown to cream, with or without bluish tinge.

Hemipenis: the completely everted hemipenis of SMF 78318 is a medium-sized, bilobed organ; sulcus spermaticus bordered by well-developed sulcal lips, bifurcating at base of apex, branches continuing to tips of lobes; asulcate processus absent; lobes strongly calyculate; truncus with transverse folds.

*Diagnosis/Similar Species. Anolis allisoni* is readily distinguished from all other Honduran anoles by its elongated snout (especially pronounced in adult males) and its longitudinally elongated ear opening with the posterior margin forming a longitudinal depression. The only other solid green anole in life in Honduras is *Norops biporcatus*, which is restricted to the mainland.

*Illustrations* (Figs. 1, 2, 79). Ruibal and Williams, 1961 (head, ear opening); Ruibal, 1964 (head, ear opening); Köhler, 1996d (adult), 1998b (adult), 2000 (adult, head and dewlap), 2003 (adult, head and dewlap), 2008 (adult, head and dewlap); Lee, 1996 (adult, ear opening), 2000 (adult, ear opening); Stafford and Meyer, 1999 (adult, ear opening); Lundberg, 2001 (adult), 2002b (adult); Monzel, 2001 (adult); McCranie et al., 2005 (adult, head and dewlap, ear opening); Diener, 2008 (adult).

*Remarks. Anolis allisoni* is a member of the *A. carolinensis* subgroup (Glor et al., 2005) or the *A. carolinensis* species group (Ruibal and Williams, 1961; Ruibal, 1964; Williams, 1976a; Nicholson et al., 2012). Ruibal and Williams (1961) suggested that the Central American island populations of *A. allisoni* might not be conspecific with those on Cuba, but a phylogenetic analysis based on mtDNA sequence data (Glor et al., 2005) showed very little divergence between an animal trade specimen purported to be from Isla de Guanaja and those from Cuba.

*Natural History Comments. Anolis allisoni* is known from near sea level to 30 m elevation in the Lowland Moist Forest formation. This species is active during sunny days on coconut palms, thorn palms, other palms, various tree types, buildings, wooden fences, and occasionally the ground. However, coconut palms seem to be the preferred habitat. It frequently perches more than 1 m above the ground. *Anolis allisoni* has been found during February, from May to July, and from September to December, so it is likely active throughout the year. The species sleeps at night on leaves and stems of low vegetation and on tree branches. Henderson and Powell (2009 and references cited therein) listed similar habitats for this species on Cuba. On Cuba, *A. allisoni* feeds primarily on insects and other invertebrates, but also feeds on nectar (Henderson and Powell, 2009, and references cited therein). On Cuba, this species lays eggs year round, but with increased egg production during the rainy season; communal egg laying has also been reported (Henderson and Powell, 2009, and references cited therein).

*Etymology.* The name *allisoni* is a patronym honoring Allison V. Armour, who assisted Thomas Barbour, the describer of the species, with transportation aboard his yacht *Utowana*.

*Specimens Examined* (413 [53] + 1 skeleton; Map 1). ATLÁNTIDA: La Ceiba, USNM 570550. ISLAS DE LA BAHÍA: Cayo Cochino Grande, lighthouse near La Ensenada, KU 220121; Cayo Cochino Pequeño, KU 220118–20; Isla de Barbareta, CM 27610, USNM 520263, 563083; Isla de Guanaja, El Bight, USNM 565425, ZMB 73626–27; Isla de Guanaja, SE shore opposite Guanaja, LACM 47775–76, LSUMZ 21436–38, 22405, UF 28566–72; Isla de Guanaja, La Playa Hotel, LSUMZ 21439; Isla de Guanaja, near Monumento a Cristóbal Colón, SMF 75979; Isla de Guanaja, North East Bight, SMF 75978; Isla de Guanaja, between Savannah Bight and East End, SMF 78320; Isla de Guanaja, between Savannah Bight and Mangrove Bight, SMF 78317–18; Isla de Guanaja, 1.5 km W of Savannah Bight, SMF 78319; Isla de Guanaja, Savannah Bight, FMNH 283643, LSUMZ 22412; Isla de Guanaja, near W end of island, TCWC 21950, 26700; "Isla de Guanaja," CM 27612 (9), FMNH 53822–27, KU 101379–88, LSUMZ 9703–9724, 10276–81; Isla de Roatán, near Corozal, SMF 80853; Isla de Roatán, Coxen Hole, AMNH

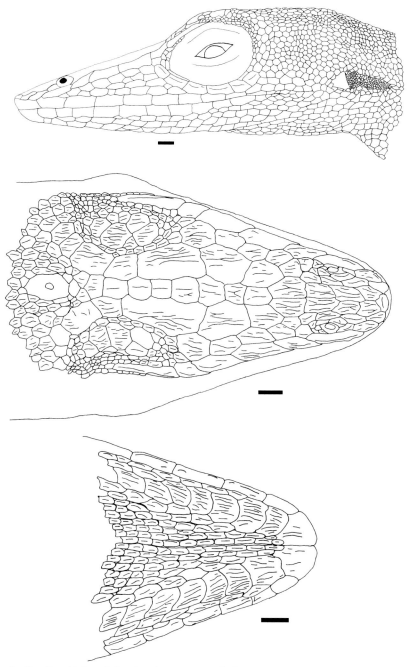

Figure 1.   *Anolis allisoni* head in lateral, dorsal, and ventral views. SMF 78319, adult female from 1.5 km W of Savannah Bight, Isla de Guanaja, Islas de la Bahía. Scale bar = 1.0 mm. Drawing by Gunther Köhler.

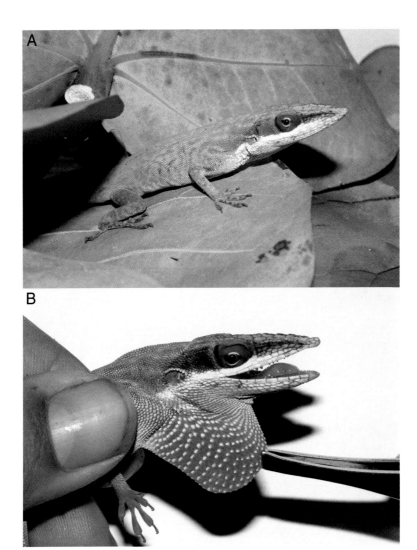

Figure 2.    *Anolis allisoni.* (A) Adult male (FMNH 282565) from Palmetto Bay, Isla de Roatán, Islas de la Bahía; (B) adult male dewlap (FMNH 282565). Photographs by James R. McCranie.

80062 (formerly part of UMMZ 67694, previously part of MCZ R-26755), ANSP 26112 (was in UMMZ collection), BMNH 1985.1095–99, CM 28991–93, FMNH 34539 (66), KU 47167 (formerly MCZ R-26737), MCZ R-26725–31, 26732 (skeleton), 26733–36, 26738–53, 26755, 171169–70, MVZ 212076, UIMNH 41497 (formerly MCZ R-26754), UMMZ 66825 (2; formerly part of MCZ R-26755), 67694 (2; formerly part of MCZ R-26755), USNM 75859; Isla de Roatán, near Diamond Rock, USNM 563084; Isla de Roatán, Fort Key, MCZ R-150946; Isla de Roatán, Flowers Bay, USNM 563085–87; Isla de Roatán, between Flowers Bay and West End Point, USNM 563089; Isla de Roatán, near Flowers Bay, USNM 563088; Isla de Roatán, near French Harbor, UF 28562; Isla de Roatán, French Harbor, LSUMZ 22393; Isla de Roatán, Jobs Bight, LSUMZ 33813–16; Isla de Roatán, Key Hole, USNM 563090; Isla de Roatán, Mudd

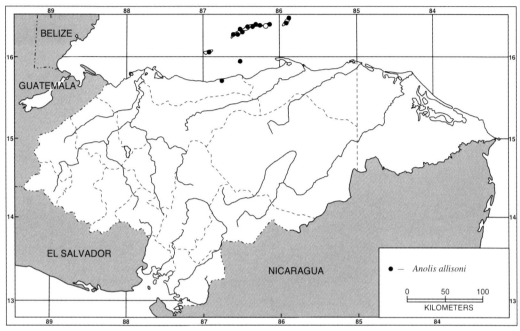

Map 1.   Localities for *Anolis allisoni*. Solid symbols denote specimens examined. The solid symbol off the coast of north-central Honduras represents the small Cayos Cochinos, and the open symbol just east of Isla de Roatán represents an accepted record for Isla Morat.

Hole Bay, SMF 80854; Isla de Roatán, near Oak Ridge, MCZ R-150943–45, UTA R-10688–98, 31975, 55233–35; Isla de Roatán, Oak Ridge, CM 27600 (20), 26701 (18), 27604 (4), 27605 (6), FMNH 53807 (27), 53808–21; Isla de Roatán, Palmetto Bay, FMNH 282565–66, MVZ 267187; Isla de Roatán, Port Royal, USNM 578740; Isla de Roatán, near Port Royal Harbor, TCWC 52414–16; Isla de Roatán, about 1.6 km N of Roatán, AMNH 149565, LSUMZ 52613; Isla de Roatán, about 2.5 km N of Roatán, LSUMZ 33809; Isla de Roatán, 0.5 km N of Roatán, LACM 47770–74, LSUMZ 21433–35, UF 28472–84, 28497–99; Isla de Roatán, 4.8 km W of Roatán, UF 28514–16, 28535; Isla de Roatán, about 3.2 km W of Roatán, CM 64584, LSUMZ 52610–12; Isla de Roatán, Roatán, LSUMZ 22312–13, 22328–32, TCWC 21946–47, 21968–70; Isla de Roatán, Rocky Point, USNM 563091–92; Isla de Roatán, Sandy Bay, KU 203134–35, 203145; Isla de Roatán, Santa Elena, BMNH 1938.10.4.2; Isla de Roatán, Turquoise Bay,

FMNH 283642; Isla de Roatán, 6.6 km E of West End, MVZ 263854; Isla de Roatán, West End, SMF 79205–06, 89205–06, TCWC 21948, USNM 578738–39; Isla de Roatán, West End Point, USNM 563093; Isla de Roatán, West End Town, USNM 563094–98; "Isla de Roatán," BMNH 1985.1099, MCZ R-191094–103; Isla de Utila, Utila Town, FMNH 283583.

*Other Records* (Map 1). ISLAS DE LA BAHÍA: Isla Morat, LSUHC 3696–98, 3703–05 (Grismer et al., 2001); Isla de Utila (MPC Herp Photo P766).

Genus *Norops* Wagler

*Norops* Wagler, 1830: 149 (type species: *Anolis auratus* Daudin, 1802: 89 [= *Lacerta aurata* Bonnaterre, 1789: 52], by monotypy).

*Geographic Distribution and Content*. This genus occurs from Sonora and Tamaulipas, Mexico, southward to northwestern

Peru west of the Andes and from southeastern Peru to southeastern Brazil east of the Andes. It also occurs on the islands in the Bahamas, Cuba (and nearby islands), Jamaica (and nearby islands), the Islas del Cisne, the Bay Islands, the Corn Islands, San Andrés, Providencia, Aruba, Curaçao, Margarita, and Trinidad and Tobago in the Caribbean and on Isla del Cocos and Isla Gorgona in the Pacific. The genus is introduced and established in extralimital populations on Hawaii and in the southeastern United States, in the Cayman Islands, Bermuda, Grenada, St. Vincent, Taiwan, and Singapore, with new introductions being regularly reported. Nicholson et al. (2012) recognized 174 named living species. We recognize an additional species not included by Nicholson et al. (2012; *N. heteropholidotus*) and consider one species recognized by those authors (*N. dariense*) not to represent a valid species. Thus, of the 174 named species of *Norops* presently recognized, 38 are known to occur in Honduras.

*Remarks.* Boulenger (1885: 73) reported *Norops ustus* (Cope) as occurring in "Honduras and Yucatan." However, that "Honduras" record is actually from Belize (specimen not examined by us). *Norops ustus* is a synonym of *N. sericeus* (Hallowell; see Köhler and Vesely, 2010), which does not occur in Honduras. Boulenger (1885: 90) also listed a specimen of *N. chrysolepis* (Duméril and Bibron) from "Honduras," but that is a South American species. Dunn and Emlen (1932) opined that Boulenger's specimen in question represented *N. copei* (see our Remarks for *N. biporcatus*). Dunn and Emlen (1932) confused *N. biporcatus* and *N. capito* in their concept of *N. copei*.

Nicholson et al. (2012) placed the 174 living species of *Norops* they recognized into three species groups. Nicholson et al. (2012) included *N. nelsoni* and *N. sagrei* in the *N. sagrei* species group, and all other *Norops* species occurring in Honduras were placed in the *N. auratus* species group. That latter group contains 150 species, 36 of which occur in Honduras, thus containing 86.2% of the total Honduran species of *Norops*. We place those 36 Honduran species into 10 subgroups of the *N. auratus* species group (five species are considered incertae sedis in that species group).

*Etymology. Norops* is Greek and means "bright, flashing, gleaming," probably in reference to either the blue dewlap found in males of the type species of *Norops* or to the pale vertebral or dorsolateral stripes found in many specimens of the otherwise drab type species of the genus (also see Nicholson et al., 2012).

## *Norops amplisquamosus* McCranie, Wilson, and Williams

*Norops amplisquamosus* McCranie, Wilson, and Williams, 1992: 209 (holotype, KU 219924; type locality: "El Cusuco [15°31′N, 88°12′W], a *finca* located 5.6 km WSW Buenos Aires, 1550 m elevation, Sierra de Omoa, Departamento de Cortés, Honduras"); Anonymous, 1994: 116; Köhler, 2000: 59, 2003: 96, 2008: 102; Köhler et al., 2001: 254; Wilson et al., 2001: 135; Wilson and McCranie, 2003: 59, 2004b: 43, 2004c: 24.

*Anolis amplisquamosus*: Townsend, 2006: 35, 2009: 298; Townsend and Wilson, 2006: 245, 2008: 148, 2009: 68; Townsend et al., 2006: 31; Köhler et al., 2007: 390; Köhler, 2014: 210.

*Geographic Distribution. Norops amplisquamosus* occurs at intermediate elevations in the Sierra de Omoa in northwestern Cortés, Honduras. It is known only from the vicinity of its type locality.

*Description.* The following is based on 11 males (KU 219924–30; SMF 77748, 77750; UF 149646; USNM 549357) and 13 females (KU 219937, 219940, 219943–45; SMF 77747, 79173; UF 149641–43, 149645, 149647–48). *Norops amplisquamosus* is a small anole (SVL 46 mm in largest male examined [KU 219929], 49 mm in largest female examined [UF 149645]); dorsal head scales smooth, rugose, or weakly keeled in internasal, prefrontal, and frontal areas, most scales smooth in parietal area; deep frontal depression present; no parietal depression; 3–6 (4.7 ± 1.0) postrostrals; anterior nasal usually divided, lower section contacting rostral and first supralabial; 3–7 (5.2 ± 1.1) internasals; canthal ridge sharply defined; scales comprising supraorbital semicircles ridged near outer edges, largest scale in semicircles varies from larger than, to smaller than, largest supraocular scale, other scales in supraocular semicircles usually smaller than largest supraocular scale; supraorbital semicircles well defined; 0–2 (0.7 ± 0.6) scales separating supraorbital semicircles at narrowest point; 1–2 (1.6 ± 0.4) scales separating supraorbital semicircles and interparietal at narrowest point; interparietal well defined, irregular in outline, longer than wide, larger than, or equal to, size of ear opening; 2 rows of about 2–5 (total number) distinctly enlarged, smooth to weakly keeled supraocular scales present; enlarged supraoculars in inner row in broad contact with supraorbital semicircles; 2 elongate superciliaries, posterior much shorter than anterior; usually 3 enlarged canthals; 4–7 (5.6 ± 0.9) scales between second canthals; 5–8 (6.9 ± 0.8) scales between posterior canthals; loreal region slightly concave, 12–26 (18.9 ± 3.4) smooth to weakly keeled loreal scales in maximum of 3–5 (4.0 ± 0.6) horizontal rows; 5–7 (6.1 ± 0.5) supralabials and 5–7 (5.9 ± 0.5) infralabials to level below center of eye; suboculars usually weakly keeled, in broad contact with supralabials; ear opening vertically oval; scales anterior to ear opening granular, similar in size to those posterior to ear opening; 4–5 (4.1 ± 0.3) postmentals, outer pair largest; largest gular scales weakly keeled, remaining gulars smooth to slightly conical; male dewlap large, extending past axilla onto chest; male dewlap with 7–8 (7.3 ± 0.49) horizontal gorgetal-sternal scale rows, about 7.5 mean number of scales per row (*n* = 7); 2 (modal number) anterior marginal pairs in adult male dewlap; female dewlap scarcely indicated; nuchal crest usually present in adult males; dorsal ridge absent; about 7–11 middorsal scale rows distinctly and abruptly enlarged, distinctly keeled, dorsal scales lateral to abruptly enlarged middorsal series grading into granular lateral scales; about 2–3 vertebral rows slightly smaller than adjacent enlarged middorsal scale rows; enlarged scales scattered among granular laterals; 30–47 (40.7 ± 5.4) dorsal scales along vertebral midline between levels of axilla and groin in males, 34–49 (41.7 ± 4.2) in females; 22–29 (24.9 ± 2.3) dorsal scales along vertebral midline contained in 1 head length in males, 18–29 (22.8 ± 2.9) in females; ventral scales on midsection smaller than largest dorsal scales; midventral body scales smooth to weakly keeled, imbricate; 34–51 (43.2 ± 5.1) ventral scales along midventral line between levels of axilla and groin in males, 39–50 (44.8 ± 3.4) in females; 24–31 (27.8 ± 1.8) ventral scales contained in 1 head length in males, 22–30 (25.1 ± 2.1) in females; 80–102 (89.9 ± 6.7) scales around midbody in males, 80–98 (91.5 ± 5.1) in females; tubelike axillary pocket absent; precloacal scales not keeled; enlarged postcloacal scales present in males; tail nearly rounded or oval, TH/TW 1.08–1.36 in 2♂; basal subcaudal scales weakly to distinctly keeled; lateral caudal scales distinctly keeled, homogeneous, although indistinct division in segments discernable; dorsal medial caudal scale row not enlarged, distinctly keeled, not forming crest; scales on anterior surface of antebrachium unicarinate; 20–25 (22.4 ± 1.3) subdigital lamellae on Phalanges II–IV of Toe IV of hind limbs; 5–8 (6.4 ± 1.1) subdigital scales on Phalanx I of Toe IV of hind limbs; SVL 37.9–45.7 (41.6 ± 2.7) mm in males, 40.0–48.8 (43.0 ± 2.7) mm in females; TAL/SVL 2.20–2.47 in seven males, 1.33–2.49 in 11 females; HL/SVL 0.27–0.29 in males, 0.25–0.28 in females; SHL/SVL 0.22–0.25 in males, 0.22–0.26 in females; SHL/HL 0.79–0.93 in males, 0.84–0.97 in females; longest toe of adpressed hind limb usually reaching between anterior to ear opening and midlength of eye.

Color in life of the male holotype (KU 219924) was described by McCranie et al. (1992: 209, 211): "dorsum dirty yellowish-tan laterally, gray-brown medially with indistinct brown chevrons; venter pale yellow; dewlap Chrome Orange (color 16) with pale yellow and dirty yellow scales." Two adult male paratypes (KU 219926, 219928) "were similar to the holotype except that KU 219928 was recorded as having some suffusion of rust and rusty gray colors dorsally and KU 219926 had a pale, olive-brown interorbital bar" (McCranie et al., 1992: 213). Another adult male paratype (KU 219930) had the dorsum heavily suffused with rust color (McCranie et al., 1992).

Color in alcohol: dorsal surfaces of head and body brown to dark brown, lateral surface of body brown; some males and most females have a dark brown middorsal stripe, dark stripe, when present, bordered below by irregular pale brown area; indistinct dark brown middorsal chevrons present in some males; two females (KU 219939, 219949) have broad cream middorsal stripes that are outlined by a thin dark brown stripe on each side; lateral surface of head brown, except most supralabials and suboculars usually cream, forming a pale labial stripe; dorsal surfaces of limbs brown with indistinct darker brown crossbars; dorsal surface of tail brown, frequently with dark brown stripe anteriorly (a continuation of middorsal stripe), otherwise without distinct markings; ventral surfaces of head and body cream, lightly flecked with brown; subcaudal surface cream proximally, lightly flecked with brown proximally, remainder of subcaudal surface becoming more heavily flecked, until distal two-thirds brown.

Hemipenis: the partially everted hemipenis of SMF 77748 is a small, bilobed organ; apex calyculate; asulcate processus present.

*Diagnosis/Similar Species. Norops amplisquamosus* is distinguished from all other Honduran *Norops*, except *N. heteropholidotus*, *N. muralla*, and *N. sminthus*, by the combination of distinctly and abruptly enlarged middorsal scale rows, heteroge-neous lateral scales, smooth to weakly keeled and imbricate ventral scales, and a pair of greatly enlarged postcloacal scales in males. *Norops amplisquamosus* differs from *N. heteropholidotus*, *N. muralla*, and *N. sminthus* by having an orange-yellow dewlap in males (male dewlap red).

*Illustrations* (Figs. 3, 4, 92–94). McCranie et al., 1992 (head scales, body scales); Köhler, 2000 (head and dewlap), 2003 (adult, head and dewlap), 2008 (adult, head and dewlap), 2014 (dorsal and ventral scales; as *Anolis*); Townsend and Wilson, 2006 (adult; as *Anolis*), 2008 (adult, dewlap; as *Anolis*).

*Remarks. Norops amplisquamosus*, along with *N. crassulus*, *N. heteropholidotus*, *N. morazani*, *N. muralla*, *N. rubribarbaris*, and *N. sminthus*, make up the Honduran segment of the *N. crassulus* species sub-group of the Nicholson et al. (2012) *N. auratus* species group. We define this subgroup as having heterogeneous lateral body scales, at least six distinctly enlarged middorsal scale rows, and the male dewlap some shade of orange or red in life (also see Remarks in *N. crassulus* account). *Norops amplisquamosus* tissues were not included in the phylogenetic analyses of Nicholson et al. (2012).

Köhler and Obermeier (1998) and Köhler et al. (1999) included *N. laeviventris* and *N. nebulosus* in the *N. crassulus* group. However, phylogenetic analyses based on molecular data by Nicholson (2002) and Nicholson et al. (2005, 2012) do not demonstrate a close relationship between *N. laeviventris–N. intermedius* (the latter considered a synonym of *N. laeviventris* by Köhler [see McCranie et al., 2000]) and two *N. crassulus* sequenced by her (one reported as *N. sminthus*). The phylogenetic analyses based on morphology in Poe (2004) also did not recover a close relationship between *N. laeviventris–N. intermedius* and *N. crassulus–N. sminthus*.

*Natural History Comments. Norops amplisquamosus* is known from 1,530 to 1,990 m elevation in the Lower Montane Wet Forest formation. McCranie et al. (1992: 214) noted, "Specimens were

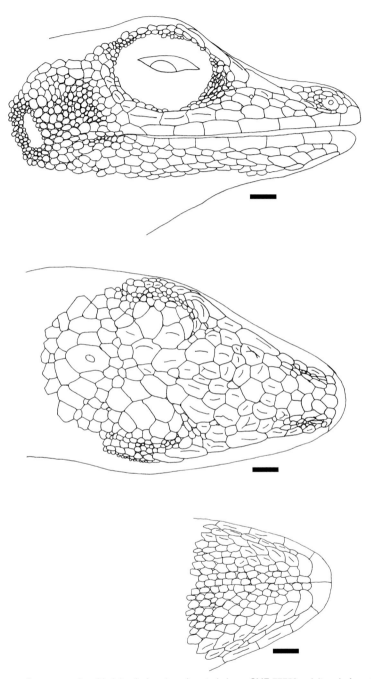

Figure 3.  *Norops amplisquamosus* head in lateral, dorsal, and ventral views. SMF 77750, adult male from Sendero El Danto, Cortés. Scale bar = 1.0 mm. Drawing by Gunther Köhler.

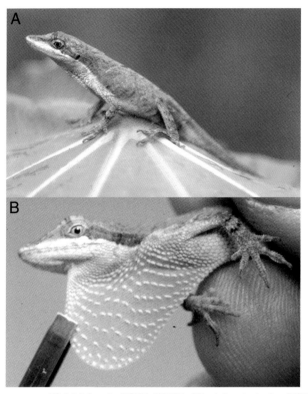

Figure 4. *Norops amplisquamosus.* (A) Adult female (USNM 549356); (B) adult male dewlap (USNM 549357), both from El Cusuco, Cortés. Photographs by James R. McCranie.

collected while active on the ground or on tree stumps during the day. At night, specimens were found sleeping with the head up on fern leaves or leaves of other vegetation, up to 2 m above the ground." Subsequently to writing the above, a few active individuals were also seen on trunks of tree ferns and low on the trunks of broadleaf trees. This species was found during April, May, July, August, and November and is probably active throughout the year during suitable weather. Trips to the area by McCranie during 2003 and 2005 did not result in finding the species along the old road leading from the visitor's center of Parque Nacional Cusuco to the private farm located above the visitor's center, where it was abundant during the 1980s and 1990s. Townsend et al. (2006) also suggested that populations of this species have severely declined in the vicinity of its type locality. Nothing has been reported on diet or reproduction in *N. amplisquamosus.*

*Etymology.* The name *amplisquamosus* is formed from the Latin *amplus* (large), *squama* (scale), and *-osus* (full of or augmented by) and alludes to the distinctive large dorsal scales of this species.

*Specimens Examined* (44 [0]; Map 2). CORTÉS: Bosque Enano, UF 149641; Cantiles Camp, UF 149642; Cerro Jilinco, UF 149643–44; El Cusuco, KU 219924–49, SMF 79173, UF 149645–46, USNM 549356–58, 578741–42; Quebrada de Cantiles, UF 149647–48; Sendero El Danto, SMF 77747–50.

*Norops beckeri* (Boulenger)
*Anolis beckeri* Boulenger, 1882: 921 (syntypes: IRSNB 2010 [1–2] [Köhler, 2010:

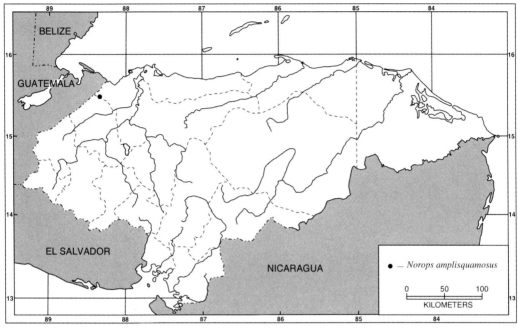

Map 2.   Localities for *Norops amplisquamosus*. The solid symbol denotes specimens examined.

9]; type locality: "Yucatan"); Townsend et al., 2012: 100.

*Anolis pentaprion*: Meyer, 1969: 228; Myers, 1971b: 37; Meyer and Wilson, 1973: 19; O'Shea, 1986: 36, 1989: 16; Poe, 2004: 63; Townsend and Wilson, 2010b: 697.

*Norops beckeri*: Campbell, 1998: 132.

*Norops pentaprion*: Espinal et al., 2001: 106; Wilson et al., 2001: 136; McCranie et al., 2006: 124; Wilson and Townsend, 2006: 105.

*Geographic Distribution. Norops beckeri* occurs at low and moderate elevations on the Atlantic versant from Tabasco and southern Quintana Roo, Mexico, to northern Nicaragua. In Honduras, it occurs across the northern portion of the country.

*Description.* The following is based on five males (CM 64619; UMMZ 58394; USNM 242056, 344805, 580300) and seven females (MCZ R-38835; SMF 91288; UF 156628–29; UMMZ 58392; USNM 570085, 578759). *Norops beckeri* is a medium-sized anole (SVL 61 mm in largest Honduran male examined [USNM 344805], 68 mm in largest Honduran female [UF 156629] examined); dorsal head scales rugose in internasal, prefrontal, and frontal areas, rugose to smooth in parietal area; shallow frontal and parietal depressions present; 5–11 (6.3 ± 1.6) postrostrals; anterior nasal divided, lower scale contacting rostral and first supralabial; 4–9 (6.1 ± 1.4) internasals; canthal ridge sharply defined; scales comprising supraorbital semicircles ridged, especially anterior ones, largest scale in semicircles larger than largest supraocular scale; supraorbital semicircles well defined; supraorbital semicircles usually broadly contacting each other medially (rarely separated by 1 scale); 1–4 (2.1 ± 0.9) scales separating supraorbital semicircles and interparietal at narrowest point; interparietal well defined, irregular in outline, longer than wide, larger than ear opening; 2–3 rows of about 3–6 (total number) enlarged, smooth to weakly keeled supraocular scales; enlarged supraoculars completely separated from supraorbital semicircles; 3 short superciliaries, posteriormost shortest; 2–4 enlarged canthals; 6–14 (8.8 ± 2.1) scales between second canthals; 6–12 (8.3 ± 1.9) scales between posterior canthals; loreal region slightly concave, 20–66 (30.9 ± 13.0) mostly smooth or rugose (some keeled) loreal scales in maximum of 3–10 (4.8 ± 2.0) horizontal rows; 7–10 (8.4 ± 0.9) supralabials and 7–11 (8.6 ± 1.1) infralabials to level below center of eye; suboculars smooth to rugose, in broad contact with supralabials; ear opening vertically oval; scales anterior to ear opening slightly larger than those posterior to ear opening; 5–7 (6.0 ± 0.4) postmentals, outer pair largest; gular scales not keeled; male dewlap moderately large, extending to beyond level of axilla onto chest; male dewlap with 6 horizontal gorgetal-sternal scale rows, about 6–9 (n = 2) scales per row; 3 (modal number) anterior marginal pairs in male dewlap; female dewlap well developed, extending to level of axilla; female dewlap with 4–6 horizontal gorgetal-sternal scale rows, with about 9–14 scales per row (n = 2); 3–4 (modal number) anterior marginal pairs in female dewlap; low nuchal crest present, but dorsal ridge absent; about 2 middorsal scale rows slightly enlarged, mostly smooth, dorsal scales lateral to middorsal series grading into granular lateral scales; no enlarged scales among laterals; 84–101 (93.8 ± 7.1) dorsal scales along vertebral midline between levels of axilla and groin in four males, 90–125 (103.5 ± 12.6) in six females; 56–80 (68.0 ± 10.3) dorsal scales along vertebral midline contained in 1 head length in males, 48–70 (58.3 ± 7.2) in females; ventral scales on midsection slightly larger than largest dorsal scales; ventral body scales smooth, obliquely conical, juxtaposed; 70–94 (83.5 ± 10.1) ventral scales along midventral line between levels of axilla and groin in four males, 65–92 (75.5 ± 11.3) in six females; 51–69 (56.2 ± 7.4) ventral scales contained in 1 head length in males, 33–58 (44.7 ± 8.8) in females; 144–178 (158.0 ± 12.9) scales around midbody in males, 146–170 (154.9 ± 7.8) in females; tubelike axillary pocket absent; precloacal scales not keeled; enlarged postcloacal scales absent; tail nearly rounded

to slightly oval; TH/TW 0.86–1.50; basal subcaudal scales smooth to faintly keeled; lateral caudal scales keeled, homogeneous, although indistinct division in segments discernable; 4 supracaudals present per caudal segment; dorsal medial caudal scale row enlarged, keeled, not forming crest; 2 median subcaudal scale rows distinctly enlarged, keeled; most scales on anterior surface of antebrachium smooth; 18–32 (26.5 ± 4.9) subdigital lamellae on Phalanges II–IV of Toe IV of hind limbs, those proximal ones only slightly widened; 7–19 (10.1 ± 4.7) subdigital scales on Phalanx I of Toe IV of hind limbs; SVL 34.9–60.6 (49.8 ± 9.5) mm in males, 50.3–68.4 (58.5 ± 7.7) mm in females; TAL/SVL 1.41–1.57 in three males, 0.87–2.47 in females; HL/SVL 0.25–0.29 in males, 0.16–0.27 in females; SHL/SVL 0.17–0.21 in males, 0.16–0.19 in females; SHL/HL 0.61–0.82 in males, 0.63–1.04 in females; longest toe of adpressed hind limb usually reaching past shoulder region.

Color in life of an adult female (USNM 570085): dorsum mottled Clay Color (26) and Brownish Olive (29) with series of Opaline Green (162D) blotches coursing down middorsum and crossing body transversely; limbs mottled Clay Color and Brownish Olive; tail mottled Olive-Gray (42) and Brownish Olive with series of Olive-Brown (28) crossbands; head Brownish Olive with Turquoise Blue (65) spots; venter Spectrum Yellow (55) with brown mottling; chin mottled Brownish Olive and Turquoise Blue; dewlap Carmine (8) with pale blue scales; iris rust red.

Color in alcohol: dorsal surface of head brown with paler brown mottling; dorsal surface of body brown with pale brown lichenlike pattern, pattern especially prominent laterally; dorsal surfaces of limbs brown with pale brown lichenlike pattern; dorsal surface of tail brown with pale brown crossbands; ventral surface of head white with brown mottling; ventral surface of body white with brown mottling and lateral lines; subcaudal surface white with brown mottling proximally, brown with dark brown crossbands distally; throat lining grayish black.

Hemipenis: unknown.

*Diagnosis/Similar Species. Norops beckeri* is distinguished from all other Honduran *Norops*, except *N. utilensis*, by having the proximal subdigital scales of the hind toes differentiated only as slightly broadened lamellae, relatively short hind legs (longest toe of adpressed hind limb usually reaching only past shoulder region), and smooth, obliquely conical, and juxtaposed ventral scales. *Norops utilensis* has five supracaudal scales per caudal segment (four supracaudals in *N. beckeri*).

*Illustrations* (Figs. 5, 6, 110). Lee, 1996 (head scales, caudal scales; as *Anolis pentaprion*), 2000 (head scales, caudal scales; as *N. pentaprion*); Stafford and Meyer, 1999 (adult; as *N. pentaprion*); McCranie et al., 2006 (adult; as *N. pentaprion*).

*Remarks. Norops beckeri* is in the *N. pentaprion* species subgroup (Köhler, 2010) of the *N. auratus* species group of Nicholson et al. (2012). Myers (1971b) and Williams (1976b) had recognized a *N. pentaprion* species group that included *N. beckeri* as a synonym of *N. pentaprion* (Cope). The only other Honduran species in this subgroup is *N. utilensis* (see Remarks for *N. utilensis*). This subgroup is characterized by having granular or only slightly widened proximal subdigital scales on the fourth digit of the hind limbs, relatively short hind limbs, a well-developed female dewlap, and granular or conical and nonoverlapping ventral scales. *Norops beckeri* tissues were not included in the Nicholson et al. (2012) study, but the seemingly closely related *N. utilensis* and *N. pentaprion* were included but, surprisingly, without recovering a close relationship between those two species.

Köhler (2010) recently provided evidence that what has traditionally been considered *Norops pentaprion* actually represents a complex of three species. *Norops pentaprion* is restricted to the Caribbean lowlands of southern Nicaragua to northwestern Colombia (also on Pacific versant in central and eastern Panama), *N. beckeri* was revalidated for the northern populations formerly referred to *N. pentaprion*, and *N. charlesmeyeri* was described as a new species for the western Costa Rican (mostly Pacific versant) and northwestern Panamanian populations.

*Natural History Comments. Norops beckeri* is known from near sea level to about 1,400 m elevation in the Lowland Moist Forest and Premontane Wet Forest formations and peripherally in the Lowland Dry Forest. This species probably occurs in gallery forest in the Lowland Dry Forest formation. One specimen of this highly arboreal and cryptic species was sleeping at night on vegetation about 2 m above the ground, and another was active on the trunk of an isolated tree about 3 m above the ground in a pasture, both during July. The two Los Pinos, Cortés, specimens were collected during April, and one of the USNM specimens from Atlántida was collected during August. Townsend et al. (2012) reported finding a specimen sleeping on a fence post at night (month of collection not given). The lichenlike color pattern of this anole probably helps keep it camouflaged to human observers. Villarreal Benítez (1997) reported that individuals from Veracruz, Mexico, are found in shady situations 3–5 m above the ground in savannas, mangrove forest, and evergreen forest. Perez-Higareda et al. (1997) reported arthropods, hatchling *N. beckeri*, and fruit of an introduced shrub in stomach contents of a population of this species from Veracruz, Mexico. *Norops beckeri* apparently is most reproductively active during the rainy season (Lee, 1996, and references cited therein).

*Etymology.* The name *beckeri* is a patronym honoring the Belgian arachnologist León Becker.

*Specimens Examined* (15 [5]; Map 3). ATLÁNTIDA: Guaymas District, UMMZ 58392–95; Jilamito Nuevo, USNM 578759; Lancetilla, MCZ R-38835. COLÓN: Cerro Calentura, CM 64619; Salamá, USNM 242056. CORTÉS: Los Pinos, UF 156628–29. GRACIAS A DIOS: Caño Awalwás, USNM 570085; Palacios, BMNH 1985. 1121. OLANCHO: Montaña del Ecuador,

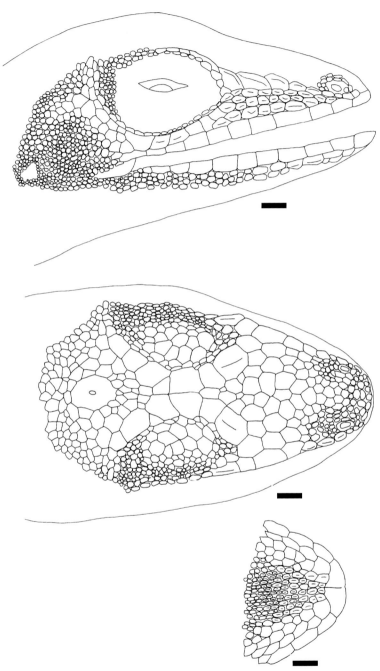

Figure 5.   *Norops beckeri* head in lateral, dorsal, and ventral views. SMF 91288, adult female from San José de Colinas, Santa Bárbara. Scale bar = 1.0 mm. Drawing by Gunther Köhler.

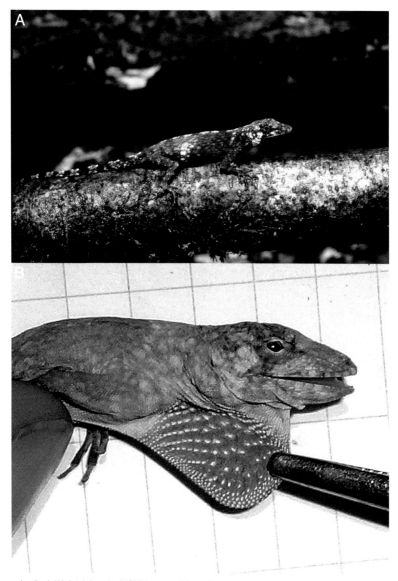

Figure 6.  *Norops beckeri.* (A) Adult female (USNM 570085) from Caño Awalwás, Gracias a Dios; (B) adult male dewlap (museum number not available) from S of Las Choapas, Veracruz, Mexico. Photographs by James R. McCranie (A) and Steven Poe (B).

USNM 344805. SANTA BÁRBARA: San José de Colinas, SMF 91288. YORO: Los Guares, USNM 580300.

*Norops bicaorum* Köhler
*Anolis lemurinus*: Meyer and Wilson, 1973: 17 (in part); Wilson and Hahn, 1973: 110 (in part).

*Norops lemurinus*: Wilson and Cruz Díaz, 1993: 17 (in part); Wilson et al., 2001: 136 (in part).
*Norops bicaorum* Köhler, 1996b: 21 (in part) (holotype, SMF 77100; type locality: "Honduras, Islas de la Bahia, Isla de Utila, on the trail to Rock Harbour, about 3 km north of the town of Utila

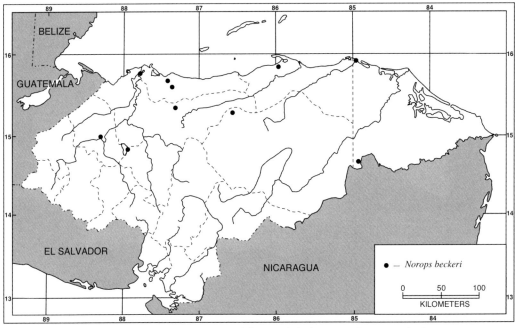

Map 3.   Localities for *Norops beckeri*. Solid symbols denote specimens examined.

[16°6.34′N; 86°53.94′W"]), 1998a: 47, 1998b: 374, 1999a: 49 (in part), 2000: 68, 2003: 96, 2008: 103; Monzel, 1998: 159 (in part), 2001: 25; Tiedemann and Grillitsch, 1999: 153; Lundberg, 2000: 3; Köhler and McCranie, 2001: 244; Köhler et al., 2001: 254; van Beest and Hartman, 2003: 3; Powell, 2003: 37; Wilson and McCranie, 2003: 59; McCranie et al., 2005: 102; Wilson and Townsend, 2006: 105; Klütsch et al., 2007: 1125 (in part); Diener, 2008: 12; Kramer, 2010: 7; Nicholson et al., 2012: 12; McCranie and Valdés Orellana, 2014: 45.

*Anolis bicaorum*: Nicholson et al., 2005: 933; Köhler et al., 2007: 390; Pyron et al., 2013: fig. 19.

*Geographic Distribution. Norops bicaorum* is known to occur only at low elevations on the eastern portion of Isla de Utila, Islas de la Bahía, Honduras.

*Description.* The following is based on 10 males (LSUMZ 22307; SMF 77100–01, 77107, 77560, 77984, 79993, 79995, 81127;

USNM 565426) and 10 females (FMNH 283602, SMF 77104, 77559, 78362, 79260, 79992, 79994, 80805–06; USNM 565427). *Norops bicaorum* is a moderately large anole (SVL 73 mm in largest male [USNM 565426] and 86 mm in largest female [SMF 77559]); dorsal head scales keeled in internasal region, smooth or tuberculate in prefrontal, frontal, and parietal areas; weak to moderate frontal and parietal depressions present; 5–9 (7.2 ± 1.3) postrostrals; anterior nasal divided, lower section contacting rostral and first supralabial; 6–10 (8.0 ± 1.4) internasals; canthal ridge sharply defined; scales comprising supraorbital semicircles keeled, largest scale in semicircles about same size as, or larger than, largest supraocular scale; supraorbital semicircles well defined; 1–2 (1.4 ± 0.5) scales separating supraorbital semicircles at narrowest point; 2–5 (3.3 ± 0.8) scales separating supraorbital semicircles and interparietal at narrowest point; interparietal well defined, greatly enlarged relative to adjacent scales, surrounded by scales of moderate size, longer than wide, usually slightly smaller than ear opening; 2–3 rows of about 5–8 (total number) enlarged, keeled supraocular scales; enlarged supraoculars completely separated from supraorbital semicircles by 1–2 rows of small scales; 2 elongate, overlapping superciliaries, posterior much shorter than anterior; 3–6 (4.2 ± 1.2) enlarged canthals; 8–18 (11.3 ± 2.9) scales between second canthals; 8–15 (11.0 ± 1.9) scales between posterior canthals; loreal region slightly concave, 42–79 (54.9 ± 10.1) mostly strongly keeled (some smooth or rugose) loreal scales in maximum of 6–9 (7.5 ± 0.8) horizontal rows; 6–8 (7.3 ± 0.7) supralabials and 6–9 (7.5 ± 0.8) infralabials to level below center of eye; suboculars strongly keeled, usually separated from supralabials by 1 row of scales; ear opening vertically oval; scales anterior to ear opening granular, similar in size to those posterior to ear opening; 6–8 (6.1 ± 0.5) postmentals, outer pair largest; keeled granular scales present on chin and throat; male dewlap large, extending to beyond level of chest; male dewlap with 7–11 horizontal gorgetal-sternal scale rows, about 9–11 scales per row (*n* = 6); 2–3 (modal number) anterior marginal scales in male dewlap; female dewlap small or absent; nuchal crest and dorsal ridge present in males; about 2 middorsal scale rows slightly enlarged, strongly keeled, dorsal scales grading into smaller granular lateral scales; no enlarged scales scattered among laterals; 64–93 (80.0 ± 11.2) dorsal scales along vertebral midline between levels of axilla and groin in males, 72–110 (92.2 ± 11.1) in females; 36–66 (45.3 ± 8.8) dorsal scales along vertebral midline contained in 1 head length in males, 38–64 (46.0 ± 7.2) in females; ventral scales on midsection much larger than largest dorsal scales; ventral body scales keeled, mucronate, imbricate; 44–60 (54.8 ± 6.1) ventral scales along midventral line between levels of axilla and groin in males, 44–59 (50.7 ± 5.1) in females; 28–42 (36.1 ± 5.4) ventral scales contained in 1 head length in males, 23–36 (28.8 ± 3.8) in females; 136–172 (149.6 ± 10.1) scales around midbody

in males, 115–172 (150.3 ± 15.4) in females; tubelike axillary pocket absent, although a shallow, scaled axillary pocket present; precloacal scales weakly to strongly keeled; no enlarged postcloacal scales in males; tail slightly compressed, TH/TW 1.17–1.50; all subcaudal scales keeled; lateral caudal scales keeled, homogeneous, although indistinct division in segments discernable; dorsal medial caudal scale row enlarged, keeled, not forming crest; scales on anterior surface of antebrachium distinctly keeled, unicarinate; 23–31 (27.9 ± 2.0) subdigital lamellae on Phalanges II–IV of Toe IV of hind limbs; 7–10 (8.5 ± 1.1) subdigital scales on Phalanx I of Toe IV of hind limbs; SVL 47.0–73.0 (64.4 ± 7.7) mm in males, 61.8–86.0 (66.2 ± 7.1) mm in females; TAL/SVL 1.66–2.27 in six males, 1.25–2.26 in eight females; HL/SVL 0.22–0.27 in males, 0.21–0.25 in females; SHL/SVL 0.28–0.31 in males, 0.23–0.31 in females; SHL/HL 1.11–1.31 in males, 1.06–1.46 in females; longest toe of adpressed hind limb usually reaching between posterior and anterior borders of eye.

Color in life of an adult male (SMF 79364): dorsal surfaces of body and limbs Clay Color (26) with Olive Brown (28) to Sepia (119) blotches; dewlap Flame Scarlet (15) with suffusion of black pigment centrally and white gorgetals. Color in life of another adult male (USNM Herp Image 2722, specimen lost): dorsal surfaces of body Smoke Gray (44) with dorsal crossbars that are Dark Grayish Brown (20) centrally and darker gray than ground color on outer edges; dorsum of head Smoke Gray with Dark Grayish Brown crossbars and markings and lines radiating outward from eyes; lateral body stripe dark brown, but paler than Smoke Gray ground color; dorsal surfaces of limbs Smoke Gray with Dark Grayish Brown crossbands; dewlap Brick Red (132A).

Color in alcohol: dorsal surfaces of head, body, and tail some shade of brown with a variety of possible dorsal markings (e.g., broad dark brown transverse bands, diamonds, pale vertebral stripe) or patternless;

dorsal surfaces of hind limbs with oblique dark crossbands; tail with faint to distinct dark brown crossbands; distinct dark brown lines radiate outward from eye; distinct dark brown interorbital bar present; dark brown lyriform nuchal mark frequently present.

Hemipenis: the completely everted hemipenis of SMF 81127 is a medium-sized, bilobed organ; sulcus spermaticus bordered by well-developed sulcal lips, bifurcating at base of apex; branches opening into broad concave area distal to point of bifurcation on each lobe; low asulcate ridge present; lobes strongly calyculate; truncus with transverse folds.

*Diagnosis/Similar Species. Norops bicaorum* is distinguished from all other Honduran *Norops*, except *N. lemurinus* and *N. roatanensis*, by the combination of a dark brown lyriform mark in the nuchal region (usually present, obscure or absent in some specimens), long hind legs (longest toe of adpressed hind limb usually reaching between posterior and anterior borders of eye), keeled, mucronate, imbricate ventral scales, about two dorsal scale rows slightly enlarged, a shallow and scaled axillary pocket, and no enlarged postcloacals in males. *Norops bicaorum* differs from *N. lemurinus* by having an orange-red male dewlap in life with suffusion of black pigment centrally and white gorgetal scales, the sulcal branches of the hemipenis opening into broad concave area distal to point of bifurcation on each lobe, and a low asulcate ridge present (male dewlap red to red-orange in life, without suffusion of black pigment, often with black or dark brown edged gorgetal scales, sulcal branches continuing to tips of lobes, and asulcate ridge absent in *N. lemurinus*). *Norops bicaorum* is most similar to *N. roatanensis* from which it differs by male dewlap coloration (orange-red in *N. bicaorum* versus pink-red in *N. roatanensis*), body size (males average about 64 mm SVL, females about 66 mm SVL in *N. bicaorum* versus males average about 56 mm SVL, females about 58 mm in *N. roatanensis*), and having sulcal branches opening into broad concave area and low asulcate processus present (sulcate branches continuing to tip of each lobe and asulcate processus absent in *N. roatanensis*).

*Illustrations* (Figs. 7, 8, 116). Köhler, 1996b (adult, head scales), 1998a (adult), 2000 (adult, head and dewlap), 2003 (adult, head and dewlap), 2008 (adult, head and dewlap); Köhler and McCranie, 2001 (head and dewlap); Monzel, 2001 (adult, head and dewlap); van Beest and Hartman, 2003 (adult, juvenile); Powell, 2003 (adult; as *Anolis*); McCranie et al., 2005 (adult, dewlap, chin scales); Diener, 2008 (adult, dewlap); Kramer, 2010 (adult, juvenile, male dewlap; as *Anolis*); Hallmen, 2011 (head and dewlap; as *Anolis*).

*Remarks. Norops bicaorum* is a member of the *N. lemurinus* species subgroup (Köhler, 1996b, Köhler and McCranie, 2001; also see Remarks for *N. lemurinus*) of the *N. auratus* species group of Nicholson et al. (2012). The other Honduran species in this subgroup are *N. lemurinus* and *N. roatanensis*. Members of this subgroup are characterized by having distinct dark lines radiating outward from the eye, a red male dewlap in life, strongly keeled midventral scales, a shallow and scaled axillary pocket present, and usually having a distinct dark-colored lyriform nuchal mark. The recovered phylogenetic analyses in Nicholson et al. (2005, 2012) showed *N. bicaorum* and *N. lemurinus* to be sister to a clade containing *N. limifrons* and *N. zeus*.

*Natural History Comments. Norops bicaorum* is known from near sea level to 20 m elevation in the Lowland Moist Forest formation. The type series was collected in the months of March, April, August, and September. Numerous other individuals were active on overcast and sunny days during October and September. Active individuals are usually on tree trunks about 0.5 to 3 m above the ground, but it was also seen on large limestone rocks or on the ground in forested areas. Specimens on the ground when first observed quickly run to the nearest tree and climbed the trunk

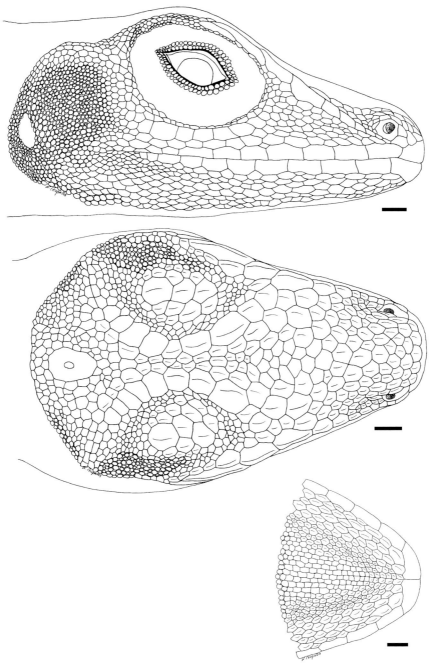

Figure 7. *Norops bicaorum* head in lateral, dorsal, and ventral views. SMF 77101, adult male from trail to Rock Harbour, Isla de Utila, Islas de la Bahía. Scale bar = 1.0 mm. Drawing by Lara Czupalla.

Figure 8.  *Norops bicaorum* (not collected). (A) Adult male; (B) adult male dewlap, both from trail N of Utila Town, Isla de Utila, Islas de la Bahía. Photographs by James R. McCranie.

when pursued. It also sleeps on low vegetation at night. This species was extremely abundant in forested areas during the 1980s, and remains common in those areas, but seemingly at lower levels than before. A successful captive breeding of wild caught individuals was reported by van Beest and Hartman (2003). Those captives were capable of reproduction year round with females burying a single egg in the terrarium substrate about every 15 days. Captives were fed "meadow plankton" and "home bred insects" (van Beest and Hartman, 2003: 9).

*Etymology.* The name *bicaorum* is derived from BICA (the initials of the Bay Islands Conservation Association) and the Latin suffix *-orum* (belonging to).

*Specimens Examined* (47 [12]; Map 4). ISLAS DE LA BAHÍA: Isla de Utila, Jakes Bight, SMF 77559, 77984; Isla de Utila, Jericó, USNM 565426; Isla de Utila, trail to Rock Harbor, NMW 35502 (2), SMF 77100–02, 77104–05, 77107, 77559–60,

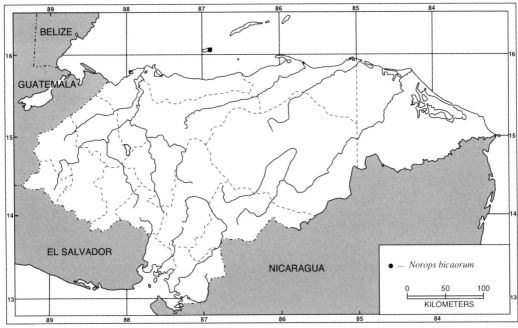

Map 4.  Localities for *Norops bicaorum*. The solid symbol denotes specimens examined.

77562, 78362, 79260, UNAH (4 unnumbered), ZFMK 63553–54; Isla de Utila, trail to Rock Harbour 1–2 km N of Utila town, SMF 79991–95, 80805–07, 81127, 82553, 83020; Isla de Utila, 2 km N of Utila, USNM 565427; Isla de Utila, Utila, LSUMZ 22272, 22295, 22305–08, UF 28396, 28404–05, 28441–43; Isla de Utila, near Utila, FMNH 283602.

*Other Records.* ISLAS DE LA BAHÍA: Isla de Utila, trail N of Utila (USNM Herp Image 2722, but specimen lost by AFE COHDEFOR employee).

*Norops biporcatus* (Wiegmann)
*Dactyloa biporcata* Wiegmann, 1834: 47 (neotype, MNHN 2426 [see Köhler and Bauer, 2001: 124; Anonymous, 2002: 230, and Remarks]; type locality: "Santa Rosa de Pansos [Guatemala]" by neotype selection).
*Anolis biporcatus*: Werner, 1896: 346; Meyer, 1969: 213; Meyer and Wilson, 1973: 15; Lieb, 1981: 295; O'Shea, 1986: 36, 1989: 16; Franklin and Franklin,

1999a: 109; Poe, 2004: 62; Wilson and Townsend, 2007: 145; Gutsche, 2012: 69.
*Anolis copei*: Dunn and Emlen, 1932: 27 (in part).
*Anolis lemurinus*: Meyer and Wilson, 1973: 15 (in part).
*Norops biporcatus*: Savage, 1973: 11; Espinal et al., 2001: 106; Wilson et al., 2001: 135; Castañeda, 2002: 41; McCranie et al., 2002a: 25, 2006: 118; Köhler and Vesely, 2003: 231; McCranie, 2005: 20; McCranie and Castañeda, 2005: 14; Castañeda and Marineros, 2006: 3–8; Wilson and Townsend, 2006: 105; McCranie and Solís, 2013: 242.
*Anolis* (*Norops*) *biporcatus*: McCranie, 2007a: 177.

*Geographic Distribution. Norops biporcatus* occurs at low and moderate elevations on the Atlantic versant from southern Veracruz, Mexico, to western Venezuela and on the Pacific versant from northwestern Costa Rica to northwestern Ecuador, including Isla Gorgona, Colombia. The species also occurs marginally on the Pacific versant in western Nicaragua (e.g., on Volcán Mombacho). In Honduras, this species occurs throughout much of the northern half of the country, with some isolated populations in the central and south-central portions of the country.

*Description.* The following is based on 11 males (SMF 79208, 80896, 86995, 86998–99; USNM 563099–100, 563108–09, 563111, 563118) and 17 females (AMNH 46981; SMF 79146–47, 79207, 80895–97; USNM 563101–02, 563107, 563112–15, 563117, 563119–20). *Norops biporcatus* is a large anole (SVL 95 mm in largest Honduran male [SMF 86998] examined, 100 mm in largest Honduran female [SMF 79146] examined; maximum reported SVL 105 mm [Henderson, 1972]); dorsal head scales strongly keeled in internasal and prefrontal regions, rugose in frontal and parietal areas; deep frontal depression present; parietal depression well developed; 5–9 (6.6 ± 1.1) postrostrals; anterior nasal divided, lower scale usually contacting first supralabial, but separated from rostral; 5–10 (7.1 ± 1.2) internasals; canthal ridge sharply defined; scales comprising supraorbital semicircles rugose, largest scale in semicircles larger than largest supraocular scale; supraorbital semicircles well defined; 2–5 (2.8 ± 0.7) scales separating supraorbital semicircles at narrowest point; 2–6 (3.8 ± 0.8) scales separating supraorbital semicircles and interparietal at narrowest point; interparietal well defined, irregular in outline, longer than wide, smaller than ear opening; 2–3 rows of about 2–8 (total number) slightly enlarged, usually keeled supraocular scales; slightly enlarged supraoculars usually completely separated from supraorbital semicircles by small supraoculars; 3–4 elongate superciliaries, posteriormost shortest; 2–5 (3.1 ± 0.7) enlarged canthals; 7–14 (9.2 ± 1.8) scales between second canthals; 8–14 (9.7 ± 1.4) scales between posterior canthals; loreal region slightly concave, 42–62 (51.8 ± 5.7) rugose to mostly smooth loreal scales in maximum of 6–9 (7.5 ± 0.8) horizontal rows; 9–13

(10.8 ± 1.0) supralabials and 9–13 (10.9 ± 1.0) infralabials to level below center of eye; suboculars distinctly keeled, usually separated from supralabials by 1 row of scales (occasionally 1 subocular in contact with supralabials); ear opening vertically oval; scales anterior to ear opening about same size as, to slightly larger than, those posterior to ear opening; 6–8 (6.3 ± 0.7) postmentals, outer pair largest; gular scales keeled; male dewlap small, extending to about level of axilla; male dewlap with about 7 horizontal gorgetal-sternal scale rows, with about 7–13 scales per row (*n* = 3); 3–4 (modal number) anterior marginal pairs in male dewlap; female dewlap absent or rudimentary; low nuchal crest present in males, low dorsal ridge present, especially in males; about 1–4 middorsal scale rows slightly enlarged, faintly keeled, dorsal scales lateral to middorsal series grading into granular lateral scales; no enlarged scales among laterals; 54–90 (74.5 ± 14.5) dorsal scales along vertebral midline between levels of axilla and groin in males, 55–90 (77.1 ± 9.9) in females; 36–52 (45.5 ± 4.8) dorsal scales along vertebral midline contained in 1 head length in males, 38–51 (43.1 ± 4.0) in females; ventral scales on midsection much larger than largest dorsal scales; ventral body scales strongly keeled, some mucronate, some imbricate; 34–54 (45.7 ± 7.0) ventral scales along midventral line between levels of axilla and groin in males, 35–56 (47.0 ± 5.4) in females; 27–34 (30.2 ± 2.7) ventral scales contained in 1 head length in males, 24–36 (29.3 ± 3.1) in females; 76–120 (99.6 ± 14.4) scales around midbody in males, 84–118 (103.4 ± 10.1) in females; tubelike axillary pocket absent; precloacal scales rugose to faintly keeled; enlarged postcloacal scales absent in males; tail slightly to distinctly compressed, TH/TW 1.13–1.65 in 25; basal subcaudal scales keeled; lateral caudal scales keeled, homogeneous, although indistinct division in segments discernable; dorsal medial caudal scale row enlarged anteriorly, keeled, forming low crest; most scales on anterior surface of antebrachium unicarinate; 30–38 (34.0 ±

2.2) subdigital lamellae on Phalanges II–IV of Toe IV of hind limbs; 9–12 (10.4 ± 1.0) subdigital scales on Phalanx I of Toe IV of hind limbs; SVL 52.0–95.0 (85.4 ± 11.8) mm in males, 83.0–100.0 (92.0 ± 4.3) mm in females; TAL/SVL 1.77–2.41 in ten males, 1.76–2.33 in 16 females; HL/SVL 0.25–0.30 in males, 0.24–0.29 in females; SHL/SVL 0.22–0.27 in males, 0.20–0.25 in females; SHL/HL 0.77–1.05 in males, 0.80–1.05 in females; longest toe of adpressed hind limb usually reaching about ear opening.

Color in life of an adult male (USNM 563109): dorsal surfaces Parrot Green (60); ventral surface of head pale green; belly cream with green and brown mottling; dewlap bluish gray centrally with reddish brown outer edge; dewlap pinkish gray between bluish gray and reddish brown areas; dewlap also pinkish gray basally; iris copper with gold rim. Color in life of an adult female (USNM 563102): dorsum Lime Green (159) with dark brown-edged Sky Blue (66) spots along dorsal midline of body and on lateral portion of body; limbs Lime Green above; venter white with pale brown streaking and spotting; dorsal and lateral portions of head Lime Green with scattered dark brown streaking and spotting; chin pale yellowish green; dewlap white with pale brown streaking; iris Cinnamon-Rufous (40).

Color in alcohol: dorsal surfaces of head and body dark brown, with pale brown mottling on lateral surface of body; dorsal surfaces of limbs dark brown with pale brown cross-reticulations; dorsal surface of tail dark brown with indistinct pale brown crossbars; ventral surface of head cream to pale brown, with varying amounts of dark brown reticulations, dark streaks in throat region sometimes present; ventral surface of body cream with varying amounts of darker brown flecking, mottling, and/or reticulations; subcaudal surface brown with darker brown crossbars.

Hemipenis: the almost completely everted hemipenis of SMF 79208 is a medium-sized, bilobed organ; sulcus spermaticus bordered by well-developed sulcal lips, bifurcating at base of apex, branches continuing to tips of lobes; no asulcate processus; asulcate surface of apex strongly calyculate; truncus with distinct transverse folds.

*Diagnosis/Similar Species. Norops biporcatus* is distinguished from all other Honduran *Norops*, except *N. petersii*, by the combination of its large size (males to 95 mm SVL, females to 100 mm SVL), short hind legs (longest toe of adpressed hind limb usually reaching to about ear opening), and 9–13 supralabials to level below center of eye. *Norops biporcatus* differs from *N. petersii* by having the anterior dorsal head scales strongly keeled (rugose or weakly keeled in *N. petersii*), the midventral scales strongly keeled and mucronate (only weakly keeled with rounded or truncated posterior margins in *N. petersii*), green coloration in life (brown or olive in *N. petersii*), 34–56 midventrals between levels of axilla and groin (68–86 in *N. petersii*), and 76–120 scales around midbody (116–133 in *N. petersii*). *Norops biporcatus* differs from *N. loveridgei*, another large anole, by having a green coloration in life (some shade of brown in *N. loveridgei*), usually fewer than 60 loreal scales (more than 60 in *N. loveridgei*), fewer than 130 scales around midbody (usually more than 130 in *N. loveridgei*), and by having shorter legs with the longest toe of the adpressed hind limb usually reaching about ear opening (usually reaching between posterior and anterior borders of eye in *N. loveridgei*).

*Illustrations* (Figs. 9, 10, 79, 90). Williams, 1966 (head; as *Anolis b. biporcatus*), 1970 (nasal-rostral and superciliary regions; as *Anolis b. biporcatus*); Myers, 1971a (adult; as *Anolis*); Álvarez del Toro, 1983 (adult; as *Anolis*); Stafford, 1991 (adult; as *Anolis*); Fläschendräger and Wijffels, 1996 (adult; as *A. b. biporcatus*); Lee, 1996 (adult, juvenile; as *Anolis*), 2000 (adult); Campbell, 1998 (adult); Stafford and Meyer, 1999 (adult); Köhler, 2000 (adult, ventral scales, head and male dewlap), 2001b (ventral scales, head and dewlap), 2003 (adult,

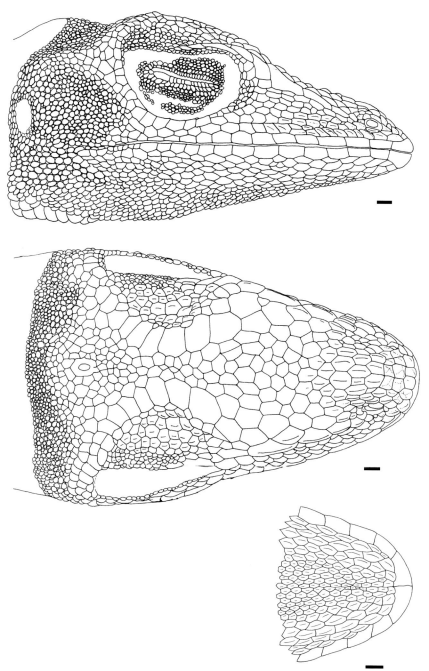

Figure 9.    *Norops biporcatus* head in lateral, dorsal, and ventral views. SMF 79208, adult male from Pataste, Olancho. Scale bar = 1.0 mm. Drawing by Gunther Köhler.

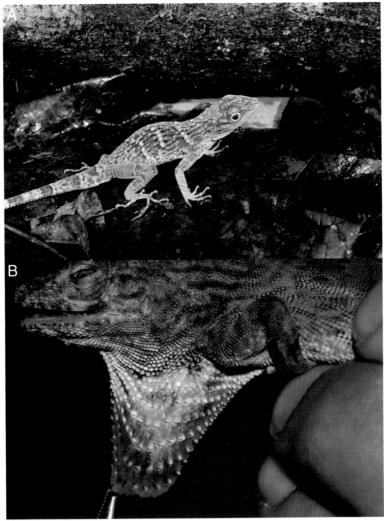

Figure 10.    *Norops biporcatus*. (A) Adult male (USNM 563111) from San San Hil Kiamp, Gracias a Dios; (B) adult male dewlap (FMNH 282555) from San Isidro, Copán. Photographs by James R. McCranie.

ventral scales, head and dewlap), 2008 (adult, ventral scales, head and dewlap), 2014 (ventral, caudal, postmental, postrostral, internasal, and supraocular scales, prefrontal and parietal depressions, cloacal region, tail tip, terminal phalanx on fourth toe; as *Anolis*); McCranie et al., 2002a (adult), 2006 (adult); Savage, 2002 (adult, male dewlap); Köhler and Vesely, 2003 (head scales); Guyer and Donnelly, 2005 (adult); McCranie, 2007a (adult; as *Anolis*);

Calderón-Mandujano et al., 2008 (adult; as *Anolis*).

*Remarks*. Köhler and Bauer (2001) discovered that the holotype (ZMB 524) of *Dactyloa biporcata* is conspecific with *N. petersii* (Bocourt). Köhler and Bauer (2001) also noted that the holotype of *A. copei* Bocourt (1873, *In* Duméril et al., 1870–1909) is conspecific with the species of large green anole occurring from southeastern Mexico to northwestern South America that has been consistently called *Anolis* (or *Norops*) *biporcatus* since the mid 1940s. To conserve this usage, Köhler and Bauer (2001: 124) petitioned The International Commission on Zoological Nomenclature "to use its plenary power to set aside all previous type fixations for *Dactyloa biporcata* Wiegmann, 1834, and to designate the holotype of *Anolis copei* Bocourt, 1873 (MNHN 2426), as the neotype" of *Dactyloa biporcata* Wiegmann. The International Commission on Zoological Nomenclature (Anonymous, 2002) ruled in favor of this designation.

*Norops biporcatus* is usually placed in the *N. biporcatus* species group (Nicholson, 2002; one exception is that Williams, 1976b placed it in a *N. petersi* [sic] species group). Other Honduran species sometimes placed in that group are *N. capito*, *N. loveridgei*, and *N. petersii*. However, phylogenetic analyses based on molecular data, including that of *N. capito*, *N. biporcatus*, and *N. loveridgei* (*N. petersii* not included) do not support this group of species as forming a monophyletic lineage (Nicholson, 2002; Nicholson et al. 2005, 2012; also see Poe, 2004, for a mostly morphological analysis). *Norops biporcatus* appears to be morphologically most similar to *N. petersii*. Poe et al. (2009) did recover a sister group relationship for *N. biporcatus* and *N. petersii*. Therefore, we include *N. biporcatus* and *N. petersii* in a *N. biporcatus* species subgroup within the *N. auratus* species group of Nicholson et al. (2012). Those two species share a large size, relatively short hind limbs, and numerous supralabials.

Franklin and Franklin (1999a) reported *N. biporcatus* as a new record for the department of Gracias a Dios. However, O'Shea (1989) had earlier reported this species from that department.

*Natural History Comments. Norops biporcatus* is known from near sea level to 1050 m elevation in the Lowland Moist Forest and Premontane Wet Forest formations and peripherally in the Premontane Moist Forest formation. This species is active on tree trunks and limbs from near the ground to the canopy level. It sleeps at night on small tree branches and leaves of vegetation overhanging streams and rivers. *Norops biporcatus* has been found from January to November, thus it is active throughout the year, but seems to be more visible during rainy weather. Specimens occasionally emit a squeak or squeaks when first captured. This species appears to have excellent vision as it is sometimes seen from a distance moving to the opposite side of a tree trunk in an effort to avoid detection. Individuals of *N. biporcatus* are deliberate in their movements and are easily captured when in reach of the collector. McCoy (1975; Guatemala) documented that females lay one egg per clutch, with multiple clutches per year. Villarreal Benítez (1997) reported reproduction occurred during the dry season in Veracruz, Mexico. Savage (2002 and references cited therein) documented that Costa Rican individuals feed on a wide variety of arthropods, with juveniles feeding mostly on beetles and adults largely on ants; there are also records of smaller anoles in their diet. Villarreal Benítez (1997) reported that a population in Veracruz, Mexico, feeds on various invertebrates such as arachnids and coleopterans and that some stomachs contained small fruits.

*Etymology.* The name *biporcatus* is formed from the Latin *bi-* (two, twice), *porca* (ridge), and *-atus* (provided with, having the nature of, pertaining to). Combined, the name means having two ridges, probably referring to "Rostri porcis duabus" (Wiegmann, 1834: 47) meaning "snout with two ridges," possibly in reference to the canthal ridges extending onto the snout in the type series.

*Specimens Examined* (67 [6] + 3 skeletons, 3 eggs; Map 5). ATLÁNTIDA: mountains S of Corozal, LSUMZ 21367–68; Estación Forestal CURLA, USNM 563099–100; Lancetilla, AMNH 46981, MCZ R-32207, 34387, SMF 79207. COLÓN: Quebrada Machín, USNM 563101; Río Guaraska, BMNH 1985.1100; Río Paulaya, BMNH 1985.1101; 5 km SW of Trujillo, LACM

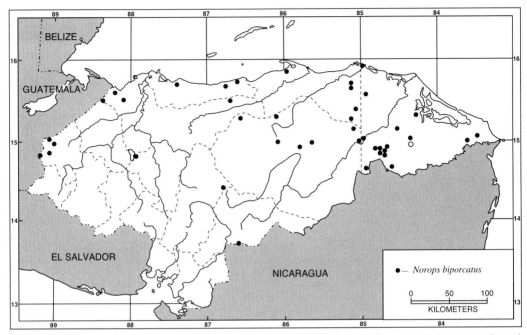

Map 5.   Localities for *Norops biporcatus*. Solid symbols denote specimens examined and the open symbol an accepted record.

47290. COPÁN: Copán, FMNH 28534; Laguna del Cerro, SMF 79147; Río Amarillo, SMF 86999, 91733; San Isidro, FMNH 282555. CORTÉS: Los Pinos, USNM 573122; Quebrada Agua Buena, SMF 79146; W of San Pedro Sula, LACM 135072, MCZ R-32205; Santa Teresa, SMF 91246. EL PARAÍSO: Las Manos, UTA R-41246. FRANCISCO MORAZÁN: El Chile, SMF 80895–96. GRACIAS A DIOS: Bodega de Río Tapalwás, SMF 86997, UF 144614, USNM 563112–14; Caño Awalwás, USNM 563102; Cerro Wampú, USNM 563118; Hiltara Kiamp, USNM 563103; Kasunta, SMF 86995; Kipla Tingni Kiamp, USNM 563104; Krahkra, USNM 563105; about 15 km S of Mocorón, UTA R-52145; Palacios, BMNH 1985.1102; Raudal Kiplatara, SMF 86011; Rawa Kiamp, SMF 91247; Rus Rus, UF 144612, USNM 563106–09, 563115; San San Hil, SMF 86996; San San Hil Kiamp, USNM 563110–11; Swabila, UF 144613; Urus Tingni Kiamp, USNM 563116; Warunta, SMF 91734; Warunta Tingni Kiamp, USNM 563117. OLANCHO: Cuaca, UTA R-53661,

53663; El Aguacatal, UTA R-53662; Pataste, SMF 79208; Quebrada Las Cantinas, USNM 342273; confluence of Quebrada Siksatara and Río Wampú, USNM 563120; between mouth of Río Wampú and Yapuwás, USNM 563119; Talgua Arriba, SMF 91735. YORO: La Libertad, SMF 86998. "HONDURAS": UF 53678 (skeleton, egg), 55467–68 (both skeletons, eggs), 90376–77, 91259, 123366, UMMZ 58379 (carries the unknown locality "Loma de Río Palo").

*Other Records* (Map 5). GRACIAS A DIOS: about 30 km S of Mocorón (Franklin and Franklin, 1999a). "HONDURAS": (Werner, 1896).

*Norops capito* (Peters)

*Anolis* (*Draconura*) *capito* Peters, 1863: 142 (two syntypes, ZMB 4684, 36298 [see Bauer et al., 1995: 58]; type locality: "Costa Rica").

*Anolis capito*: Werner, 1896: 346; Meyer, 1969: 216; Meyer and Wilson, 1973: 16; Cruz Diaz, 1978: 25; O'Shea, 1986: 36, 1989: 16; Franklin and Franklin, 1999b:

109; Townsend, 2006: 35; Townsend et al., 2006: 32; Townsend and Wilson, 2008: 150; Gutsche, 2012: 70; Köhler, 2014: 210.

*Anolis copei*: Dunn and Emlen, 1932: 27 (in part).

*Norops capito*: Savage, 1973: 11; Espinal, 1993: table 3; Köhler et al., 2000: 425, 2005a: 130; Nicholson et al., 2000: 30; Espinal et al., 2001: 106; McCranie and Köhler, 2001: 232; Wilson et al., 2001: 136; Castañeda, 2002: 15; McCranie et al., 2002a: 27, 2006: 119; McCranie, 2005: 20; Castañeda, 2006: 29; Castañeda and Marineros, 2006: 3–8; Wilson and Townsend, 2006: 105.

*Geographic Distribution. Norops capito* occurs at low and moderate elevations on the Atlantic versant from Tabasco, Mexico, to eastern Panama and from southeastern Costa Rica to central Panama on the Pacific versant. It is also found marginally on the Pacific versant in northwestern Costa Rica. In Honduras, this species is known from across most of the northern portion of the country.

*Description.* The following is based on 15 males (SMF 79094, 80822, 86969, 86972–73; UF 142454, 144616; USNM 563125, 563130–32, 563146, 563152, 563155, 563167) and 13 females (SMF 79093, 86968, 86970; UF 144617–19, USNM 563122, 563124, 563138, 563157, 563159, 563172, 563174). *Norops capito* is a moderately-large anole (SVL 83 mm in largest Honduran male [UF 142454] examined, 100 mm in largest Honduran female examined [SMF 91248, not included in the following description]); dorsal head scales rugose to keeled in internasal, prefrontal, frontal, and parietal areas; deep frontal depression present; parietal depression shallow; 5–9 (7.0 ± 1.0) postrostrals; anterior nasal divided, lower scale contacting rostral or separated from rostral by single scale, lower scale usually separated from first supralabial; 6–10 (8.4 ± 1.1) internasals; canthal ridge sharply defined; scales comprising supraorbital semicircles keeled, largest scale in semicircles larger than largest supraocular scale; supraorbital

semicircles well defined; 2–3 (2.6 ± 0.5) scales separating supraorbital semicircles at narrowest point; 1–4 (2.9 ± 0.7) scales separating supraorbital semicircles and interparietal at narrowest point; interparietal well defined, irregular in outline, slightly longer than wide, smaller than ear opening; 2–3 rows of about 5–9 (total number) enlarged, keeled supraocular scales; enlarged supraoculars varying from completely separated from supraorbital semicircles by 1 row of small scales to 1–2 enlarged supraoculars in broad contact with supraorbital semicircles; 2–3 elongate superciliaries, posteriormost shortest; usually 2 (occasionally 3 or 4) enlarged canthals; 7–15 (10.3 ± 1.6) scales between second canthals; 9–13 (10.7 ± 1.2) scales between posterior canthals; loreal region slightly concave, 27–54 (40.2 ± 6.6) mostly keeled (some smooth or rugose) loreal scales in a maximum of 4–9 (6.7 ± 1.0) horizontal rows; 8–11 (9.6 ± 0.9) supralabials and 10–13 (11.4 ± 0.9) infralabials to level below center of eye; suboculars keeled, separated from supralabials by 1 row of scales; ear opening vertically oval; scales anterior to ear opening distinctly larger than those posterior to ear opening; 4–6 (5.0 ± 0.8) postmentals, outer pair largest; gular scales not keeled; male dewlap small, extending to about level of anterior portion of forelimb insertion; male dewlap with 7–10 horizontal gorgetal-sternal scale rows, about 7–13 scales per row ($n = 7$); 2–4 (modal number) anterior marginal pairs in male dewlap; female dewlap absent or rudimentary; no nuchal crest or dorsal ridge; about 8–14 middorsal scale rows slightly enlarged, smooth, juxtaposed, most pentagonal or hexagonal, dorsal scales lateral to middorsal series grading into granular lateral scales; no enlarged scales among laterals; 56–76 (66.9 ± 6.2) dorsal scales along vertebral midline between levels of axilla and groin in males, 57–74 (67.4 ± 5.8) in females; 16–45 (33.7 ± 9.8) dorsal scales along vertebral midline contained in 1 head length in males, 17–40 (33.7 ± 6.6) in females; ventral scales on midsection much larger than largest dorsal scales; ventral body scales keeled, most imbricate; 49–65 (56.5 ± 4.0) ventral scales along midventral line between levels of axilla and groin in males, 46–60 (53.5 ± 4.0) in females; 14–36 (28.9 ± 7.4) ventral scales contained in 1 head length in males, 16–35 (27.8 ± 5.8) in females; 96–120 (107.0 ± 6.7) scales around midbody in males, 100–114 (107.6 ± 4.9) in females; tubelike axillary pocket absent; precloacal scales rugose to faintly keeled; enlarged postcloacal scales absent in males; tail oval to slightly compressed, TH/TW 1.05–1.44; basal subcaudal scales keeled; lateral caudal scales keeled, homogeneous; dorsal medial caudal scale row not enlarged, keeled, not forming crest; most scales on anterior surface of antebrachium rugose to unicarinate; 23–34 (26.6 ± 2.1) subdigital lamellae on Phalanges II–IV of Toe IV of hind limbs; 8–14 (11.6 ± 1.4) subdigital scales on Phalanx I of Toe IV of hind limbs; SVL 66.0–83.4 (77.7 ± 4.5) mm in males, 67.0–89.1 (81.5 ± 6.5) mm in females; TAL/SVL 1.79–2.08 in males, 1.58–1.84 in 12 females; HL/SVL 0.22–0.26 in males, 0.23–0.27 in females; SHL/SVL 0.31–0.36 in males, 0.30–0.33 in females; SHL/HL 1.24–1.53 in males, 1.14–1.41 in females; longest toe of adpressed hind limb usually reaching beyond tip of snout.

Color in life of an adult male (USNM 342275): dorsal surface of head Yellowish Olive-Green (50) with dark brown markings along lip; dorsal surface of body Yellowish Olive-Green with lichenose dark brown marbling in form of vague crossbands; dorsal surfaces of limbs Yellowish Olive-Green with indistinct darker crossbars; belly pale olive-yellow; dewlap Buff-Yellow (53) with white scales; iris copper. An adult female (USNM 563168) was as follows: dorsum Dark Brownish Olive (129) laterally with darker brown and yellowish tan stripes; broad middorsal stripe dirty Robin Rufous (340), outlined below by black anteriorly, middorsal stripe extending onto tail; dorsal surfaces of limbs Dark Brownish Olive with darker brown and yellowish tan crossbars; belly and subcaudal surface yellowish brown with Dark Brownish Olive stripes; chin yellowish brown with Dark Brownish Olive

blotches and mottling; dewlap Cinnamon (39) with dark gray scales; iris Pratt's Rufous (140).

Color in alcohol: dorsal surface of head brown to grayish brown, usually with narrow, darker brown crosslines on snout and V-shaped interorbital band; dorsal surface of body grayish brown to brown, with darker brown reticulations or mottling, or darker brown middorsal diamond-shaped markings, some females with broad, pale brown middorsal stripe; dorsal surfaces of limbs grayish brown to brown, with darker brown reticulations or crossbars; dorsal surface of tail grayish brown to brown, with darker brown reticulations; lateral surface of head with dark brown reticulations or supraorbital vertical lines, eyelids with darker brown vertical medial bar; ventral surface of head cream to pale brown, with brown cross-reticulations, usually pale cream to white crossbar present at level anterior to eye (crossbar more prominent in juveniles to young adults); ventral surface of body pale cream to pale brown, with or without brown reticulations or scattered flecking; subcaudal surface pale cream to pale brown, with or without brown reticulations, brown crossbands sometimes present on distal half.

Hemipenis: the almost completely everted hemipenis (SMF 85969) is a large stout bilobed organ, with well-developed lobes; sulcus spermaticus bordered by well-developed lips, bifurcating at base of apex, continuing to tips of lobes; flaplike processus present on asulcate side; slight "lateral" ridge evident on lobes, ridge increasing in size on each side of truncus; lobes strongly calyculate; truncus with a few transverse folds.

*Diagnosis/Similar Species.* *Norops capito* can be distinguished from all other Honduran *Norops* by the combination of its large adult size (adults average about 80 mm SVL), long hind legs (longest toe of adpressed hind limb usually reaching to beyond tip of snout), a conspicuously broad and stout head with a rather blunt snout in lateral aspect, and a pale crossband on the

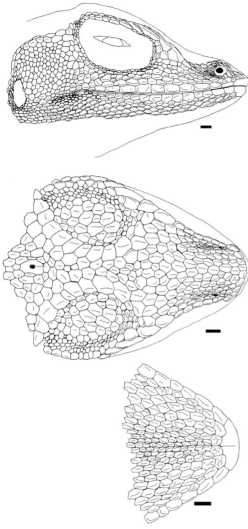

Figure 11. *Norops capito* head in lateral, dorsal, and ventral views. SMF 79894, adult male from Parque Nacional Saslaya, Atlántico Norte, Nicaragua. Scale bar = 1.0 mm. Drawing by Gunther Köhler.

chin that is especially evident in juveniles and subadults.

*Illustrations* (Figs. 11, 12, 85, 86). Duméril et al., 1870–1909 (head scales; as *Anolis*); Köhler, 1991 (adult), 1999c (adult), 1999d (adult), 2000 (adult, caudal scales, head and dewlap), 2001b (adult), 2003 (adult, head and dewlap, caudal scales), 2008 (adult, head and dewlap, caudal scales), 2014 (ventral and caudal scales, tail

tip, subdigital pad, terminal phalanx of fourth toe, axillary region; as *Anolis*); Schmidt and Henkel, 1995 (adult; as *Anolis*); Fläschendräger and Wijffels, 1996 (adult); Lee, 1996 (adult; as *Anolis*), 2000 (adult, head and dewlap); Campbell, 1998 (adult); Stafford and Meyer, 1999 (adult); Savage, 2002 (adult, dewlap); D'Cruze, 2005 (adult); Guyer and Donnelly, 2005 (adult, juvenile); Köhler et al., 2005a (adult, head scales, hemipenis); McCranie et al., 2006 (adult); Townsend and Wilson, 2008 (adult, chin: as *Anolis*).

*Remarks. Norops capito* is purported to be a member of the *N. biporcatus* species group (Nicholson, 2002, also see Remarks for *N. biporcatus*). Based on external morphology, *N. capito* does not appear to us to be closely related to the other species usually associated with the *N. biporcatus* group. Also, phylogenetic analyses based on molecular data by Nicholson (2002) and Nicholson et al. (2005, 2012) do not support a close relationship between *N. capito* and *N. biporcatus* (also see Poe, 2004, for a largely morphological analysis). The short but extremely stout head of *N. capito* is unique among the species of *Norops* occurring in Honduras. The phylogenetic analyses of Nicholson et al. (2005, 2012) placed *N. capito* in a subclade with the morphologically much different *N. tropidonotus*, whereas that of Nicholson (2002) placed it in a sister clade to three small species of southern Mexico and northern Central America (*N. isthmicus*, *N. rodriguezii*, and *N. unilobatus* [as *N. sericeus*]). Therefore, we leave *N. capito* as incertae sedis within the *N. auratus* species group of Nicholson et al. (2012).

It seems strange that *N. capito* has not been collected in the well-studied broadleaf forests in Atlántida in northern Honduras. Perhaps it occurs in low population densities in that area.

*Natural History Comments. Norops capito* is known from near sea level to 1,300 m elevation in the Lowland Moist Forest and Premontane Wet Forest formations. It is active on lower portions of tree trunks,

Figure 12.  *Norops capito*. (A) Adult male (FMNH 282577) from Bachi Kiamp, Gracias a Dios; (B) adult male dewlap (USNM 342275) from Quebrada El Pinol, Olancho. Photographs by James R. McCranie.

usually with the head facing downwards, and frequently in the shade. We have seen individuals jump to the ground after spotting a prey item but return to the tree perch after grasping its prey. The species sleeps at night on tree trunks and leaves of low vegetation. This arboreal species was found in every month of the year. Individuals appear to rely on camouflage to escape notice because they rarely move until touched by the collector. Vitt and Zani (2005: 36), in an ecological study in

southeastern Nicaragua, indicated that this is a cryptic species "that live[s] low on trunks in shaded rainforest and [is] active throughout the day but appear[s] to spend most of their time in shade." Savage (2002 and references cited therein) also documented that this is a shade-tolerant species in Costa Rica that perches on tree trunks at usually 0.25–2.00 m above the ground. One Honduran specimen (USNM 563157) was collected with a frog (*Craugastor lauraster* [Savage, McCranie, and Espinal]; USNM

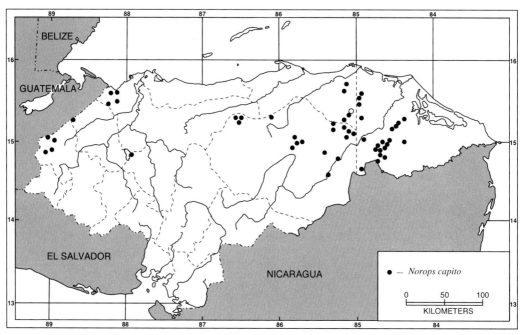

Map 6.  Localities for *Norops capito*. Solid symbols denote specimens examined and the open symbol an accepted record.

563293) in its mouth. *Norops capito* feeds on a variety of arthropods, with individuals generally eating few and relatively large prey items (Vitt and Zani, 2005). It is also known to eat other lizards and slugs (Savage, 2002; Vitt and Zani, 2005) and small frogs (*Pristimantis ridens* [Cope], Mora et al., 2012). Reproduction appears to occur throughout the year (see references in Savage, 2002) but is likely highest during the wet season (Vitt and Zani, 2005). Vitt and Zani (2005: 41) predicted that this rainforest dependent species "will likely disappear with elimination of the rainforest canopy." Our observations support that prediction.

*Etymology.* The name *capito* is Latin meaning "one with a large head" and refers to the relatively broad and stout head found in this species.

*Specimens Examined* (144 [3] + 3 skeletons; Map 6). COLÓN: El Ocotillal, SMF 85976–78; Quebrada Machín, USNM 536489 (skeleton), 563121–33; Río Guaraska, BMNH 1985.1103; Tulito, BMNH 1985.1104–07. COPÁN: Laguna del Cerro, UF 142454; Río Amarillo, SMF 91736–38; San Isidro, FMNH 282574, 282576; SMF 91248, UF 142455–57; Santa Rita, SMF 86968. CORTÉS: 6.0 km SE of Agua Azul, UF 135772; El Paraíso, UF 144723; Guanales Camp, UF 144745; Quebrada Agua Buena, SMF 79136; W of San Pedro Sula, MCZ R-31624. GRACIAS A DIOS: Auka Kiamp, USNM 563134; Awasbila, USNM 563140; Bachi Kiamp, FMNH 282577, SMF 92841–42; Bodega de Río Tapalwás, UF 144619–20, USNM 559595 (skin and skeleton), 563147–59; Cabeceras de Río Rus Rus, SMF 86969; Caño Awalwás, UF 144615–18, USNM 563135–39; Cerro Wahatingni, UF 144621; Crique Unawás, SMF 85979; Dos Bocas, FMNH 282575; Hiltara Kiamp, USNM 563141–43; Kipla Tingni Kiamp, USNM 563144, 573123–24; Leimus (Río Warunta), USNM 565432; about 30 km S of Mocorón, UTA R-42656; Pomokir, SMF 85972–75; Raudal Kiplatara, SMF 85961–64, 85966–67; Rawa Kiamp, SMF 91249; Río Cuyamel, SMF 85968–71; Río

Sutuwala, USNM 563146; Sachin Tingni Kiamp, USNM 563145, 563526; Sadyk Kiamp, SMF 91250–51, USNM 565433; San San Hil Kiamp, SMF 86971–72, USNM 563527–28; Urus Tingni Kiamp, SMF 86970, 86973, USNM 563160–64, 565430–31; Warunta Tingni Kiamp, SMF 86974, USNM 563165–67, 565428–29. OLANCHO: Caobita, SMF 85524; about 40 km E of Catacamas, TCWC 23629; Cuaca, UTA R-53192; Piedra Blanca, SMF 91739; Planes de San Esteban, SMF 91740; Quebrada de Agua, SMF 91741; Quebrada de Las Escaleras, SMF 79094; Quebrada de Las Marías, USNM 563171; Quebrada El Guásimo, SMF 80822–23, USNM 563169; Quebrada El Pinol, USNM 342274–78; Quebrada Las Cantinas, SMF 79093; confluence of Ríos Aner and Wampú, USNM 563168; Río Kosmako, USNM 563170; confluence of Ríos Sausa and Wampú, USNM 563172–74; Río Wampucito, LACM 45152; confluence of Ríos Yanguay and Wampú, USNM 563175. SANTA BÁRBARA: La Cafetalera, SMF 91252. "HONDURAS": UF 60346 (skeleton).

*Other Records* (Map 6). COLÓN: Cañon del Chilmeca, UNAH 5450, Empalme Río Chilmeca, UNAH 5452, 5510 (Cruz Díaz, 1978; specimens now lost).

*Norops carpenteri* (Echelle, Echelle, and Fitch)

*Anolis carpenteri* Echelle, Echelle, and Fitch, 1971: 355 (holotype, KU 132506; type locality: "from east bank of Río Reventazón, 500+ m elev, about 7 km ESE Turrialba, where Highway 10 crosses the river, Cartago Province, Costa Rica").

*Norops carpenteri*: Savage, 1973: 11; McCranie and Köhler, 2012: 103.

*Geographic Distribution. Norops carpenteri* occurs at low and moderate elevations on the Atlantic versant from northeastern Honduras to northwestern Panama. In Honduras, this species is known only from two nearby, low-elevation localities in the northeastern portion of the country.

*Description.* The following is based on two males (FMNH 282560; SMF 91746). *Norops carpenteri* is a small anole (SVL 41 mm in largest Honduran male [SMF 91746]; maximum reported SVL 46 mm [Myers, 1971a]); dorsal head scales weakly keeled in internasal region, rugose in prefrontal, frontal, and parietal areas; deep frontal depression present; parietal depression absent; 5–7 (6.0) postrostrals; anterior nasal entire, contacting rostral and first supralabial on three sides, contacting only rostral on one side; 4–6 (5.0) internasals; canthal ridge sharply defined; scales comprising supraorbital semicircles without keels, largest scale in semicircles about same size as largest supraocular scale; supraorbital semicircles well defined; 1–3 (2.0) scales separating supraorbital semicircles at narrowest point; 3 scales separating supraorbital semicircles and interparietal at narrowest point; interparietal well defined, greatly enlarged relative to adjacent scales, surrounded by scales of moderate size, longer than wide, larger than ear opening; 2 rows of about 5–6 (total number) enlarged, keeled supraocular scales; no enlarged supraoculars in broad contact with supraorbital semicircles; 2 elongate superciliaries, posterior about same length as anterior; 3 enlarged canthals; 10–12 (11.0) scales between second canthals; 12 scales between posterior canthals; loreal region concave, 61–67 (64.5 ± 3.0) mostly keeled (some smooth or rugose) loreal scales in maximum of 8–10 (9.0 ± 1.2) horizontal rows; 8–9 (8.5 ± 0.6) supralabials and 7–9 (8.0 ± 0.6) infralabials to level below center of eye; suboculars weakly keeled, 2 suboculars in broad contact with supralabials; ear opening vertically oval; scales anterior to ear opening granular, similar in size to those posterior to ear opening; 5–7 (6.0) postmentals, outer pair largest; granular scales present on chin and throat; male dewlap small, extending to level anterior to axilla; male dewlap with 8 horizontal gorgetal-sternal scale rows, about 95–97 total scales (*n* = 2); 2 (modal number) anterior marginal pairs in male dewlap; no nuchal crest or dorsal ridge; no middorsal scale rows distinctly enlarged; dorsal scales smooth to rugose, grading into smaller granular lateral scales; no enlarged scales scattered among laterals; 107–120 (113.5) dorsal scales along vertebral midline between levels of axilla and groin; 58–80 (69.0) dorsal scales along vertebral midline contained in 1 head length; ventral scales on midsection about same size as middorsal scales; most ventral body scales smooth (a few midventral scales weakly keeled), subimbricate; 71–96 (83.5) ventral scales along midventral line between levels of axilla and groin; 46–49 (47.5) ventral scales contained in 1 head length; 180 scales around midbody in one; tubelike axillary pocket absent; precloacal scales rugose; no enlarged postcloacal scales; tail slightly compressed, TH/TW 1.20–1.25; basal subcaudal scales smooth; lateral caudal scales rugose, homogeneous, although indistinct division in segments discernable; dorsal medial caudal scale row not enlarged, not keeled, not forming crest; most scales on anterior surface of antebrachium rugose; 22–24 (22.5 ± 1.0) subdigital lamellae on Phalanges II–IV of Toe IV of hind limbs; 6–9 (7.8 ± 1.5) subdigital scales on Phalanx I of Toe IV of hind limbs; SVL 31.9–41.3 (36.6) mm; TAL/SVL 1.28 in one; HL/SVL 0.26–0.29; SHL/SVL 0.21–0.22; SHL/HL 0.74–0.81; longest toe of adpressed hind limb not reaching ear opening.

Color in life of an adult male (FMNH 282560): dorsum greenish brown with numerous paler greenish brown and pale brown spots on lateral surfaces of body; tail same except for dark brown crossbands; chin and belly dirty white with dark brown flecking; ventral surfaces of limbs brown with dark brown mottling; subcaudal surface pale brown with dark brown crossbands; iris brown; dewlap Spectrum Orange (17) with pale brown to cream scales. Color in life of another adult male (SMF 91746): dorsum greenish brown with paler green spots middorsally and white spots laterally on body.

Color in alcohol: all dorsal surfaces, except supraoculars and parietal area, pale brown with numerous dark brown scales; supraocular scales dark brown; scales in parietal area dark brown with some slightly paler brown scales; all ventral surfaces pale brown with many dark brown scales.

Hemipenis: the partially everted hemipenis of SMF 81821 (from Bartola, Río San Juan, Nicaragua) is a stout, bilobed organ; sulcus spermaticus bordered by well-developed sulcal lips, bifurcating at base of apex; low asulcate ridge present; lobes extensively calyculate; truncus with transverse folds.

*Diagnosis/Similar Species. Norops carpenteri* is distinguished from all other Honduran *Norops*, except *N. limifrons, N. ocelloscapularis, N. rodriguezii, N. yoroensis,* and *N. zeus,* by having a single elongated prenasal scale, smooth and subimbricate ventral scales, and slender habitus. *Norops carpenteri* differs from *N. limifrons* and *N. zeus* by having short hind legs (longest toe of adpressed hind limb not reaching eye, usually not reaching beyond ear opening versus usually reaching between anterior border of eye and tip of snout in *N. limifrons* and *N. zeus*). *Norops carpenteri* also differs from *N. limifrons* and *N. zeus* by having an orange male dewlap in life and a greenish brown dorsum with pale spots in life (male dewlap dirty white, with or without basal orange-yellow spot in life, and dorsum brown without pale spots in life in *N. limifrons* and *N. zeus*). *Norops carpenteri* differs from *N. rodriguezii, N. ocelloscapularis,* and *N. yoroensis* by having a greenish brown dorsum with pale spots in life and in lacking a pale lateral stripe (dorsum brown without pale spots in life and pale lateral stripe usually present in those three species). *Norops carpenteri* also differs from *N. ocelloscapularis* by lacking an ocellated shoulder spot (ocellated shoulder spot usually present in *N. ocelloscapularis*).

*Illustrations* (Figs. 13, 14). Echelle et al., 1971 (head scales); Myers, 1971a (adult, head scales; as *Anolis procellaris*); Savage, 2002 (adult); Köhler, 2003 (adult, head and

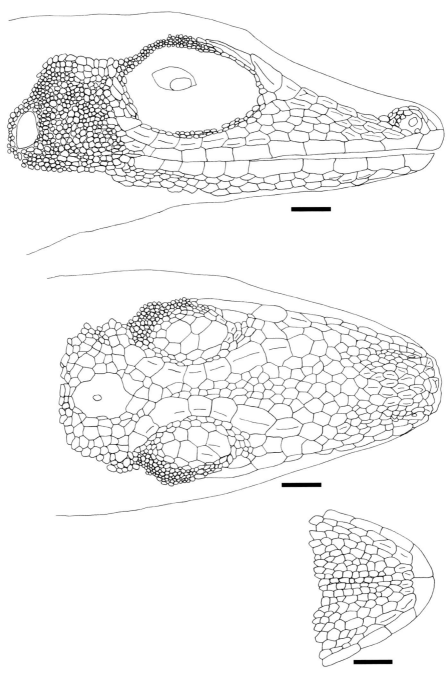

Figure 13.   *Norops carpenteri* head in lateral, dorsal, and ventral views. SMF 91746, adult male from Bachi Kiamp, Gracias a Dios. Scale bar = 1.0 mm. Drawing by Gunther Köhler.

Figure 14.   *Norops carpenteri*. (A) Adult male (FMNH 282560) from Leimus (Río Warunta), Gracias a Dios; (B) adult male dewlap (FMNH 282560). Photographs by James R. McCranie.

dewlap), 2008 (adult, head and dewlap); Guyer and Donnelly, 2005 (adult, juvenile).

*Remarks. Norops carpenteri* is included in the *N. fuscoauratus* species subgroup (see Remarks for *N. ocelloscapularis*). *Norops carpenteri* was recovered as a sister species to *N. ocelloscapularis* in the molecular and morphological study of Nicholson et al. (2012).

*Natural History Comments. Norops carpenteri* is known from 30 to 40 m elevation in the Lowland Moist Forest formation. One Honduran specimen was active in disturbed vegetation on a riverbank during July and another was sleeping at night on leaves of a small tree about 2 m above the ground in an abandoned crop field during May. Both areas were originally covered with broadleaf rainforest. Broadleaf rainforest still occurs in most places away from the river in those areas. The type series of this species was collected on lichen-covered

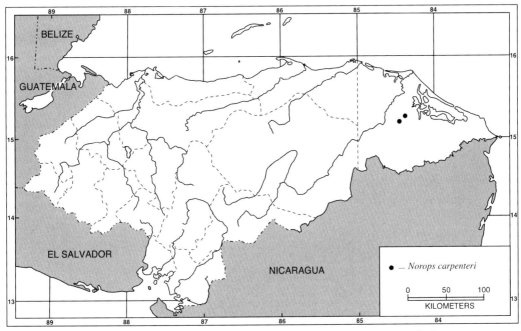

Map 7.   Localities for *Norops carpenteri*. Solid symbols denote specimens examined.

rocks, shrubs, and tree trunks along a well-shaded riverbank in Costa Rica (Echelle et al., 1971). The holotype of the synonym *N. procellaris* was found "in a dense-river swamp forest that is dominated by an understory of large palms" (Myers, 1971a: 7). Guyer and Donnelly (2005) reported that at La Selva, Costa Rica, this species occurs in a wide variety of habitats. Savage (2002), based on data in an unpublished dissertation, stated that this species might spend more time higher up on tree trunks and might be a tree crown species in Costa Rica. Although known only from two low-elevation localities in Honduras, Savage (2002) reported the species has been collected up to 1,100 m elevation in Nicaragua. The diet of *N. carpenteri* in Costa Rica consists of small arthropods, dominated by homopterans, orthopterans, and araneids, and reproduction occurs throughout the year (see Guyer and Donnelly, 2005). Juveniles of this species are said to feed on small snails (see Savage, 2002).

*Etymology.* The name *carpenteri* is a patronym for Charles C. Carpenter, who was a professor at the University of Oklahoma at the time the type series was collected.

*Specimens Examined* (2 [0]; Map 7). GRACIAS A DIOS: Bachi Kiamp, SMF 91746; Leimus (Río Warunta), FMNH 282560.

## Norops crassulus (Cope)

*Anolis crassulus* Cope, 1864: 173 (five syntypes, ANSP 8023–27 [see Stuart, 1942: 2, who designated these as "lectocotypes"]; type locality restricted to "Central Guatemala" by Stuart, 1942: 2); Meyer, 1969: 216; Hahn, 1971: 111 (in part); Meyer and Wilson, 1972: 108, 1973: 16 (in part); Wilson et al., 1991: 70; Köhler et al., 2007: 391; Wilson and Townsend, 2007: 145; Townsend and Wilson, 2009: 63; Gutsche, 2012: 71; Köhler, 2014: 210.

*Norops crassulus*: Villa et al., 1988: 48; McCranie et al., 1992: 208; Cruz et al.,

1993: 28; Köhler et al., 1999: 297, 2001: 254; Wilson et al., 2001: 136; Köhler, 2003: 108, 2008: 116; Wilson and McCranie, 2004b: 43.

*Anolis sminthus*: Wilson et al., 1991: 70; Nicholson et al., 2005: 933.

*Norops* sp. (*crassulus* group): McCranie et al., 1992: 208.

*Norops* cf. *sminthus*: Köhler and Obermeier, 1998: 136 (in part); Köhler et al., 1999: 298 (in part).

*Norops sminthus*: Nicholson, 2002: 120; Nicholson et al., 2012: 12.

*Geographic Distribution.* *Norops crassulus* occurs at moderate and intermediate elevations from southeastern Chiapas, Mexico, to central Honduras on the Atlantic versant and in southern Guatemala, western El Salvador, and southwestern Honduras on the Pacific versant (but see Remarks). In Honduras, this species occurs in the mountainous southwestern region on both versants and in the Sierra de Agalta in the central portion of the country.

*Description.* The following is based on 10 males (KU 219950, 219955–58; SMF 78829–30, 86963; UF 103410–11) and 13 females (KU 219951–54, 219959–60; SMF 78092, 78799–800, 86964–67). *Norops crassulus* is a medium-sized anole (SVL 57 mm in largest Honduran male [UF 103410] examined, 54 mm in largest Honduran female [KU 219954] examined); dorsal head scales rugose to strongly keeled in internasal, prefrontal, frontal, and parietal areas; deep frontal depression present; parietal depression absent; 4–6 (5.1 ± 0.5) postrostrals; anterior nasal usually divided, occasionally entire, lower scale contacting rostral and first supralabial; 5–7 (5.6 ± 0.7) internasals; canthal ridge sharply defined; scales comprising supraorbital semicircles smooth, rugose or weakly keeled, largest scale in semicircles larger than largest supraocular scale; supraorbital semicircles well defined; 0–2 (0.8 ± 0.5) scales separating supraorbital semicircles at narrowest point; 2–3 (2.5 ± 0.5) scales separating supraorbital semicircles and in-

terparietal at narrowest point; interparietal well defined, slightly to distinctly enlarged relative to adjacent scales, surrounded by scales of moderate size, longer than wide, smaller than ear opening; 2–3 rows of about 5–8 (total number) enlarged, smooth to strongly keeled supraocular scales; 1–3 enlarged supraoculars in broad contact with supraorbital semicircles; 2–3 elongate superciliaries, posteriormost only slightly shorter than anteriormost; usually 3 enlarged canthals; 5–7 (5.4 ± 0.6) scales between second canthals; 5–9 (6.6 ± 1.1) scales between posterior canthals; loreal region slightly concave, 14–29 (19.2 ± 4.3) mostly strongly keeled (some smooth or rugose) loreal scales in 20, in maximum of 3–6 (4.3 ± 0.8) horizontal rows; 5–7 (6.2 ± 0.6) supralabials and 4–7 (6.1 ± 0.7) infralabials to level below center of eye; suboculars weakly to strongly keeled, usually 2–3 suboculars in broad contact with supralabials; ear opening vertically oval; scales anterior to ear opening not granular, slightly larger than those posterior to ear opening; 3–7 (4.3 ± 0.8) postmentals, outer pair usually largest; weakly keeled granular scales present on chin and throat; male dewlap moderately large, extending to level of axilla; male dewlap with 5–8 horizontal gorgetal-sternal scale rows, about 5–8 scales per row (*n* = 4); 2 (modal number) anterior marginal pairs in male dewlap; female dewlap relatively well developed, but smaller than male dewlap; low nuchal crest and low dorsal ridge present in males; about 14–18 middorsal scale rows distinctly enlarged, keeled; dorsal scales lateral to middorsal series grading into keeled, nongranular lateral scales; flank scales heterogeneous, solitary enlarged keeled or elevated scales scattered among laterals; 34–46 (41.5 ± 4.6) dorsal scales along vertebral midline between levels of axilla and groin in males, 34–57 (44.3 ± 7.8) in females; 22–33 (26.8 ± 3.2) dorsal scales along vertebral midline contained in 1 head length in males, 21–33 (26.5 ± 3.9) in females; ventral scales on midsection slightly larger than largest dorsal scales; ventral body scales strongly keeled, mucronate, imbricate, 31–37 (34.7 ± 2.4) ventral scales along midventral line between levels of axilla and groin in males, 25–36 (30.7 ± 3.7) in females; 20–27 (23.9 ± 2.4) ventral scales contained in 1 head length in males, 16–22 (18.9 ± 2.2) in females; 54–114 (100.1 ± 19.2) scales around midbody in males, 70–120 (99.7 ± 12.6) in females; tubelike axillary pocket absent; precloacal scales strongly keeled; a pair of greatly enlarged postcloacal scales in males; tail slightly to distinctly compressed, TH/TW 1.09–2.00 in 20; all subcaudal scales keeled; lateral caudal scales keeled, homogeneous, although indistinct division in segments discernable; dorsal medial caudal scale row enlarged, keeled, not forming crest; scales on anterior surface of antebrachium distinctly keeled, unicarinate; 22–26 (24.5 ± 1.0) subdigital lamellae on Phalanges II–IV of Toe IV of hind limbs; 6–8 (6.9 ± 0.5) subdigital scales on Phalanx I of Toe IV of hind limbs; SVL 34.5–56.5 (44.9 ± 7.9) mm in males, 36.0–54.2 (48.2 ± 5.3) mm in females; TAL/SVL 1.35–2.46 in four males, 1.20–2.60 in eight females; HL/SVL 0.27–0.31 in males, 0.25–0.29 in females; SHL/SVL 0.22–0.28 in males, 0.22–0.26 in females; SHL/HL 0.82–0.98 in males, 0.79–0.93 in females; longest toe of adpressed hind limb usually reaching between ear opening and posterior margin of eye.

Color in life of an adult male (SMF 78829): dorsal surface of head Buff (24) with Drab (27) interorbital crossbar; dorsal surface of body Buff with Drab chevrons, chevrons with Sepia (219) spots; dorsolateral surface of body Yellow Ocher (123C); middorsal and dorsolateral coloration separated by pale yellowish-brown stripe; dorsal surface of forelimbs Yellow Ocher; dorsal surface of hind limbs Raw Umber (23) with Sepia mottling and crossbands; dorsal surface of tail Buff with Sepia spots on vertebral area; dewlap Flame Scarlet (15) with yellowish brown scales; iris Buff with dark brown reticulations. Dewlap color of another adult male (SMF 86963): Chrome Orange (16) with dirty white scales. Dewlap color of a third adult male (FMNH 283625):

Chrome Orange (16) with pale brown scales. Color in life of an adult female (SMF 86964): middorsal pale stripe Pale Horn Color (92); dorsolateral field below middorsal stripe Sepia (119) above and Mikado Brown (121C) below; lateral stripe present, paler brown than middorsal stripe; lateral field Drab-Gray (119D); dorsal surface of head Drab (27); lateral surface of head with continuation of pale lateral stripe and similarly colored postorbital stripe that connects with middorsal stripe; iris coppery brown; venter pale brown; dewlap Spectrum Orange (17). Color in life of another adult female (SMF 78799): dorsal surface of head Cinnamon (123A) with dark brownish gray markings; dorsal surface of body Cinnamon, infused with dark brownish gray middorsally; dorsal surfaces of limbs Mahogany Red (132B) with paler crossbars; two Brick Red (132A) spots at base of tail; dorsal surface of tail otherwise brown with Mahogany Red chevrons; belly Flesh Ocher (132D); dewlap Spectrum Orange (17) with Flesh Ocher scales; iris bronze with darker smudging. Color in life of another adult female (KU 219960): dorsal surface of head coppery bronze, occipital region outlined below by pale bronze temporal stripe; sides of body gray-brown; dorsolateral stripe above gray-brown sides cream anteriorly, grading to brown-gray posteriorly, fading into color of side just posterior to midbody, joining anteriorly with creamy yellow lip stripe; middorsum coppery gray-brown with middorsal stripe grading from pale bronze on nape to gray on most of body, middorsal stripe flanked by linear series of pale, outlined crescents that are dark brown on edges, coppery brown in center; dorsal surface of forelimbs pale coppery brown with darker brown band on forearm; dorsal surface of hind limbs coppery brown with silvery gray crossbands; tail coppery brown with brown-gray middorsal stripe; chest pale gold, grading to white with gray and orange punctations on belly; chin white with some pale gray smudging anteriorly; dewlap orange-red with scales that are white at tip and golden yellow at base; eyelids with antique gold scales; iris gold.

Color in alcohol: dorsal and lateral surfaces of head and body brown to grayish brown; dark brown interorbital bar usually present; sides of body usually with pale brown longitudinal lateral stripe; ventral surfaces of head and body pale grayish brown or dirty white; females show a wide degree of variation in dorsal pattern including narrow pale brown middorsal stripe that is bordered by dark brown and scalloped or triangular blotches in some specimens.

Hemipenis: the completely everted hemipenis of SMF 78100 (from Departamento Santa Ana, El Salvador) is a medium-sized, stout, slightly bilobed organ; sulcus spermaticus bordered by well-developed sulcal lips, opening into single broad concave area at base of apex; lobes calyculate; truncus with transverse folds; single, undivided asulcate processus present.

*Diagnosis/Similar Species. Norops crassulus* is distinguished from all other Honduran *Norops*, except *N. morazani* and *N. rubribarbaris*, by the combination of having distinctly enlarged middorsal scale rows, heterogeneous lateral scales, strongly keeled, mucronate, imbricate ventral scales, and a pair of greatly enlarged postcloacal scales in males. *Norops crassulus* differs from *N. rubribarbaris* by having about 14–18 rows of enlarged dorsal scales (8–11 in *N. rubribarbaris*) and by having a well-developed female dewlap (no female dewlap to female dewlap small in *N. rubribarbaris*). *Norops crassulus* differs from *N. morazani* by having a hemipenis with an undivided asulcate processus (divided in *N. morazani*).

*Illustrations* (Figs. 15, 16, 95, 96, 98). Mertens, 1952b (adult); Köhler, 2003 (adult, head and dewlap), 2008 (adult, head and dewlap), 2014 (dorsal and flank scales; as *Anolis*); Leenders and Watkins-Colwell, 2004 (dewlap); Köhler et al., 2005b (adult, dewlap, head scales).

*Remarks. Norops crassulus* is a member of the *N. crassulus* species subgroup (Köhler et al., 1999; Nicholson, 2002) of the *N. auratus* species group of Nicholson et al.

(2012). The Honduran members of this group are *N. amplisquamosus, N. crassulus, N. heteropholidotus, N. morazani, N. muralla, N. rubribarbaris,* and *N. sminthus* (see Remarks for *N. amplisquamosus*). It needs to be noted that the *N. sminthus* sequenced in Nicholson (2002) and Nicholson et al. (2005, 2012) has strongly keeled ventral scales and is thus more closely related to *N. crassulus* (see *N. sminthus* account). Nicholson (2002, personal communication 8 March 2011), using nuclear DNA data, recovered a 4% difference between *N. crassulus* samples from Chiapas, Mexico, and Olancho, Honduras. The Olancho population of *N. crassulus* also lacks pale lateral stripes, whereas those stripes are distinct in the southwestern Honduran populations (Intibucá and La Paz) of *N. crassulus*. Thus, based on molecular and morphological data, the Olancho population of *N. crassulus* appears to represent an undescribed species.

*Norops haguei* (Stuart) of Guatemala is sometimes listed as a distinct species (Lieb, 1981; Savage and Guyer, 1989; Köhler and Obermeier, 1998; Köhler et al., 1999; Köhler, 2000; Nicholson et al., 2012) or as a subspecies of *N. crassulus* (Stuart, 1948, 1955; Smith et al., 1968). McCranie et al. (1992: 214) opined that the Honduran material of *N. crassulus* that they examined indicated that *N. haguei* "may not even be a recognizable race of *crassulus.*" *Norops anisolepis* (Smith, Burley, and Fritts, 1968) of Chiapas, Mexico, was listed as a member of the *N. crassulus* species group by Savage and Guyer (1989) but was not listed as a valid species by Nicholson et al. (2012). As pointed out by McCranie et al. (1992), a study of the *N. crassulus*-like populations with strongly keeled ventral scales from throughout their range from western Chiapas, Mexico, to northwestern El Salvador and southwestern and central Honduras is needed.

Leenders and Watkins-Colwell (2004) discussed sexual dimorphism in dewlap color in this species from El Salvador.

*Natural History Comments. Norops crassulus* is known from 1,200 to 2,285 m

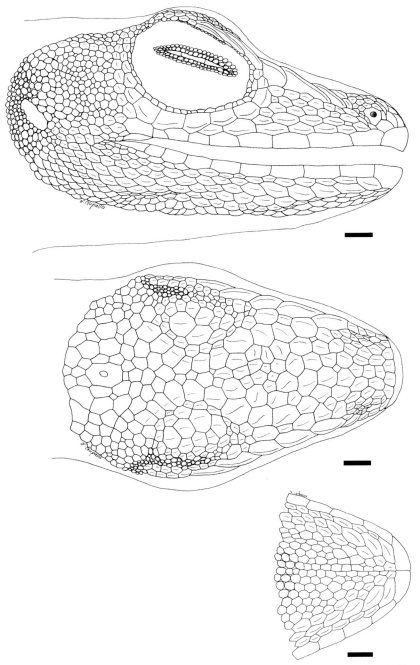

Figure 15.    *Norops crassulus* head in lateral, dorsal, and ventral views. SMF 78800, adult female from Pico La Picucha, Olancho. Scale bar = 1.0 mm. Drawing by Lara Czupalla.

Figure 16. *Norops crassulus*. (A) Adult male (FMNH 283625) from Santa Catarina, Intibucá; (B) adult male dewlap (FMNH 283625). Photographs by James R. McCranie.

elevation in the Premontane Moist Forest, Lower Montane Wet Forest, and Lower Montane Moist Forest formations. This species is usually found in low vegetation, brush piles, or on the ground, frequently in the vicinity of streams, or in forest edge situations such as along road banks. Individuals on the ground usually retreat to and climb nearby tree trunks. *Norops crassulus* individuals are usually seen in sunny areas and disappear during cloudy periods, only to reemerge when sunny conditions return. Apparently, the areas where this species occurs are too cold for it to remain exposed when sunrays are not available. Its actions are deliberate, and it is easily captured. It has only been found during April, June, and August but is probably active throughout the year during sunny periods. Nothing has been reported on its diet or reproduction. Stuart (1942: 4) reported that females of the closely related *N. haguei* (see Remarks

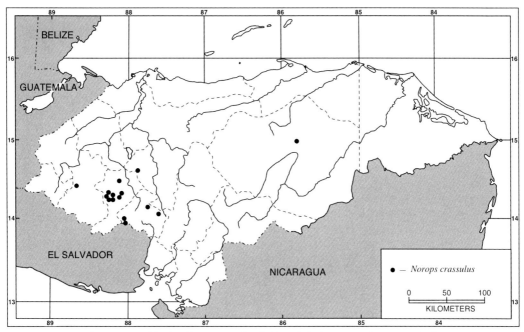

Map 8.    Localities for *Norops crassulus*. Solid symbols denote specimens examined.

above) collected in Guatemala during May "contained several very large eggs."

*Etymology.* The name *crassulus* might be derived from the Latin *crassus* meaning "thick, fat, stout." The name might allude to the "thick" anterior head scales mentioned by Cope (1864). The species also has a rather thick, stout body.

*Specimens Examined* (91 [9]; Map 8). COMAYAGUA: 9.8 km SW of Siguatepeque, KU 219977. INTIBUCÁ: Cerro El Pelón, UF 166192; near El Rodeo, SMF 86963–67; 18.1 km NW of La Esperanza, KU 219950–53; 11 km NW of La Esperanza, KU 194271; 17 km NE of La Esperanza, UF 103410–11; 2.4 km NE of La Esperanza, LACM 47683–85; 4.8 km ENE of La Esperanza, LACM 47686–90; 12.9 km ENE of La Esperanza, LACM 47691; about 10 km SE of La Esperanza, UF 121773; La Esperanza, CM 59118–19, LSUMZ 38816–24; San Pedro La Loma, KU 219954; Santa Catarina, FMNH 283624–27; Zacate Blanco, KU 194268–70, UF 166190–91. LA PAZ: Cantón Palo Blanco, KU 184084–87; Cantón El Zancudo, KU 184072–77; near Guajiquiro, UF 166185–89; Opatoro, SMF 78092; mountains S of San Pedro de Tutule, KU 219955–56, 219978–81; about 5 km S of Santa Elena, KU 194300–04. LEMPIRA: Naranjos, FMNH 236387, KU 209323–26, 219957–60. OLANCHO: Pico La Picucha, SMF 78799–800, 78829–30, 91752–53, USNM 578763–66.

*Other Records.* "HONDURAS": (Werner, 1896).

*Norops cupreus* (Hallowell)
*Anolis cupreus* Hallowell, 1861: 481 (14 syntypes, MCZ R-17631–32, UIMNH 40733, USNM 12211 (11) [see Barbour and Loveridge, 1929b: 218; Cochran, 1961: 86; Smith et al., 1964: 38; also see Remarks]; type locality: "Nicaragua"); Meyer, 1969: 219; Meyer and Wilson, 1973: 16; Fitch and Seigel, 1984: 4; Poe, 2004: 62; Köhler et al., 2007: 391; Lovich et al., 2010: 113; Townsend and Wilson, 2010b: 697.

*Norops cupreus*: Savage, 1973: 11; Wilson and McCranie, 1998: 16; Köhler and Kreutz, 1999: 65; Köhler et al., 2000: 425, 2001: 254; Nicholson et al., 2000: 30; Espinal et al., 2001: 106; McCranie and Köhler, 2001: 232; Wilson et al., 2001: 136; Castañeda, 2002: 15; McCranie et al., 2002a: 27, 2002b: 472; Lovich et al., 2006: 13.

*Norops* especie A: Espinal, 1993: table 3.

*Norops dariense*: McCranie et al., 2006: 120; Wilson and Townsend, 2006: 105; McCranie 2011: 260, 493.

*Anolis dariense*: Wilson and Townsend, 2007: 145; Townsend and Wilson, 2010b: 697.

*Geographic Distribution.* *Norops cupreus* occurs at low and moderate elevations on the Atlantic versant from north-central Honduras to south-central Costa Rica and on the Pacific versant from southern Honduras to southwestern Costa Rica. *Norops cupreus* is widespread in central, eastern, and southern Honduras.

*Description.* The following is based on 26 males (SDSNH 72731–39, 72741; SMF 80820–21, 80885–86, 80888, 86980–81; USNM 342279–80, 342283–84, 342286, 541035–36, 541041, 563221) and 23 females (SDSNH 72729–30; 72749, 72754; SMF 77314, 80887, 86975, 86977–79, 86986–88, 86990; USNM 541031, 541033, 541037–40, 541042, 563223, 563225). *Norops cupreus* is a moderate-sized anole (SVL 53 mm in largest Honduran male [SDSNH 72732] and 52 mm SVL in largest Honduran female [USNM 541042] examined; maximum reported SVL 57 mm [Fitch et al., 1972]); dorsal head scales rugose to keeled in internasal, prefrontal, and frontal areas, rugose to smooth in parietal area; deep frontal depression present; parietal depression absent; 4–9 (5.9 ± 1.1) postrostrals; anterior nasal divided, lower scale contacting rostral and usually first supralabial; 5–9 (6.4 ± 0.9) internasals; canthal ridge sharply defined; scales comprising supraorbital semicircles usually keeled, largest scale in semicircles usually larger than largest supraocular scale; supraorbital semicircles well defined; 1–3 (2.1 ± 0.3) scales separating supraorbital semicircles at narrowest point; 2–4 (2.9 ± 0.5) scales separating supraorbital semicircles and interparietal at narrowest point; interparietal well defined, irregular in outline, longer than wide, larger than, or equal to, size of ear opening; 2–3 rows of about 3–9 (total number) enlarged, keeled supraocular scales; enlarged supraoculars varying from completely separated from supraorbital semicircles by 1 row of small scales to 1–2 enlarged supraoculars in broad contact with supraorbital semicircles; 2–3 elongate superciliaries, posteriormost shortest; usually 3 enlarged canthals; 7–12 (9.2 ± 1.3) scales between second canthals; 7–12 (9.2 ± 1.2) scales between posterior canthals; loreal region slightly concave, 29–61 (44.0 ± 6.7) mostly keeled to rugose loreal scales in maximum of 4–8 (6.9 ± 0.8) horizontal rows; 5–11 (7.3 ± 0.9) supralabials and 6–10 (7.7 ± 0.8) infralabials to level below center of eye; suboculars distinctly to weakly keeled, separated from supralabials by 1 row of scales; ear opening vertically oval; scales anterior to ear opening slightly larger than those posterior to ear opening; 5–9 (6.4 ± 0.8) postmentals, outer pair largest; gular scales keeled; male dewlap large, extending well onto venter nearly to level of elbows when forelimbs extended alongside body; male dewlap with 9–10 horizontal gorgetal-sternal scale rows, 13.5 mean number of scales per row (*n* = 5); 4–5 (modal number) anterior marginal pairs in male dewlap; female dewlap absent; no nuchal crest or dorsal ridge; about 0–2 middorsal scale rows slightly enlarged, faintly keeled, dorsal scales lateral to middorsal series grading into granular lateral scales; no enlarged scales among laterals; 59–93 (77.3 ± 8.8) dorsal scales along vertebral midline between levels of axilla and groin in males, 64–98 (81.2 ± 8.5) in females; 40–67 (49.4 ± 6.9) dorsal scales along vertebral midline contained in 1 head length in males, 34–76 (48.4 ± 10.2) in females; ventral scales on midsection about 2–3 times larger than largest dorsal scales; ventral body scales weakly to strongly keeled, subimbricate, or imbricate; 36–82 (60.1 ± 9.1) ventral scales along midventral line between levels of axilla and groin in males, 40–78 (60.2 ± 10.6) in females; 38–53 (42.8 ± 4.0) ventral scales contained in 1 head length in males, 26–58 (37.2 ± 7.6) in females; 124–184 (162.2 ± 13.8) scales around midbody in males, 104–186 (155.5 ± 21.6) in females; tubelike axillary pocket absent; precloacal scales not keeled; enlarged postcloacal scales absent in males; tail slightly to distinctly compressed, TH/TW 1.08–1.60 in 41; basal subcaudal scales keeled; lateral caudal scales keeled, homogeneous, although indistinct division in segments discernable; dorsal medial caudal scale row slightly enlarged, keeled, not forming crest; most scales on anterior surface of antebrachium rugose to strongly unicarinate; 21–26 (24.0 ± 1.2) subdigital lamellae on Phalanges II–IV of Toe IV of hind limbs; 6–11 (8.3 ± 1.2) subdigital scales on Phalanx I of Toe IV of hind limbs; SVL 29.0–53.3 (45.7 ± 5.7) mm in males, 33.0–51.6 (46.5 ± 4.6) mm in females; TAL/SVL 1.79–2.00 in nine males, 1.48–1.95 in 14 females; HL/SVL 0.24–0.30 in males, 0.24–0.29 in females; SHL/SVL 0.26–0.33 in males, 0.27–0.32 in females; SHL/HL 1.00–1.22 in males, 0.96–1.15 in females; longest toe of adpressed hind limb usually reaching between anterior border of eye and beyond tip of snout.

Color in life of an adult male (USNM 541036): dorsal surface of body with middorsal swath of Antique Brown (37) extending from occiput to end of dilated proximal portion of tail, coursed by series of seven Sepia (219) spots, bounded by Sepia dorsolateral stripe, in turn by Sulphur Yellow (57) lateral stripe; lateral stripe bounded below by Cinnamon-Brown (33) ventrolateral stripe; dorsal surface of head pale brown with darker brown interocular band; front limbs dirty Straw Yellow (56); hind limbs Straw Yellow with slightly darker crossbands; ventral surfaces of head and body Straw Yellow; dewlap Deep Vinaceous (4)

with brown basal spot; distal portion of tail coppery brown with slightly darker crossbands; iris pale copper. Color in life of another adult male (USNM 541035): dorsal surfaces of head and body chocolate brown with slight bronze sheen, series of bronze-brown middorsal blotches separated by dark chocolate brown crossbars; lateral stripe pale gold anteriorly, grading to pale bronze posteriorly; limbs pale olive-tan; head brown with rust wash; tail pale olive brown with rust wash and indistinct brown crossbands; ventral surface of body pale yellow; chin red-orange; dewlap Maroon (31) with brown basal spot. Color in life of a third adult male (USNM 541032): dorsal surfaces of head and body pale brown with darker brown dorsolateral blotches; hind limbs pale brown with pale orange crossbands; ventral surfaces pale brown with paler brown reticulations; dewlap Warm Sepia (221A) with darker brown basal spot. Color in life of another adult male (USNM 541033): dorsal surfaces brown with dark brown middorsal chevrons; yellowish tan middorsal stripe extending onto tail; belly cream with brown mottling; dewlap brown with darker brown basal spot; dewlap scales white. Another adult male (USNM 563221) had a chocolate brown dewlap with a darker brown basal spot. Color in life of an adult female (USNM 541031): dorsal surface of body yellowish brown with gold to reddish orange middorsal stripe; dorsal surface of head yellowish brown; front limbs pale brown; hind limbs rust red with scattered yellow flecking; middorsal stripe continuing onto proximal half of tail, distal portion of tail banded pale and dark brown; ventral surface of body pale brown; chin and throat white with scattered pale brown mottling; iris bronze. Another adult female (USNM 563226): dorsal surfaces brown; chin pale brown; belly pale brown with midventral white area. Color in life of a juvenile (USNM 563222): dorsal surface of body olive with bronze cast; indistinct bronze middorsal stripe present; dorsum crossed by four dark brown X-shaped crossbands; narrow gold lateral stripe present; upper

portion of head copper with brown interocular bar; ventral surface of body cream
with brown mottling; chin and throat white
with brown longitudinal lines.

Color in alcohol: dorsal surfaces of head
and body pale brown to dark brown;
middorsal pattern variable, some specimens
lack pattern, some have dark brown spots or
chevrons (most prominent above groin
region and on base of tail), most females
have a thin vertebral pale line, occasional
females have a broad pale middorsal stripe
bordered by dark brown; pale lateral stripe
present in many individuals (stripes most
prominent above shoulder region, lateral
stripe reaching groin region and base of tail
in occasional specimens); lateral surface of
head pale brown; dorsal surfaces of limbs
brown, usually with indistinct paler brown
crossbands; dorsal surface of tail brown,
some have darker brown chevrons or crossbands, these markings more evident anteriorly; scales of ventral surfaces of head and
body white, flecked with brown, flecking
sparse in some, others moderately to heavily
flecked with brown; juveniles and some
females with brown streaks or spots on chin
and throat region; subcaudal surface lightly
to moderately flecked with brown proximally, brown flecking becoming heavier distally
until distal two-thirds mostly brown; male
dewlap brown.

Hemipenis: the completely everted hemipenis of KU 112986 (from Departamento
Chontales, Nicaragua) is a medium-sized,
bilobed organ; sulcus spermaticus bordered
by well-developed sulcal lips, bifurcating at
base of apex, its branches continuing to tip
of lobes; lobes strongly calyculate; truncus
with transverse folds; asulcate processus
absent.

*Diagnosis/Similar Species.* *Norops cupreus* is distinguished from all other Honduran *Norops*, except *N. bicaorum*, *N.
lemurinus*, and *N. roatanensis*, by the
combination of having long hind legs
(longest toe of adpressed hind limb usually
reaching between anterior border of eye
and beyond tip of snout), weakly to strongly
keeled ventral scales, about 0–2 dorsal scale

rows slightly enlarged, homogeneous lateral
scales, no tubelike axillary pocket, and no
enlarged postcloacals in males. *Norops
cupreus* differs from *N. bicaorum*, *N.
lemurinus*, and *N. roatanensis* by having
nonmucronate ventral scales (mucronate in
those three species), no dark brown lyriform
mark in nuchal region (such a mark usually
present in those three species), a brown
male dewlap in life (dewlap some shade of
red in those three species), and smaller
body size (average adult size about 45 mm
SVL in *N. cupreus* versus greater than
55 mm SVL in *N. bicaorum*, *N. lemurinus*,
and *N. roatanensis*).

*Illustrations* (Figs. 17, 18, 113). Fitch et
al., 1972 (dorsal color pattern variation; as
*Anolis*); Fitch, 1973 (adult; as *Anolis*), 1975
(adult; as *Anolis*); Schmidt and Henkel,
1995 (adult; as *Anolis cupreus dariense*);
Fläschendräger and Wijffels, 1996 (adult; as
*Anolis cupreus dariense*); Köhler, 1999c
(adult), 2000 (head and dewlap), 2001b
(head and dewlap), 2003 (adult; head and
dewlap), 2008 (adult; head and dewlap,
hemipenis), 2014 (male dewlap; as *Anolis*);
Köhler and Kreutz, 1999 (dewlap, hemipenis); Savage, 2002 (adult, dewlap);
McCranie et al., 2006 (adult; as *N. dariense*).

*Remarks.* Hallowell (1861) stated that 17
specimens were available to him for his
species description. We have been able to
account for only 14 syntypes in the literature (see species synonymy).

Savage (2002) suggested elevating *N.
cupreus dariense* (Fitch and Seigel) to a full
species, a suggestion followed by McCranie
et al. (2006). According to this classification,
*N. cupreus* would be restricted in Honduras
to the southern portion of the country,
whereas those populations from north-
central and eastern Honduras would be
referred to as *N. dariense*. An analysis of 27
morphometric and scalation characters from
*N. cupreus* from the southern portion of the
country (10 males, 5 females) with specimens from north-central and eastern Honduras (16 males, 18 females) did not reveal
any significant differences (see Table 4).

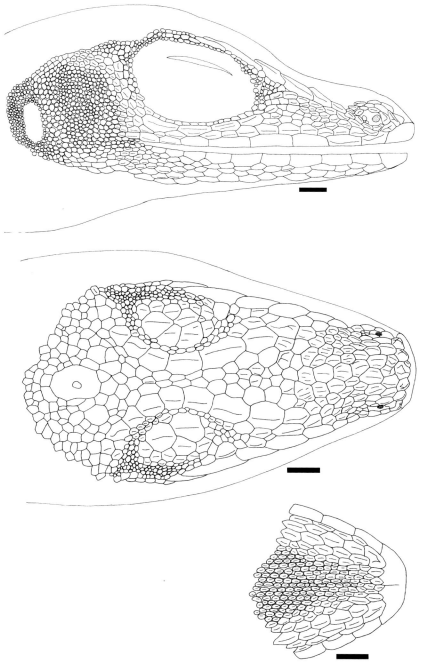

Figure 17.  *Norops cupreus* head in lateral, dorsal, and ventral views. SMF 86988, adult female from San San Hil, Gracias a Dios. Scale bar = 1.0 mm. Drawing by Gunther Köhler.

Figure 18.   *Norops cupreus.* (A) Adult male (USNM 580699) from El Rodeo, El Paraíso; (B) adult male dewlap (USNM 580699). Photographs by James R. McCranie.

Although differences in mean values in some characters were documented, there was still large overlap between the two nominal forms in all characters examined. McCranie et al. (2006: 121) stated "*Norops dariense* has smaller scales, longer hind limbs, and a different colored dewlap than typical *N. cupreus.*" At this point we cannot sufficiently address the geographic variation in male dewlap coloration in *N. cupreus* (*sensu lato*). However, the postulated dif-

ferences in dorsal scale size are only weakly supported by our data (dorsals in one HL 34–76, $x = 50.2 \pm 9.4$ in "*dariense*" versus 42–49, $x = 45.4 \pm 1.9$ in "*cupreus*"). Although, on average, specimens referable to *N. dariense* have slightly longer legs relative to SVL length compared with "true" *N. cupreus* (SHL/SVL 0.27–0.33 in male and female "*dariense*" versus 0.26–0.30 in male and 0.26–0.27 in female "*cupreus*"), there is still slight overlap between the

TABLE 4.   Morphological comparison between Honduran populations of *Norops dariense* and *Norops cupreus* (see text for explanation).

| | *Norops dariense* (16 males, 18 females) | *Norops cupreus* (10 males, 5 females) |
|---|---|---|
| **Snout-vent length (SVL)** | | |
| Both sexes combined | 29.0–52.0 (44.9 ± 5.5) | 45.0–53.3 (49.3 ± 2.4) |
| Males | 29.0–52.0 (43.4 ± 6.1) | 45.8–53.3 (49.3 ± 2.3) |
| Females | 33.0–51.6 (46.3 ± 4.7) | 45.0–51.2 (49.2 ± 2.7) |
| **Ventrals in 1 head length** | | |
| Both sexes combined | 26–58 (40.1 ± 7.2) | 35–46 (40.9 ± 2.9) |
| Males | 38–53 (43.2 ± 4.8) | 40–46 (42.3 ± 2.0) |
| Females | 26–58 (37.3 ± 8.0) | 35–40 (38.0 ± 2.0) |
| **Ventrals between axilla-groin** | | |
| Both sexes combined | 36–82 (59.8 ± 10.9) | 50–72 (61.4 ± 5.7) |
| Males | 36–82 (59.5 ± 11.0) | 50–67 (61.0 ± 5.3) |
| Females | 40–78 (60.0 ± 11.1) | 52–72 (62.2 ± 7.4) |
| **Dorsals in 1 head length** | | |
| Both sexes combined | 34–76 (50.2 ± 9.4) | 42–49 (45.4 ± 1.9) |
| Males | 40–67 (51.7 ± 7.8) | 42–49 (45.7 ± 2.3) |
| Females | 34–76 (48.8 ± 10.7) | 44–46 (44.8 ± 0.8) |
| **Dorsals between axilla-groin** | | |
| Both sexes combined | 59–98 (78.1 ± 9.0) | 60–93 (80.1 ± 9.1) |
| Males | 59–92 (75.3 ± 8.7) | 68–93 (80.7 ± 8.2) |
| Females | 64–98 (80.6 ± 8.7) | 60–91 (78.8 ± 11.5) |
| **Fourth toe lamellae** | | |
| Both sexes combined | 21–26 (23.9 ± 1.3) | 23–26 (24.0 ± 0.9) |
| Males | 22–26 (24.3 ± 1.1) | 24–26 (24.5 ± 0.7) |
| Females | 21–26 (23.6 ± 1.4) | 23–23 (23.0 ± 0.0) |
| **Fourth finger lamellae** | | |
| Both sexes combined | 6–11 (8.3 ± 1.3) | 7–9 (8.3 ± 0.7) |
| Males | 6–10 (8.3 ± 1.2) | 7–9 (8.4 ± 0.8) |
| Females | 6–11 (8.4 ± 1.4) | 8–9 (8.2 ± 0.4) |
| **Loreal rows** | | |
| Both sexes combined | 4–8 (6.9 ± 0.9) | 6–8 (6.7 ± 0.7) |
| Males | 6–8 (7.2 ± 0.6) | 6–8 (6.8 ± 0.8) |
| Females | 4–8 (6.6 ± 1.0) | 6–7 (6.6 ± 0.5) |
| **Total loreals** | | |
| Both sexes combined | 29–61 (45.4 ± 7.0) | 30–46 (40.8 ± 4.3) |
| Males | 38–61 (46.8 ± 6.1) | 36–46 (40.8 ± 3.5) |
| Females | 29–57 (44.0 ± 7.7) | 30–44 (40.8 ± 6.1) |
| **Supralabials to level below center of eye** | | |
| Both sexes combined | 5–8 (7.0 ± 0.7) | 7–11 (8.1 ± 1.0) |
| Males | 6–8 (7.2 ± 0.5) | 7–11 (8.2 ± 1.2) |
| Females | 5–8 (6.8 ± 0.7) | 7–8 (7.8 ± 0.4) |
| **Infralabials to level below center of eye** | | |
| Both sexes combined | 6–9 (7.4 ± 0.8) | 8–10 (8.4 ± 0.6) |
| Males | 6–9 (7.7 ± 0.7) | 8–10 (8.5 ± 0.7) |
| Females | 6–8 (7.2 ± 0.7) | 8–9 (8.2 ± 0.4) |
| **Scales separating supraorbital semicircles (SOS)** | | |
| Both sexes combined | 1–3 (2.1 ± 0.4) | 2–3 (2.1 ± 0.3) |
| Males | 2–3 (2.2 ± 0.4) | 2–2 (2.0 ± 0.0) |
| Females | 1–3 (2.0 ± 0.3) | 2–3 (2.2 ± 0.4) |

TABLE 4.  CONTINUED.

| | *Norops dariense* (16 males, 18 females) | *Norops cupreus* (10 males, 5 females) |
|---|---|---|
| Scales between interparietal and SOS | | |
| Both sexes combined | 2–4 (2.8 ± 0.5) | 2–4 (3.0 ± 0.4) |
| Males | 2–4 (2.8 ± 0.5) | 2–4 (2.9 ± 0.3) |
| Females | 2–4 (2.8 ± 0.5) | 2–4 (3.3 ± 0.6) |
| Scales between posterior canthals | | |
| Both sexes combined | 7–12 (9.8 ± 1.1) | 9–12 (10.3 ± 0.8) |
| Males | 9–12 (10.3 ± 0.9) | 9–12 (10.4 ± 0.8) |
| Females | 7–11 (9.3 ± 1.1) | 9–11 (10.0 ± 0.7) |
| Scales between second canthals | | |
| Both sexes combined | 7–12 (9.3 ± 1.3) | 7–10 (8.9 ± 0.9) |
| Males | 8–12 (9.7 ± 1.4) | 7–10 (8.9 ± 1.0) |
| Females | 7–11 (8.9 ± 1.2) | 8–10 (8.8 ± 0.8) |
| Postmentals | | |
| Both sexes combined | 5–9 (6.5 ± 0.9) | 6–8 (6.1 ± 0.5) |
| Males | 5–8 (6.3 ± 0.8) | 6–8 (6.2 ± 0.6) |
| Females | 6–9 (6.7 ± 1.0) | 6–6 (6.0 ± 0.0) |
| Postrostrals | | |
| Both sexes combined | 4–9 (5.9 ± 1.3) | 5–7 (6.0 ± 0.4) |
| Males | 4–9 (5.6 ± 1.4) | 6–7 (6.1 ± 0.3) |
| Females | 5–9 (6.1 ± 1.1) | 5–6 (5.8 ± 0.4) |
| Internasals | | |
| Both sexes combined | 5–9 (6.6 ± 1.0) | 6–6 (6.0 ± 0.0) |
| Males | 5–8 (6.3 ± 1.1) | 6–6 (6.0 ± 0.0) |
| Females | 5–9 (6.8 ± 0.9) | 6–6 (6.0 ± 0.0) |
| Scales around midbody | | |
| Both sexes combined | 104–186 (155.9 ± 18.7) | 148–186 (168.3 ± 11.7) |
| Males | 124–170 (158.0 ± 14.1) | 156–184 (169.0 ± 10.7) |
| Females | 104–186 (154.1 ± 22.2) | 148–186 (167.0 ± 14.7) |
| Tail length/SVL | | |
| Both sexes combined | 1.48–2.00 (*n* − 15) | 1.46–1.95 (*n* = 8) |
| Males | 1.79–2.00 (*n* = 5) | 1.80–1.95 (*n* = 4) |
| Females | 1.48–1.95 (*n* = 10) | 1.46–1.63 (*n* = 4) |
| Head length/SVL | | |
| Both sexes combined | 0.24–0.30 | 0.24–0.27 |
| Males | 0.25–0.30 | 0.24–0.27 |
| Females | 0.24–0.29 | 0.24–0.26 |
| Shank length/SVL | | |
| Both sexes combined | 0.27–0.33 | 0.26–0.30 |
| Males | 0.27–0.33 | 0.26–0.30 |
| Females | 0.27–0.33 | 0.26–0.27 |
| Head length/head width | | |
| Both sexes combined | 1.47–2.02 | 1.55–1.78 |
| Males | 1.54–1.88 | 1.55–1.78 |
| Females | 1.47–2.02 | 1.64–1.69 |
| Tail height/tail width | | |
| Both sexes combined | 1.07–1.60 (*n* = 32) | 1.08–1.40 (*n* = 12) |
| Males | 1.07–1.57 (*n* = 15) | 1.15–1.31 (*n* = 7) |
| Females | 1.08–1.60 (*n* = 17) | 1.08–1.40 (*n* = 5) |

TABLE 4. CONTINUED.

| | *Norops dariense*<br>(16 males, 18 females) | *Norops cupreus*<br>(10 males, 5 females) |
|---|---|---|
| Head width/SVL | | |
| Both sexes combined | 0.13–0.19 | 0.14–0.16 |
| Males | 0.13–0.18 | 0.14–0.16 |
| Females | 0.14–0.19 | 0.15–0.15 |
| Axilla-groin distance/SVL | | |
| Both sexes combined | 0.28–0.51 | 0.40–0.50 |
| Males | 0.28–0.47 | 0.40–0.50 |
| Females | 0.36–0.51 | 0.40–0.50 |
| Shank length/head length | | |
| Both sexes combined | 0.96–1.22 | 1.00–1.14 |
| Males | 1.00–1.22 | 1.02–1.14 |
| Females | 0.96–1.15 | 1.00–1.12 |

ranges. Given the small sample size (only five females in one group), we can expect a greater range of variation and even more overlap once additional specimens are added to this analysis. We prefer to maintain *N. dariense* in the synonymy of *N. cupreus*.

*Norops cupreus* is a member of the *N. cupreus* species subgroup (Savage and Guyer, 1989) of the *N. auratus* species group of Nicholson et al. (2012). It is the only member of that subgroup in Honduras. However, a morphological analysis by Poe (2004) did not recover a monophyletic *N. cupreus* species group as recognized by Savage and Guyer (1989). *Norops cupreus* formed a clade with *N. polylepis* (Peters) of Costa Rica and Panama in Nicholson et al. (2005, 2012). Savage and Guyer (1989) had placed *N. polylepis* in their *N. fuscoauratus* species group.

*Natural History Comments. Norops cupreus* is known from near sea level to 1,300 m elevation in the Lowland Moist Forest, Lowland Dry Forest, Lowland Arid Forest, Premontane Wet Forest, and Premontane Moist Forest formations. This species is both terrestrial and arboreal and is unique among Honduran anoles in that it occurs in areas varying from closed canopy humid broadleaf rainforest to open subhumid sunny areas with little leafy vegetation available. Individuals in closed canopy forest are active in both sunny and shaded

situations, whereas those at the other extreme in open subhumid areas are usually active in shaded areas. Those active on the ground frequently retreat to the nearest tree or boulder and climb when pursued. Others can be found in and near rock piles, where they retreat when pursued. One was active in a mixed shady-sunny spot in a banana plant during late afternoon in an area formerly of mixed dry forest and gallery forest. The species also sleeps on low vegetation at night. It has been collected during every month of the year. A specimen of *N. cupreus* from Honduras (UF 144636) was removed from the stomach of the pit viper *Bothriechis schlegelii* (UF 141057), and another is in the stomach of a small snake *Coniophanes fissidens* (USNM 561022). The ecology of *N. cupreus* has been extensively studied in the hot and dry lowlands of northwestern Costa Rica. The following is taken from Savage (2002), who summarized the literature resulting from those studies. Individuals tend to hide during the day, are active near sunset during the hot times of the year, but bask in the sun during less-hot periods. They spend much of their active period foraging in leaf litter but also forage on a wide variety of elevated areas. They feed on a variety of arthropods, but predominately on ants, lepidopteran larvae, and spiders. Reproduction is essentially restricted to the height of the rainy

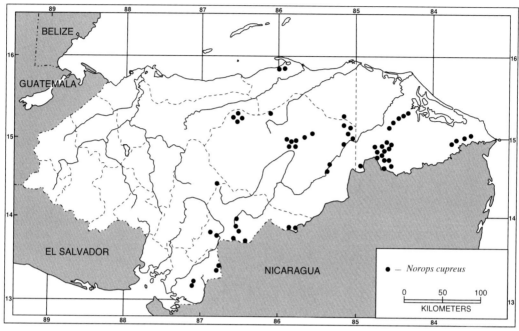

Map 9.   Localities for *Norops cupreus*. Solid symbols denote specimens examined.

season, with females laying a single egg every 7–10 days.

*Etymology.* The name *cupreus* is Latin (of copper, coppery) and alludes to the copper dorsal coloration described for the type series by Hallowell (1861).

*Specimens Examined* (221 [28]; Map 9). CHOLUTECA: Cerro Guanacaure, SDSNH 72735; Finca El Rubí, SDSNH 72733; La Fortuna, SDSNH 72729–32, 72734, 72737–41, 72743–50; Quebrada de La Florida, SDSNH 72754; San Marcos de Colón, SMF 77314–15; Tres Pilas, SDSNH 72736. COLÓN: Cerro Calentura, LSUMZ 33675, USNM 541030–31; Quebrada Machín, USNM 541040–42; 1 km SSW of Trujillo, KU 101394–95; Trujillo, LSUMZ 22429. EL PARAÍSO: Agua Fría, AMNH 70355; Arenales, LACM 16855 (listed as LACM 16856 by Meyer and Wilson, 1973); Boca Español, SMF 86986; Danlí, BYU 18192–94; El Rodeo, USNM 580699–700; Las Manos, UTA R-52141–42; La Viña de San Marcos, USNM 541036–39; Mapachín, SMF 91758; near Yuscarán, AMNH 70381. FRANCISCO

MORAZÁN: El Chile, SMF 80885–88. GRACIAS A DIOS: Awasbila, USNM 563185; Bachi Kiamp, SMF 91253–55, 91760, USNM 565441; Bodega de Río Tapalwás, UF 144630–34, USNM 563204–12; Cabeceras de Río Rus Rus, SMF 86978; Caño Awalwás, UF 144622–26, USNM 563176–84; Caño Awawás, USNM 563186–90; Cerros de Sabaní, SMF 86979; Cerro Wahatingni, UF 144636; Cerro Wisplini, SMF 91766; Crique Wahatingni, SMF 86990, USNM 563213; near Cueva de Leimus, SMF 91334; Hiltara Kiamp, SMF 86977, USNM 563191; SW end of Isla de Venado, SMF 91256–58; Kakamuklaya, USNM 573125; Karasangkan, USNM 563192–94; Kaska Tingni, USNM 563195; Kasunta, USNM 563196–97; Kipla Tingni Kiamp, SMF 86976; Krahkra, SMF 91259, USNM 563198; Leimus (Río Warunta), FMNH 282578–80, USNM 565440; Rawa Kiamp, SMF 91260; Rus Rus, USNM 563199–200, 573126; Sadyk Kiamp, SMF 91261–63, 92847; San San Hil, SMF 86987–89, USNM 563201–03; San San Hil Kiamp,

SMF 86981; Sisinbila, SMF 91761; Swabila, UF 144627–29; Tikiraya, UF 144635; Warunta Tingni Kiamp, SMF 86980, USNM 563214–18, 565436–39. OLANCHO: 0.5–1.0 km WNW of Catacamas, LSUMZ 21371–72; 6 km NW of Catacamas, UF 90222; 5 km E of Catacamas, LACM 45160–64, 45227–31, 47667–70, 47674–75; about 9 km E of Catacamas, LACM 45133; 1–3 km NW of Catacamas, LACM 47671–73; LSUMZ 21373–75; 4.5 km SE of Catacamas, LSUMZ 21376–79; between Catacamas and Dulce Nombre de Culmí, UTA R-54003–05; Cuaca, UTA R-53906, 53909, 53918, 54001–02; between El Díctamo and Parque Nacional La Muralla Centro de Visitantes, USNM 342279, 563219; El Murmullo, SMF 91754–55; Montaña de Liquidambar, USNM 342282; Montaña del Ecuador, USNM 342280–81; Piedra Blanca, SMF 91757, 91759; Quebrada de Agua, SMF 91763; Quebrada de Las Escaleras, USNM 342283–84; between Quebrada de Las Escaleras and Quebrada Las Cantinas, USNM 342286–87; Quebrada El Guásimo, SMF 80820–21, 86975, USNM 563220; Quebrada Las Cantinas, USNM 342285; Quebrada Salitre Lajas, USNM 342288–89; Qururia, USNM 563225; Raudal La Caldera, USNM 541035; confluence of Ríos Aner and Wampú, USNM 541032–33; Río Cuaca, SMF 91333; Río Kosmako, USNM 563221–24; Río Seco, USNM 563226; 4 km E of San José del Río Tinto, LACM 45146; Talgua, SMF 91756; Terrero Blanco, USNM 342290; Yapuwás, USNM 541034.

*Norops cusuco* McCranie, Köhler, and Wilson
*Anolis laeviventris*: Wilson et al., 1991: 70; Köhler et al., 2007: 391 (in part).
*Norops laeviventris*: Anonymous, 1994: 116; McCranie et al., 2000: 223 (in part); Köhler et al., 2001: 254 (in part); Wilson et al., 2001: 136 (in part); McCranie, 2005: 20.
*Norops cusuco* McCranie, Köhler, and Wilson, 2000: 214 (holotype, SMF 78842; type locality: "Parque Nacional El Cusuco Centro de Visitantes [15°29.92′N, 88°12.88′W], 1550 m elevation, Departamento de Cortés, Honduras"); Köhler, 2003: 98, 2008: 105; Wilson and McCranie, 2003: 59, 2004b: 43, 2004c: 24.
*Anolis cusuco*: Townsend, 2006: 35, 2009: 298; Townsend et al., 2006: 35; Köhler et al., 2007: 391; Townsend and Wilson, 2008: 154 (in part).

*Geographic Distribution.* *Norops cusuco* occurs at intermediate elevations in the environs of Parque Nacional Cusuco in Cortés and Santa Bárbara, and in Parque Nacional Cerro Azul in Copán, in northwestern Honduras.

*Description.* The following is based on eight males (KU 194278–79, 194285; SMF 78842, 79170, 79179, 93368; USNM 578744) and 12 females (KU 194276–77, 194280–82, 194284; SMF 79171, 79182; USNM 532567–68, 565434–35). *Norops cusuco* is a small anole (SVL 46 mm in largest male [SMF 78842], 43 mm in largest females [USNM 532567–68]); dorsal head scales keeled in internasal and prefrontal areas, smooth in frontal area, most scales smooth in parietal area; deep frontal depression present; parietal depression shallow; 3–7 ($4.7 \pm 1.0$) postrostrals in 15; anterior nasal divided or single, lower edge contacting rostral and first supralabial when single; 3–6 ($5.1 \pm 0.9$) internasals; canthal ridge sharply defined; scales comprising supraorbital semicircles smooth or faintly keeled, largest scale in semicircles larger than largest supraocular scale; supraorbital semicircles well defined; supraorbital semicircles usually in broad contact medially, occasionally 1 scale preventing contact; 1–3 ($1.8 \pm 0.6$) scales separating supraorbital semicircles and interparietal at narrowest point; interparietal well defined, irregular in outline, larger than ear opening; 2–3 rows of about 3–7 (total number) enlarged, smooth, or faintly keeled supraocular scales; enlarged supraoculars completely separated from supraorbital semicircles by 1 row of small scales; 2 elongate superciliaries, posterior shortest; usually 3 enlarged canthals; 7–10 ($8.2 \pm 0.7$) scales between second canthals; 8–11

(8.9 ± 0.9) scales between posterior canthals; loreal region slightly concave, 17–43 (30.1 ± 6.4) keeled loreal scales in maximum of 5–6 (5.4 ± 0.5) horizontal rows; 5–8 (6.3 ± 0.6) supralabials and 4–7 (5.6 ± 0.7) infralabials to level below center of eye in 15; suboculars distinctly keeled, 1–2 suboculars usually in contact with supralabials (rarely 1 scale separating suboculars from supralabials); ear opening vertically oval; scales anterior to ear opening not granular, slightly larger than those posterior to ear opening; 4–6 (5.3 ± 0.8) postmentals, outer pair usually largest; gular scales faintly keeled; male dewlap large, extending onto chest well posterior to level of axilla; male dewlap with about 10–11 horizontal gorgetal-sternal scale rows, about 8 mean number of scales per row ($n = 2$); 2 (modal number) anterior marginal pairs in male dewlap; female dewlap absent; no nuchal crest or dorsal ridge; about 0–4 middorsal scale rows slightly enlarged, keeled, dorsal scales lateral to middorsal series grading into lateral scales (lateral scales slightly enlarged in some, granular in others); a few or no enlarged scales scattered among laterals; 69–96 (80.4 ± 9.4) dorsal scales along vertebral midline between levels of axilla and groin in seven males, 69–99 (81.4 ± 10.3) in eight females; 42–61 (48.8 ± 7.7) dorsal scales along vertebral midline contained in 1 head length in males, 36–54 (42.4 ± 4.2) in females; ventral scales on midsection larger than largest dorsal scales; ventral body scales keeled, imbricate; about 45–67 (56.0 ± 7.9) ventral scales along midventral line between levels of axilla and groin in six males, 43–57 (49.3 ± 4.5) in eight females; 29–43 (35.4 ± 5.0) ventral scales contained in 1 head length in males, 24–36 (28.6 ± 3.3) in females; 130–164 (141.1 ± 13.4) scales around midbody in seven males, 120–154 (134.8 ± 10.5) in eight females; tubelike axillary pocket absent; precloacal scales not keeled; enlarged postcloacal scales present in males; tail rounded to slightly compressed, TH/TW 1.00–1.50 in 19; basal subcaudal scales smooth to keeled; lateral caudal scales keeled, homogeneous; dorsal medial caudal scale row slightly enlarged,

keeled, not forming crest; most scales on anterior surface of antebrachium unicarinate; 22–28 (25.6 ± 1.6) subdigital lamellae on Phalanges II–IV of Toe IV of hind limbs; 5–9 (6.6 ± 1.0) subdigital scales on Phalanx I of Toe IV on 28 sides of hind limbs; SVL 37.0–46.2 (41.3 ± 3.5) mm in males, 38.5–43.3 (41.2 ± 1.7) mm in females; TAL/SVL 1.43–1.67 in three males, 1.29–1.74 in seven females; HL/SVL 0.25–0.29 in males, 0.25–0.28 in females; SHL/SVL 0.20–0.23 in males, 0.19–0.22 in females; SHL/HL 0.71–0.88 in males, 0.72–0.87 in females; longest toe of adpressed hind limb usually reaching ear opening.

Color in life of the male holotype (SMF 78842) was described by McCranie et al. (2000: 217): "dorsal and lateral surfaces of head and body pale grayish brown with some dark brown shadings; a dirty white supralabial stripe and a dark brown interorbital bar present; venter dirty white with some pale gray shadings; dewlap dirty white." Color in life of a series from the type locality: (KU 194278; adult male) dorsum pale grayish tan with cream stripe extending posteriorly from shoulder region; belly dirty white; dewlap pale yellow; (KU 194279; adult male) same as that of KU 194278, except with yellowish tinge on ventral surfaces of hind limbs and base of tail; (KU 194280; adult female) dorsum pale yellowish tan; middorsal stripe pale golden rust with faintly defined brown borders; tail colored as middorsal stripe; belly and dewlap white with pale yellow tinge; (KU 194281; adult female) same as that for KU 194280, except middorsal stripe pale tan; (KU 194282; adult female) same as that for KU 194280, except lacking middorsal stripe. Dewlap color of another adult male (SMF 93368) was Pale Horn Color (92) with similarly colored scales. Color in life of another adult male (SMF 79179): dorsal surface of head rust gray with brown interocular bar; dorsal surface of body golden rust with darker markings on middorsal area; dorsal surfaces of limbs golden yellow; supralabials and loreal region cream; chin cream, heavily mottled with brown;

venter cream; tail tan above, rust orange below, pale yellow at base; dewlap pale yellow, some scales with small brown spots. Color in life of an adult female (USNM 565450): dorsum Straw Yellow (56) with middorsal series of Hair Brown (119A) chevrons on Fawn Color (25) swath; top of head Straw Yellow with brown interocular bar; side of head Straw Yellow; forelimbs Straw Yellow with slightly darker transverse bars; hind limbs Straw Yellow with brown-outlined Spectrum Yellow (55) spots on shanks; dorsal surface of tail with series of diamond-shaped pale markings separated along length of tail by brown interspaces on Straw Yellow background; venter Cream Color (54). Color in life of another adult female (FMNH 236388): head pale rust red with gray spots on cream-colored supraoculars and dark brown spotting on temporal and interocular areas; dorsal surface of body golden tan with golden brown middorsal band narrowing anteriorly to golden stripe; dark brown lateral stripe present, stripe widening somewhat posteriorly, located between axilla and groin; dorsal surfaces of limbs tan with brown crossbars; chin cream with brown punctations; belly cream.

Color in alcohol: dorsal surfaces of head and body grayish-brown with some brown mottling along dorsal midline of body; dorsal surfaces of limbs grayish brown with indistinct brown crossbars; dorsal surface of tail brown with indistinct darker brown crossbands; ventral surfaces of head and body grayish white with some brown flecking; subcaudal surface brown.

Hemipenis: the fully everted hemipenis of SMF 78842 is a rather stout organ with only rudimentary lobes; sulcus spermaticus bordered by well-developed sulcal lips, bifurcating at base of apex; large asulcate processus present; no particular surface structure discernible on either truncus or apex.

*Diagnosis/Similar Species. Norops cusuco* is distinguished from all other Honduran *Norops*, except *N. laeviventris* and *N. kreutzi*, by the combination of having a few to no slightly enlarged middorsal scale rows, heterogeneous lateral scales in some, keeled ventral scales, male dewlap dirty white, pale yellow, or pale gray in life, and a pair of enlarged postcloacal scales in males. *Norops cusuco* differs from *N. laeviventris* by having a larger male dewlap (extending well onto chest posterior to level of axilla in *N. cusuco* versus to about level of axilla in *N. laeviventris*). *Norops cusuco* differs from *N. kreutzi* by having a dirty white, pale yellow, or pale gray male dewlap in life, with gorgetal scales of same color (dewlap pale yellow with purple gorgetal scales in life in *N. kreutzi*). Those two species also occur in isolated mountain ranges with about 55 km of unsuitable habitat separating each other's closest known localities.

*Illustrations* (Figs. 19, 20). McCranie et al., 2000 (adult, head scales, head and dewlap, hemipenis); Köhler, 2003 (adult), 2008 (adult); Townsend and Wilson, 2008 (adult; as *Anolis*).

*Remarks. Norops cusuco* is a member of the *N. laeviventris* species subgroup (McCranie et al., 2000) of the *N. auratus* species group of Nicholson et al. (2012). The other Honduran members of this subgroup are *N. kreutzi* and *N. laeviventris*. Köhler and Obermeier (1998) and Köhler et al. (1999) believed *N. laeviventris* to be part of the *N. crassulus* species group. However, molecular and morphological data in Nicholson (2002), Poe (2004), and Nicholson et al. (2005) do not support that relationship. The Honduran members of the *N. laeviventris* subgroup are characterized by having white to pale yellow male dewlaps, strongly keeled ventral scales, and a small size. *Norops cusuco* was not included in the phylogenetic analysis of Nicholson et al. (2012).

*Natural History Comments. Norops cusuco* is known from 1,350 to 1,990 m elevation in the Lower Montane Wet Forest formation and peripherally in the Premontane Wet Forest formation. It is usually active on tree branches and tree trunks about 50–150 cm above the ground, and one was in a sunny spot on a small log at about 10:00 a.m. Townsend and Wilson (2008)

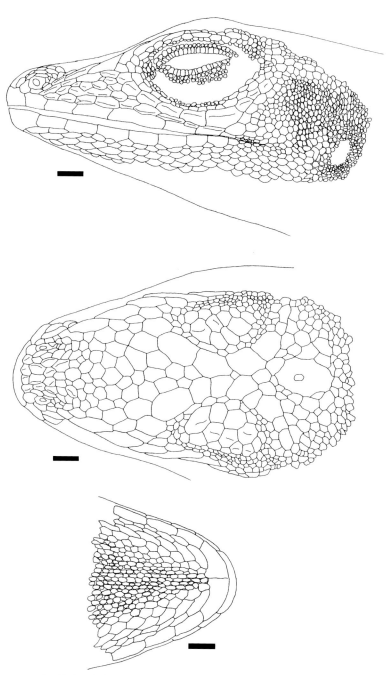

Figure 19.   *Norops cusuco* head in lateral, dorsal, and ventral views. SMF 78842, adult male from El Cusuco, Cortés. Scale bar = 1.0 mm. Drawing by Gunther Köhler.

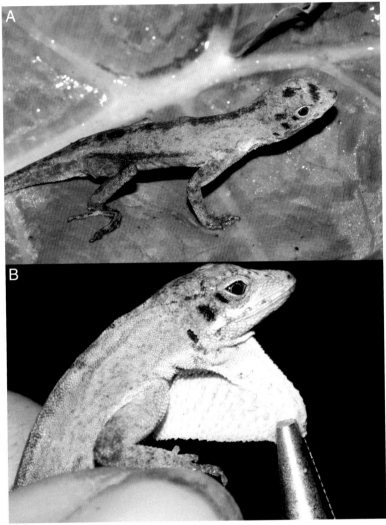

Figure 20. *Norops cusuco*. (A) Adult male (SMF 93368) from Quebrada Las Piedras, Copán; (B) adult male dewlap (SMF 93368). Photographs by James R. McCranie.

reported finding it on vegetation up to 2.5 m above the ground. This species sleeps on leaves and stems of low vegetation at night. *Norops cusuco* was collected from March to September and during November and is probably active throughout the year under favorable conditions. Nothing has been reported on its diet or reproduction.

*Etymology.* The name *cusuco* refers to Parque Nacional Cusuco, where its type locality is located.

*Specimens Examined* (36 [0]; Map 10). COPÁN: Quebrada Grande, FMNH 236388, SMF 79179, USNM 532569, 565450; Quebrada Las Piedras, SMF 93368. CORTÉS: El Cusuco, KU 194275–85, SMF 78842, 79170–71, 79182, UF 142740, 144213, 149649–50, 166193, USNM 532567–68, 565434–35, 578743–45; Guanales Camp, UF 149651, 149653; Quebrada de Cantiles, UF 149655. SANTA BÁRBARA: La Fortuna Camp, UF 149652.

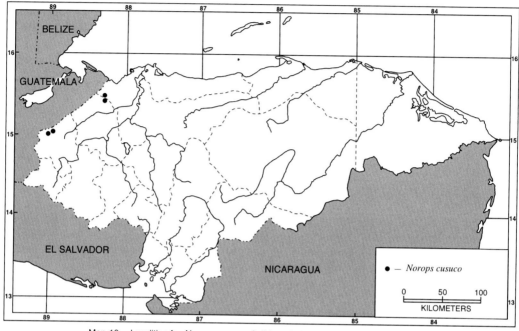

Map 10.   Localities for *Norops cusuco*. Solid symbols denote specimens examined.

*Norops heteropholidotus* (Mertens)

*Anolis heteropholidotus* Mertens, 1952a: 89 (holotype, SMF 43041; type locality: "Oberhalb Hacienda Los Planes am Miramundo, 2000 m H., Dept. Santa Ana, El Salvador"); Köhler et al., 2007: 391; Townsend and Wilson, 2009: 63; Köhler, 2014: 210.

*Anolis sminthus*: Meyer, 1969: 232 (in part); Meyer and Wilson, 1972: 108 (in part), 1973: 20 (in part); Wilson et al., 1979b: 62.

*Norops heteropholidotus*: Villa et al., 1988: 49; Köhler and McCranie, 1998: 12; Köhler and Obermeier, 1998: 136; Köhler et al., 1999: 298, 2001: 254; Köhler, 2003: 101, 2008: 107; Wilson and McCranie, 2003: 59 (in part), 2004b: 43; Cruz et al., 2006: 3–6.

*Norops sminthus*: McCranie et al., 1992: 208 (in part); Wilson et al., 2001: 136 (in part); Wilson and McCranie, 2003: 59 (in part).

*Norops* cf. *sminthus*: Köhler and Obermeier, 1998: 136 (in part); Köhler et al., 1999: 298 (in part).

*Anolis* cf. *sminthus*: Townsend and Wilson, 2009: 63 (in part).

*Geographic Distribution.* *Norops heteropholidotus* occurs at intermediate elevations in northwestern El Salvador, southeastern Guatemala, and southwestern Honduras on the Pacific versant and in southwestern Honduras on the Atlantic versant. In Honduras, this species occurs along both sides of the Continental Divide in the mountains of the southwestern portion of the country from western Ocotepeque to eastern Intibucá.

*Description.* The following is based on 14 males (KU 219969–71, 219974–75; MVZ 39974; SMF 78013–14, 78017, 78020–21, 78024–25, 78384) and 15 females (KU 219972–73, 219976; SMF 77741, 78015–16, 78018–19, 78022–23, 78026–30). *Norops heteropholidotus* is a medium-sized anole (SVL 50 mm in largest Honduran male [SMF 78017] examined, 58 mm in largest Honduran female [KU 219976] examined; Townsend and Wilson [2009] reported a maximum SVL 51 mm in males and 59 mm in females); dorsal head scales rugose or keeled in internasal, prefrontal, frontal, and parietal regions; frontal depression present; parietal depression absent; 3–7 (5.1 ± 1.1) postrostrals; anterior nasal divided, lower section contacting rostral and first supralabial, or only rostral; 3–7 (5.4 ± 1.0) internasals; canthal ridge sharply defined; scales comprising supraorbital semicircles rugose or weakly keeled, largest scale in semicircles larger than largest supraocular scale; supraorbital semicircles well defined; 0–2 (1.0 ± 0.6) scales separating supraorbital semicircles at narrowest point in 21; 2–4 (2.6 ± 0.6) scales separating supraorbital semicircles and interparietal at narrowest point; interparietal well defined, slightly to distinctly enlarged relative to adjacent scales, surrounded by scales of moderate size, longer than wide, smaller than ear opening; 2–3 rows of about 5–8 (total number) enlarged, smooth to faintly keeled supraocular scales; 1–3 enlarged supraoculars in broad contact with supraorbital semicircles; 2 elongate superciliaries, posterior only slightly shorter than anterior; usually 3 enlarged canthals; 5–8 (5.9 ± 0.8) scales between second canthals; 6–11 (7.8 ± 1.2) scales between posterior canthals; loreal region slightly concave, 16–30 (22.8 ± 4.2) mostly strongly keeled loreal scales in maximum of 3–6 (4.3 ± 0.6) horizontal rows; 6–8 (6.8 ± 0.6) supralabials and 6–9 (7.1 ± 0.9) infralabials to level below center of eye; suboculars weakly to strongly keeled, usually 2 suboculars in broad contact with supralabials; ear opening vertically oval; scales anterior to ear opening not granular, slightly larger than those posterior to ear opening; 4–6 (4.2 ± 0.5) postmentals, outer pair usually largest; keeled granular scales present on chin and throat; male dewlap moderately large, extending past level of axilla onto chest; male dewlap with 5–8 horizontal gorgetal-sternal scale rows, about 5–9 scales per row (*n* = 7); 1–2 (modal number) anterior marginal pairs in male dewlap; female dewlap relatively well developed, but smaller than male dewlap; low nuchal crest and low dorsal ridge present in males; about 8–10 middorsal scale rows distinctly enlarged, keeled, without small scales irregularly interspersed among the enlarged dorsal scales; dorsal scales lateral to middorsal series abruptly larger than granular lateral scales; flank scales heterogeneous, solitary enlarged keeled or elevated scales scattered among laterals; 32–46 (41.7 ± 4.7) dorsal scales along vertebral midline between levels of axilla and groin in males, 35–65 (49.2 ± 8.2) in females; 24–40 (29.9 ± 5.3) dorsal scales along vertebral midline contained in 1 head length in males, 22–36 (29.2 ± 4.3) in females; ventral scales on midsection about same size as largest dorsal scales; ventral body scales smooth, flat, 28–54 (40.6 ± 7.6) ventral scales along midventral line between levels of axilla and groin in males, 26–43 (36.0 ± 4.8) in females; 22–42 (30.9 ± 5.2) ventral scales contained in 1 head length in males, 20–28 (23.7 ± 2.7) in females; 90–138 (116.6 ± 13.4) scales around midbody in males, 84–136 (119.1 ± 12.4) in females; tubelike axillary pocket absent; precloacal scales not keeled; pair of greatly enlarged postcloacal scales in males; tail rounded to distinctly compressed, TH/TW 1.00–1.62 in 28; all subcaudal scales keeled; lateral caudal scales keeled, homogeneous, although indistinct division in segments discernable; dorsal medial caudal scale row enlarged, keeled, not forming crest; scales on anterior surface of antebrachium distinctly keeled, unicarinate; 23–29 (24.4 ± 1.3) subdigital lamellae on Phalanges II–IV of Toe IV of hind limbs; 5–9 (6.7 ± 0.9) subdigital scales on Phalanx I of Toe IV of hind limbs; SVL 38.0–50.0 (44.4 ± 3.7) mm in males, 44.0–58.0 (50.6 ± 3.3) mm in females; TAL/SVL 2.23–2.44 in four males, 1.86–2.37 in six females; HL/SVL 0.26–0.32 in males, 0.25–0.28 in females; SHL/SVL 0.23–0.27 in males, 0.22–0.26 in females; SHL/HL 0.80–0.92 in males, 0.80–0.94 in females; longest toe of adpressed hind limb usually reaching between ear opening and mideye.

Color in life of an adult male (KU 219975): dorsal surface of head brown with dark brown interocular bar; dorsal surface of body pale brown with brownish gray middorsal line, on either side of this line is series of three brown triangular blotches; lateral surface of body mottled brown and gray; dorsal surfaces of forelimbs brown with gray crossbars; dorsal surfaces of hind limbs rust brown with gray crossbars; proximal third of tail gray middorsally, mottled gray and brown laterally, the two areas separated by rust brown line, remainder of tail rust brown with gray crossbars; temporal region with thin brown lines radiating posteriorly; chin white, heavily mottled with gray; belly cream; dewlap bright red-orange with white scales; iris bronze. Color in life of an adult female (SMF 91264): middorsum with Cinnamon (39) stripe with darker brown mottling; dorsolateral field Light Drab (119C); lateral field darker brown than dorsolateral field; dorsal and lateral surfaces of head Light Drab with pale yellow subocular spot; iris pale brown on anterior half, darker brown on posterior half; chin and belly pale brown; dewlap Chrome Orange (16) with cream scales.

Color in alcohol: dorsal surfaces of head and body some shade of brown; middorsal pattern variable, some specimens lack pattern, some have dark brown chevrons; most females have a thin vertebral pale line, occasional females have a broad pale middorsal stripe bordered by dark brown, or a pattern of dark brown diamonds; lateral surface of head pale grayish brown; dorsal surfaces of limbs brown, usually with indistinct dark brown crossbands; dorsal surface of tail brown, some have darker brown chevrons or crossbands, those markings more evident proximally.

Hemipenis: the completely everted hemipenis of SMF 78024 is a medium-sized, slightly bilobed organ; sulcus spermaticus bordered by well-developed sulcal lips, opening into single broad concave area at base of apex; lobes calyculate; truncus with transverse folds; single asulcate processus present.

*Diagnosis/Similar Species. Norops heteropholidotus* is distinguished from all other Honduran *Norops*, except *N. muralla* and *N. sminthus*, by the combination of having about 8–10 distinctly enlarged middorsal scale rows, heterogeneous lateral scales, smooth, imbricate ventral scales, a red male dewlap in life, and a pair of greatly enlarged postcloacal scales in males. *Norops heteropholidotus* differs from *N. muralla* by having the medial dorsal scales uniform in size, without interspersed small scales (small scales irregularly interspersed among enlarged medial dorsal scales in *N. muralla*). *Norops heteropholidotus* differs from *N. sminthus* by having smooth midventral scales (weakly keeled in *N. sminthus*). *Norops amplisquamosus* shares most of the characters listed above with *N. heteropholidotus*, but differs most notably in having an orange-yellow male dewlap in life (red dewlap in *N. heteropholidotus*).

*Illustrations* (Figs. 21, 22, 91, 100, 101). Köhler, 1996c (body scales), 2000 (head and dewlap), 2003 (adult, head and dewlap), 2008 (adult, head and dewlap), 2014 (dorsal and flank scales; as *Anolis*); Köhler and McCranie, 1998 (head and dewlap); Köhler and Obermeier, 1998 (hemipenis); Leenders and Watkins-Colwell, 2004 (adult, dewlap); Köhler et al., 2005b (adult, head scales).

*Remarks.* Meyer and Wilson (1972) placed *Norops heteropholidotus* in the synonymy of *N. sminthus* (as *Anolis*). Köhler (1996c) demonstrated that *N. heteropholidotus* is a species distinct from *N. sminthus*. This species is a member of the *N. crassulus* species subgroup (Köhler and Obermeier, 1998; Köhler et al., 1999, 2001; also see Remarks for *N. amplisquamosus*) of the *N. auratus* species group of Nicholson et al. (2012). Nicholson et al. (2012) did not include *N. heteropholidotus* as a valid species of *Norops*, thus apparently considering it synonymous with *N. sminthus*. *Norops heteropholidotus*, *N. amplisquamosus*, *N. crassulus*, *N. muralla*, and *N. sminthus* are also

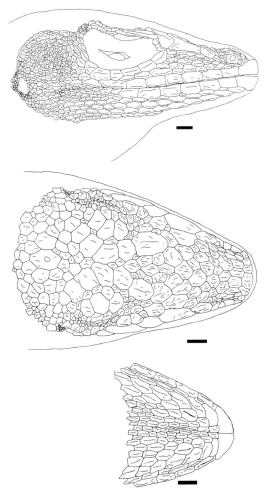

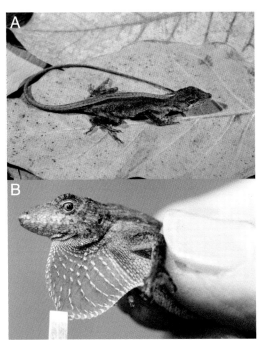

Figure 22. *Norops heteropholidotus*. (A) Adult female (SMF 91264) from El Rodeo, Intibucá; (B) adult male dewlap (from series SMF 78020–30) from Quebrada La Quebradona, Ocotepeque. Photographs by James R. McCranie (A) and Gunther Köhler (B).

Figure 21. *Norops heteropholidotus* head in lateral, dorsal, and ventral views. SMF 78015, adult female from Sumpul, Ocotepeque. Scale bar = 1.0 mm. Drawing by Gunther Köhler.

similar to each other in being rather slow in their efforts to escape human pursuit.

Leenders and Watkins-Colwell (2003, 2004) discussed sexual dimorphism in dewlap color in this species in El Salvador. The specimen listed as *N. sminthus* from El Salvador (MCZ R-57280) by Poe (2004) actually represents *N. heteropholidotus*.

*Natural History Comments. Norops heteropholidotus* is known from 1,860 to 2,200 m elevation in the Lower Montane Moist Forest formation. This species is active on the ground, low on tree trunks, on piles of lumber, and in low vegetation. It appears to be most common in forest edge situations that receive periods of direct sunlight. At one site, (Sumpul, Ocotepeque) it was particularly common in low vegetation along a dirt road. *Norops heteropholidotus* basks in the sun and is also active in shady areas, and can remain active during short cloudy periods during midday under cooler conditions. Individuals of *N. heteropholidotus* are deliberate in their movements and are easily captured. The species was found during January, February, May, June, and September and is probably active throughout the year under favorable conditions. Nothing has been reported on its diet or reproduction.

*Etymology.* The name *heteropholidotus* is formed from the Greek *heteros* (different) and *pholidotos* (clad in scales), and refers to the enlarged scales scattered among the granular lateral scales in this species.

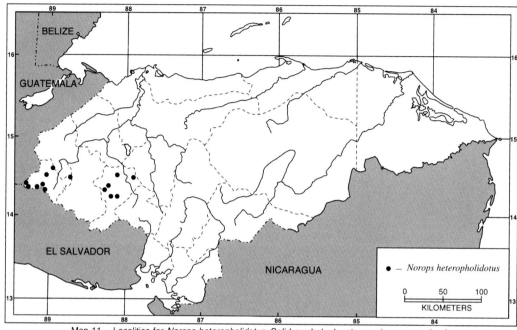

Map 11.   Localities for *Norops heteropholidotus*. Solid symbols denote specimens examined.

*Specimens Examined* (82 [4]; Map 11). INTIBUCÁ: El Rodeo, SMF 91264; 20 km NNE of Jesús de Otoro, LSUMZ 33683; 2.4 km NE of La Esperanza, LACM 47682 (listed as LACM 47862 by Meyer and Wilson, 1973); San Pedro La Loma, KU 219969, SMF 78384, UF 166279–81, 166283–88; Segua, SMF 91265; Zacate Blanco, KU 194299, LSUMZ 38825. OCO-TEPEQUE: Cerro El Pital, MVZ 39971, 39974; El Chagüitón, KU 219970–73; El Portillo de Cerro Negro, SMF 91267–72; 3.6 km S of El Portillo de Ocotepeque, MVZ 263869; near El Portillo de Ocotepeque, UF 166194–210; El Portillo de Ocotepeque, KU 219974–76, LACM 47679–81, SMF 77741; Las Hojas, FMNH 283600; Quebrada El Comatal, SMF 91273–74; near Quebrada La Quebradona, SMF 78020–30, USNM 573127; Sumpul, SMF 78013–19, 91266, USNM 573128–31; "no other data," UTA R-52146.

*Norops johnmeyeri* (Wilson and McCranie)
*Anolis johnmeyeri* Wilson and McCranie,
    1982: 133 (holotype, LSUMZ 37834; type

locality: "El Cusuco [15°30′N, 88°13′W], 5.6 km WSW Buenos Aires [the latter locality about 19 km N Cofradía], 1580 me-ters elevation, Sierra de Omoa, Depto. Cortés, Honduras"); McCranie et al., 1984: 337; Vanzolini, 1986: 4; Rossman and Good, 1993: 9; Anonymous, 1994: 116; Nieto-Montes de Oca, 1994a: 334, 1994b: 113, 1996: 26, 2001: 52; Townsend, 2006: 35, 2009: 298; Townsend and Wilson, 2006: 245, 2008: 156; Townsend et al., 2006: 35; Köhler et al., 2007: 391; Köhler, 2014: 210.
*Norops johnmeyeri*: Villa et al., 1988: 49; Campbell et al., 1989: 235; McCranie et al., 1993a: 386, 1993b: 394; Köhler, 2000: 69, 2003: 100, 2008: 107; Köhler et al., 2001: 255; Wilson et al., 2001: 136; Wilson and McCranie, 2003: 59, 2004b: 43, 2004c: 24; McCranie, 2005: 20; Köhler and Smith, 2008: 223; Snyder, 2011: 8.

*Geographic Distribution. Norops john-meyeri* occurs at moderate and intermediate elevations and is known from the vicinity of

Quebrada Grande, Copán, and in Parque Nacional Cusuco, Cortés, on the Atlantic versant of northwestern Honduras.

*Description.* The following is based on 10 males (SMF 77757, 77761, 78824, 86983; USNM 322903, 322905, 322908, 549363, 549365–66) and 14 females (KU 192623; LSUMZ 37834; SMF 77755–56, 77758, 77760, 78825–26, 86982, 86984; USNM 322902, 322904, 322907, 549364). *Norops johnmeyeri* is a relatively large anole (SVL 73 mm in largest male [SMF 86983], 68 mm in largest female [SMF 78825]); dorsal head scales keeled in internasal, prefrontal, frontal, and parietal areas; moderate frontal and parietal depressions present; 4–10 (6.2 ± 1.6) postrostrals; anterior nasal divided, lower section usually contacting rostral and first supralabial or only first supralabial, occasionally lower anterior nasal separated from rostral and first supralabial by row of scales; 5–8 (6.4 ± 0.8) internasals; canthal ridge sharply defined; scales comprising supraorbital semicircles keeled, largest scale in semicircles about same size as, or larger than, largest supraocular scale; supraorbital semicircles well defined; 1–5 (2.9 ± 0.9) scales separating supraorbital semicircles at narrowest point; 3–7 (4.9 ± 1.1) scales separating supraorbital semicircles and interparietal at narrowest point; interparietal well defined, moderately to greatly enlarged relative to adjacent scales, surrounded by scales of moderate size, longer than wide, smaller than ear opening; 3–4 rows of about 3–8 (total number) enlarged, keeled supraocular scales; enlarged supraoculars in broad contact with supraorbital semicircles; 2 elongate superciliaries, posterior much shorter than anterior; usually 3 enlarged canthals; 7–10 (8.1 ± 1.1) scales between second canthals; 7–12 (9.7 ± 1.2) scales between posterior canthals; loreal region slightly concave, 36–66 (47.7 ± 8.7) mostly strongly keeled (some smooth or rugose) loreal scales in maximum of 5–10 (6.8 ± 1.3) horizontal rows; 7–10 (8.6 ± 0.9) supralabials and 7–10 (8.7 ± 0.7) infralabials to level below center of eye; suboculars strongly keeled, usually separated from

supralabials by 1 row of scales, or 1 subocular narrowly in contact with 1 supralabial below eye; ear opening vertically oval; scales anterior to ear opening not granular, larger than those posterior to ear opening; 4–8 (6.1 ± 0.7) postmentals, outer pair usually largest; keeled granular scales present on chin and throat; male dewlap moderately large, extending posteriorly to level of axilla onto chest; male dewlap with 6–8 horizontal gorgetal-sternal scale rows, about 7–9 scales per row (*n* = 3); 3 (modal number) anterior marginal pairs in male dewlap; female dewlap moderately large, although slightly smaller than male dewlap, extending to level of axilla; nuchal crest and dorsal ridge present in adult males; about 2 middorsal scale rows slightly enlarged, keeled, subimbricate, dorsal scales lateral to middorsal series gradually larger than granular lateral scales; no enlarged scales scattered among laterals; 41–64 (51.7 ± 6.9) dorsal scales along vertebral midline between levels of axilla and groin in males, 38–59 (51.0 ± 5.8) in females; 31–52 (38.4 ± 6.3) dorsal scales along vertebral midline contained in 1 head length in males, 29–44 (34.2 ± 4.8) in females; ventral scales on midsection larger than largest dorsal scales; ventral body scales smooth, flat, imbricate; 38–58 (48.9 ± 6.2) ventral scales along midventral line between levels of axilla and groin in males, 34–56 (43.1 ± 6.2) in females; 33–47 (37.5 ± 3.8) ventral scales contained in 1 head length in males, 26–38 (31.1 ± 3.7) in females; 122–154 (138.2 ± 9.4) scales around midbody in males, 108–156 (134.1 ± 13.2) in females; tubelike axillary pocket absent; precloacal scales conical, not keeled; a pair of slightly enlarged postcloacal scales in males; tail slightly to distinctly compressed, TH/TW 1.10–2.00; basal subcaudal scales smooth; lateral caudal scales weakly keeled, homogeneous, although indistinct division in segments discernable; dorsal medial caudal scale row not enlarged, keeled, not forming crest; most scales on anterior surface of antebrachium distinctly keeled, unicarinate; 23–28 (25.1 ± 1.5) subdigital lamellae on Phalanges II–IV of Toe IV of hind limbs; 8–

12 (10.4 ± 1.4) subdigital scales on Phalanx I of Toe IV of hind limbs; SVL 61.0–72.5 (67.2 ± 3.8) mm in males, 39.0–68.0 (58.7 ± 8.8) mm in females; TAL/SVL 1.69–1.76 in three males, 1.44–1.82 in 11 females; HL/SVL 0.28–0.30 in males, 0.26–0.32 in females; SHL/SVL 0.26–0.28 in males, 0.25–0.27 in females; SHL/HL 0.90–0.95 in males, 0.85–0.98 in females; longest toe of adpressed hind limb usually reaching between posterior and anterior borders of eye.

Color in life of an adult male (USNM 565442): dorsum closest to Grayish Olive (43) with series of four brown lateral spots edged narrowly posteriorly with Pale Horn Color (92); limbs Grayish Olive with slightly paler crossbands; top of head Grayish Olive; top of tail Cinnamon (39); dewlap Chrome Orange (16) peripherally, Cyanine Blue (74) centrally, gorgetal scales Straw Yellow (56), except on posterior half of central blue spot where scales same color as surrounding skin; belly Cream Color (54); underside of tail pale cinnamon with pale yellow crossbands; inner edge of eyelids Sulfur Yellow (157); iris rust red. Color in life of another adult male (KU 192624) was described by McCranie et al. (1984: 337): "dorsum pale brown with cream-outlined brown blotches along and on either side of the dorsal midline; head and limbs yellowish tan, latter with dark-outlined cream-colored bands; tail yellowish tan with cream-outlined brown bands; venter pale yellowish tan; iris rust red; dewlap red-orange with large central royal blue spot, scales across middle pale yellow, those of periphery yellow." Color in life of a third adult male (USNM 322903): dorsal surface of body with olive green swath and olive brown chevrons on middorsum; lateral surface of body brown; dorsal and lateral surfaces of head olive brown; dorsal surfaces of limbs olive green with rust tinge and pale olive crossbars; dorsal surface of tail olive green with pale olive chevrons; chin olive yellow; ventral surfaces of body, limbs, and tail olive yellow; dewlap red-orange with cobalt blue central spot and yellow scales (blue spot occupies about half of area of dewlap);

perimeter of eyelid yellow; iris copper red. Color in life of a fourth adult male (USNM 322905): dorsal surface of body pale olive; lateral surface of body with alternating rust red, olive gray, and dark olive green bars; dorsal surfaces of limbs pale olive with pale rust crossbars; chin yellow; belly yellow with scattered pale rust orange spotting; dewlap red-orange with cobalt blue central spot and yellow scales (blue spot occupies about half of area of dewlap). Color in life of the adult female holotype (LSUMZ 37834) was described by Wilson and McCranie (1982: 135): "dorsum pale olive-brown with small, scattered irregular paler spots; a thin tan line (dark-bordered below) extending posteromedially from behind the eye toward the midline of the body where it joins with the line on the other side of the body to form a narrow pale line extending along the dorsal midline, the latter of which broadens somewhat on the base of the tail and continues along the middorsal region of the tail and grades into the ground color of the tail; the pale middorsal stripe broken at intervals by about ten narrow, short diffuse dark brown marks; head pale olive brown with a narrow, dark interocular line; limbs pale olive-brown with narrow pale bands on the upper and lower legs and feet; venter pale yellowish cream; dewlap canary yellow with a large central dark blue spot crossed by oblique rows of white scales; iris red." Color in life of another adult female (USNM 322904): dorsal surface of head rust red with olive tinge; dorsal surface of body rust red with scant indication of darker band on sides; forelimbs rust red dorsally, grading to olive rust red ventrally; dorsal surfaces of hind limbs rust red with paler crossbars; dorsal surface of tail olive with rust red tinge, grading to rust red towards tip; lateral surface of head orange-rust; chin olive yellow with orange-rust spots; dewlap yellow with cobalt blue central spot and yellow scales; belly yellow with rust spotting and brown line down midventer; ventral surfaces of limbs pale olive yellow. Color in life of a third adult female (USNM 322907): dorsal surfaces of head and body dark olive green with scattered paler short narrow bars irregularly arranged on sides; dorsal surfaces of limbs dark olive green with pale olive gray crossbars; dorsal surface of tail dark olive green with olive gray crossbars, tail becoming increasingly rust colored distally; belly pale gray with olive tinge; dewlap pale yellow with blue central spot and scattered orange flecks; iris red-orange. Color in life of a fourth adult female (KU 192623): dorsum olive brown, except middorsal scales olive tan; dorsal surfaces of limbs olive brown with tan crossbars; tail pale brown with tan-outlined brown markings; belly pale olive brown; dewlap golden yellow with blue central spot. A juvenile male (USNM 322906) had a red-orange dewlap with a cobalt blue central spot and yellow scales.

Color in alcohol: dorsal and lateral surfaces of head and body pale brown to dark brown; middorsal color can be uniform, or darker brown blotches or paler brown stripe with zigzag edges can be present; dorsal surfaces of limbs pale brown to dark brown with indistinct to distinct darker brown crossbars; dorsal surface of tail similar basally to middorsal pattern, becoming paler brown with darker brown crossbands for about distal half; ventral surfaces of head and body cream, lightly to heavily flecked with brown; subcaudal surface cream with brown flecking proximally, brown flecking increasing distal to that point until subcaudal surface dark brown with pale brown crossbands for about distal half.

Hemipenis: unknown.

*Diagnosis/Similar Species. Norops johnmeyeri* is distinguished from all other Honduran *Norops*, except *N. pijolense* and *N. purpurgularis*, by the combination of smooth, flat, and distinctly imbricate ventral scales, keeled supraocular scales, long hind limbs (longest toe of adpressed hind limb usually reaching between posterior and anterior borders of eye), and a pair of slightly enlarged postcloacal scales in males. *Norops johnmeyeri* differs from *N. pijolense* and *N. purpurgularis* by its larger body size

(SVL to 73 mm in males, 68 mm in females of *N. johnmeyeri* versus to about 59 mm in males, about 60 mm in females of *N. pijolense* and *N. purpurgularis*), and a orange-red male dewlap with a large central blue blotch in life, female dewlap well developed and yellow with a large central blue blotch in life (male dewlap rose with purple spot or uniformly purple with female dewlaps small and usually same color as that of male dewlap in *N. pijolense* and *N. purpurgularis*).

*Illustrations* (Figs. 23, 24, 108). Wilson and McCranie, 1982 (adult, female dewlap; as *Anolis*), 2004c (adult); Nieto-Montes de Oca, 1994b (head scales; as *Anolis*); Köhler, 2000 (adult, head and dewlap), 2003 (adult, head and dewlap), 2008 (adult, head and dewlap), 2014 (ventral scales; as *Anolis*); Townsend and Wilson, 2006 (adult; as *Anolis*), 2008 (adult, dewlap; as *Anolis*); Snyder, 2011 (adult).

*Remarks. Norops johnmeyeri*, along with two other Honduran species (*N. pijolense* and *N. purpurgularis*), has been placed in the *N. schiedii* species group (McCranie et al., 1993b). For differing opinions on the relationships of these three species, see Campbell et al. (1989) and Nieto-Montes de Oca (2001). The latter author considered the three Honduran species of the "*schiedii* group" of McCranie et al. (1993b) as an uncertain group status within *Norops*. However, there seems little doubt that these three Honduran species are closely related to each other based on external morphology. However, for the purpose of this study, we keep *N. johnmeyeri*, *N. pijolense*, and *N. purpurgularis* in the *N. schiedii* species subgroup of the *N. auratus* species group of Nicholson et al. (2012). We also add *N. loveridgei* to this subgroup. *Norops loveridgei* forms a clade with *N. purpurgularis* in both molecular phylogenetic analyses done to date (Nicholson et al., 2005, 2012). *Norops loveridgei* is also similar to the other Honduran members of the *N. schiedii* subgroup in having a well-developed female dewlap, long hind limbs, and smooth ventral scales.

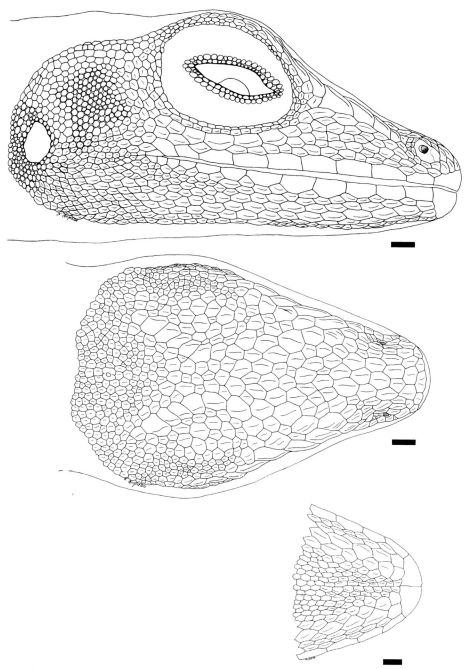

Figure 23.    *Norops johnmeyeri* head in lateral, dorsal, and ventral views. SMF 77757, adult male from Sendero El Danto, Cortés. Scale bar = 1.0 mm. Drawing by Lara Czupalla.

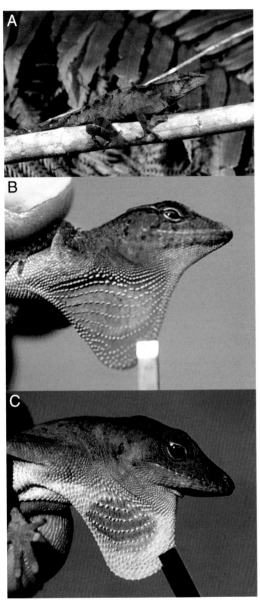

Figure 24. *Norops johnmeyeri.* (A) Adult male (USNM 549363); (B) male dewlap (from series SMF 77755-61); (C) adult female dewlap (from series SMF 77755–61) from El Cusuco, Cortés. Photographs by James R. McCranie (A) and Gunther Köhler (B, C).

*Natural History Comments. Norops johnmeyeri* is known from 1,300 to 2,000 m elevation in the Lower Montane Wet Forest formation and peripherally in the Premon-

tane Wet Forest formation. It is active in low vegetation, low on tree trunks, and on the ground near tree trunks. It sleeps at night on low tree branches and on trunks and leaves of tree ferns. The species is usually found in closed canopy forest well away from streams and forest edge situations. Its escape behavior is rather deliberate and is easily captured. *Norops johnmeyeri* has been found during March and April and from July to September and is probably active throughout the year during sunny conditions. Nothing has been reported on its diet or reproduction other than the female holotype (collected during April) contains two eggs "one of which appears ready for deposition; the other is less well-developed" (Wilson and McCranie, 1982: 138).

*Etymology.* The name *johnmeyeri* is a patronym honoring John R. Meyer, who began a review of the Honduran herpetofauna that culminated in his dissertation (Meyer, 1969).

*Specimens Examined* (41 [0]; Map 12). COPÁN: Quebrada Grande, KU 192624–25, SMF 86982–83, USNM 565442. CORTÉS: Bosque Enano, UF 149551; Cantiles Camp, UF 149552, 149554–61; El Cusuco, KU 192623, LSUMZ 37834, SMF 77755, 86984, USNM 322902–08, 549363–66; Guanales Camp, UF 149553, 149592; Montaña San Ildefonso, SMF 78827; Quebrada de Cantiles, SMF 78824–26; Sendero de Cantiles, SMF 77760–61; Sendero El Danto, SMF 77756–58.

**Norops kreutzi** McCranie, Köhler, and Wilson
Norops kreutzi McCranie, Köhler, and Wilson, 2000: 218 (holotype, SMF 78844; type locality: "2.5 airline km NNE La Fortuna [15°26′N, 87°18′W], 1670–1690 m elevation, Departamento de Yoro, Honduras"); Köhler, 2003: 100, 2008: 108; Wilson and McCranie, 2003: 59, 2004b: 43.
Anolis kreutzi: Köhler et al., 2007: 391; Townsend et al., 2010: 12, 2012: 100; Wilson et al., 2013: 66.

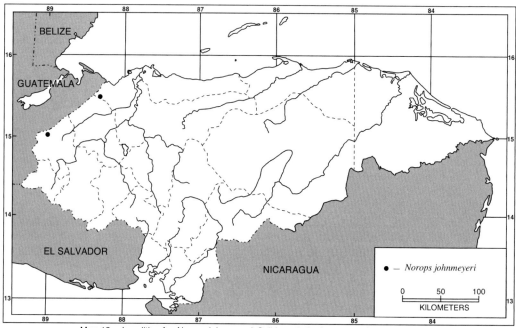

Map 12.    Localities for *Norops johnmeyeri*. Solid symbols denote specimens examined.

*Geographic Distribution. Norops kreutzi* occurs at moderate and intermediate elevations in the western portion of the Cordillera Nombre de Dios in Honduras.

*Description.* The following is based on five males (SMF 78844; USNM 532565, 578747, 578819; UTA R-53998) and eight females (SMF 79172; UF 166211–12; USNM 532571, 565443, 578746, 578748–49). *Norops kreutzi* is a small anole (SVL 48 in largest male [SMF 78844], 51 mm in largest female [USNM 532571]); most dorsal head scales in internasal and prefrontal areas rugose or faintly keeled, most scales in frontal area smooth; most scales in parietal area smooth; deep frontal depression present; parietal depression absent; 3–8 (5.2 ± 1.6) postrostrals; anterior nasal single, usually contacting rostral and first supralabial; 4–8 (5.3 ± 1.0) internasals; canthal ridge sharply defined; scales comprising supraorbital semicircles smooth or faintly keeled, largest scale in semicircles larger than largest supraocular scale; supraorbital semicircles well defined, with 1–2 scales in broad contact medially with semicircle on other side; 1–3 (1.9 ± 0.8) scales between supraorbital semicircles and interparietal; interparietal well defined, irregular in outline, longer than wide, larger than ear opening; 2–3 rows of about 4–7 (total number) enlarged keeled supraocular scales; enlarged supraoculars completely separated from supraorbital semicircles by 1 row of small scales; 2 elongate superciliaries, posterior shortest; 3 enlarged canthals; 6–10 (7.0 ± 1.3) scales between second canthals; 6–10 (8.0 ± 1.3) scales between posterior canthals; 14–37 (25.2 ± 5.8) mostly keeled loreal scales in maximum of 4–6 (4.8 ± 0.6) horizontal rows; 5–7 (6.1 ± 0.6) supralabials and 5–7 (5.7 ± 0.6) infralabials to level below middle of eye; suboculars keeled, 2–3 in contact with supralabials; ear opening vertically oval; scales anterior to ear opening not granular, slightly larger than those posterior to ear opening; 4–6 (4.7 ± 0.9) postmentals, outer pair largest; gular scales faintly keeled; male dewlap large, extending onto chest well posterior to level of axilla; male dewlap with about 9 horizontal gorgetal-sternal scale rows, about 6.7 mean number of scales per row ($n = 1$); 2 (modal number) anterior marginal pairs in male dewlap; female dewlap absent; no nuchal crest or dorsal ridge; about 4–5 middorsal scale rows slightly enlarged, keeled, dorsal scales lateral to middorsal series grading into granular lateral scales; a few or no enlarged scales scattered among laterals; 64–102 (83.6 ± 16.5) dorsal scales along vertebral midline between levels of axilla and groin in males, 62–104 (86.9 ± 13.9) in females; 42–67 (54.4 ± 11.3) dorsal scales along vertebral midline contained in 1 head length in males, 36–60 (46.4 ± 8.1) in females; ventral scales on midsection larger than largest dorsal scales; ventral body scales keeled, imbricate; 44–83 (66.8 ± 17.5) ventral scales along midventral line between levels of axilla and groin in males, 48–95 (68.1 ± 16.7) in females; 27–46 (40.2 ± 7.7) ventral scales contained in 1 head length in males, 29–53 (35.4 ± 8.4) in females; 115–160 (138.2 ± 21.1) scales around midbody in males, 130–162 (144.3 ± 13.1) in females; tubelike axillary pocket absent; precloacal scales not keeled; pair of enlarged postcloacal scales present in males; tail slightly to distinctly compressed, TH/TW 1.11–2.33; basal subcaudal scales smooth to keeled; lateral caudal scales keeled, homogeneous; dorsal medial caudal scale row slightly enlarged, keeled, not forming crest; most scales on anterior surface of antebrachium unicarinate; 23–28 (25.6 ± 1.5) subdigital lamellae on Phalanges II–IV of Toe IV of hind limbs; 5–9 (6.8 ± 1.1) subdigital scales on Phalanx I of Toe IV of hind limbs; SVL 35.7–47.5 (41.2 ± 5.4) mm in males, 35.4–51.2 (43.0 ± 5.2) mm in females; TAL/SVL 1.48–1.81 in three males, 1.11–1.89 in five females; HL/SVL 0.27–0.29 in males, 0.18–0.27 in females; SHL/SVL 0.21–0.25 in males, 0.18–0.24 in females; SHL/HL 0.75–0.79 in males, 0.72–0.79 in females; longest toe of adpressed hind limb reaching about shoulder region.

Color in life of the adult male holotype (SMF 78844) was described by McCranie et al. (2000: 221–222): "dorsal surface of body pale grayish brown with dark brown chevrons on middorsal line; head pale grayish brown with brown interocular bar; dorsal surface of forelimbs pale grayish brown with dark brown crossbars; dorsal surfaces of forefeet pale yellow with brown markings; dorsal surface of hind limbs grayish brown with rust-colored crossbars and darker brown smudging; dorsal surface of hind feet pale yellow with rust-colored and dark brown markings; sacrum with rust-red markings; tail pale brown with brown crossbars; ventral surface of body yellow medially, grading to pale yellow laterally; chin mottled cream and brown; dewlap pale yellow with purple gorgetal scales; area around vent, including underside of legs and base of tail, golden yellow; iris dark brown." Color in life of an adult female (SMF 79172): dorsal surface of head Clay Color (26) with Cinnamon-Rufous (40) markings; dorsal surface of body Clay Color with scalloped Cinnamon (39) middorsal markings; dorsal surface of forelimbs Clay Color with Cinnamon-Rufous patch near elbow and crossbar on forearms and hands; dorsal surface of hind limbs Chestnut (32); dorsal surface of tail Chestnut with Cinnamon middorsal line; chin white with Cinnamon-Rufous smudging; belly dirty white; iris dark brown.

Color in alcohol: dorsal surfaces of head and body grayish brown, dark brown chevrons usually present, one female with broad pale brown, dark brown–outlined middorsal stripe on body; dorsal surfaces of limbs grayish brown to brown, with indistinct, narrow pale brown crossbars; dorsal surface of tail brown with dark brown, narrow crossbands; ventral surfaces of head and body white to grayish white, with some brown flecking; subcaudal surface brown.

Hemipenis: unknown.

*Diagnosis/Similar Species.* *Norops kreutzi* is distinguished from all other Honduran *Norops*, except *N. cusuco* and *N. laeviven-*

*tris*, by the combination of having about 4–5 slightly enlarged middorsal scale rows, heterogeneous lateral scales in some, keeled ventral scales, male dewlap pale yellow in life, and a pair of enlarged postcloacal scales in males. *Norops kreutzi* differs from *N. cusuco* by having a pale yellow male dewlap in life with purple gorgetal scales (dirty white, pale yellow, or pale gray in life with gorgetal scales of same color in *N. cusuco*). *Norops kreutzi* differs from *N. laeviventris* by having a larger male dewlap (extending well onto chest posterior to level of axilla in *N. kreutzi* versus to about level of axilla in *N. laeviventris*).

*Illustrations* (Figs. 25, 26). McCranie et al., 2000 (adult, head scales, head and dewlap); Wilson and McCranie, 2004a (adult).

*Remarks.* *Norops kreutzi* belongs to the *N. laeviventris* species subgroup (McCranie et al., 2000) of the *N. auratus* species group of Nicholson et al. (2012, also see the Remarks for *N. cusuco*). *Norops kreutzi* was not included in the phylogenetic analyses of Nicholson et al. (2005, 2012).

*Natural History Comments.* *Norops kreutzi* is known from 980 to 1,690 m elevation in the Premontane Wet Forest and Lower Montane Wet Forest formations. Specimens are active on tree trunks about 1–2 m above the ground. One slowly crawled about 10 m down a tree trunk where it stopped in a sunny spot about 2 m above the ground. Most captured individuals were in forest edge situations such as along an old dirt road and along a river. It also sleeps on tree branches and leaves at night. The species was collected during April and from June to September and is probably active throughout the year under sunny conditions. Nothing has been reported on its diet or reproduction.

*Etymology.* The specific name is a patronym honoring Joerg Kreutz, a German biologist who has drawn several *Norops* hemipenes for our work on anoles.

*Specimens Examined* (13 [0]; Map 13). ATLÁNTIDA: La Liberación, USNM 578746–48, 578819. YORO: 2.5 airline km NNE of La Fortuna, SMF 78844, 79172, UF 166211–12, USNM 532565–66, 532571, 565443; near Yuqüela, UTA R-53998.

*Norops laeviventris* (Wiegmann)
*Dactyloa laeviventris* Wiegmann, 1834: 47 (holotype, ZMB 525 [see Smith and Follett, 1960: 78]; type locality not stated, but "Mexico" inferred from title of paper).
*Anolis laeviventris*: Meyer, 1969: 222; Hahn, 1971: 111; Meyer and Wilson, 1972: 107, 1973: 17; Wilson et al., 1991: 69, 70; Poe, 2004: 63; Köhler et al., 2007: 391 (in part); Townsend et al., 2007: 10; Wilson and Townsend, 2007: 145.
*Norops laeviventris*: Villa et al., 1988: 49; McCranie et al., 2000: 223 (in part), 2006: 217; Köhler et al., 2001: 254 (in part); Wilson et al., 2001: 136 (in part); Köhler, 2003: 109, 2008: 117; Wilson and McCranie, 2004b: 43; Wilson and Townsend, 2006: 105.
*Norops* especie B: Espinal, 1993: table 3.
*Norops* "*laeviventris*": Espinal et al., 2001: 106.

*Geographic Distribution.* *Norops laeviventris* occurs at moderate and intermediate elevations on the Atlantic versant in disjunct populations from eastern Hidalgo, northeastern Puebla, and central Veracruz, Mexico, southward to central Honduras (see Remarks). In Honduras, this species is known from widely scattered localities in the western two-thirds of the country.

*Description.* The following is based on 10 males (KU 193099; MCZ R-191106, 19110, 19112, SMF 78119, 78843, 79174, 79181; USNM 532570, 565448) and 13 females (SMF 78118, 79139, 79143, 79175–78, 79180; USNM 565444–47, 565449). *Norops laeviventris* is a small anole (SVL 42 mm in largest Honduran male examined [USNM 565448], 45 mm in largest Honduran female examined [SMF 79175]; maximum reported SVL 49 mm [McCranie et al., 2000]); most dorsal head scales rugose or faintly keeled in internasal and prefrontal areas, most scales smooth in frontal and parietal areas; deep

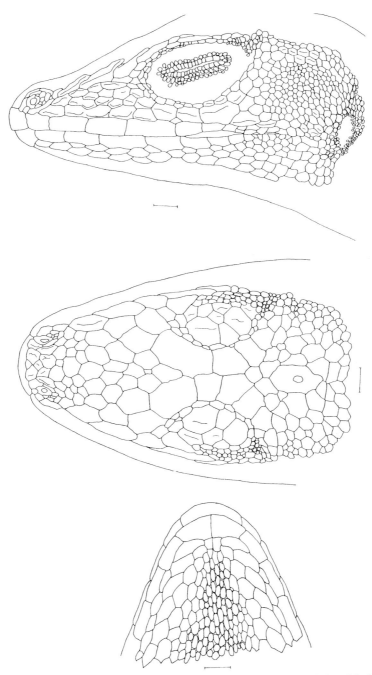

Figure 25.   *Norops kreutzi* head in lateral, dorsal, and ventral views. SMF 78844, adult male from 2.5 airline km NNE of La Fortuna, Yoro. Scale bar = 1.0 mm. Drawing by Gunther Köhler.

Figure 26.   *Norops kreutzi.* (A) Adult male (SMF 78844) from 2.5 airline km NNE of La Fortuna, Yoro; (B) adult male dewlap (SMF 78844). Photographs by James R. McCranie.

frontal depression present; parietal depression absent; 3–6 (4.0 ± 0.8) postrostrals; anterior nasal single, usually contacting rostral and first supralabial; 3–6 (4.5 ± 0.8) internasals; canthal ridge sharply defined; scales comprising supraorbital semicircles smooth or faintly keeled, largest scale in semicircles larger than largest supraocular scale; supraorbital semicircles well defined, usually with 1–2 scales in broad contact medially with semicircle on other side (occasionally 1 scale separates supraor-

bital semicircles); 1–3 (1.9 ± 0.7) scales separating supraorbital semicircles and interparietal at narrowest point; interparietal well defined, irregular in outline, longer than wide, larger than ear opening; 2–3 rows of about 3–6 (total number) enlarged keeled supraocular scales; enlarged supraoculars completely separated from supraorbital semicircles by 1 row of small scales; 2 elongate superciliaries, posterior much shorter than anterior; usually 3 enlarged canthals; 4–8 (6.3 ± 1.1) scales between

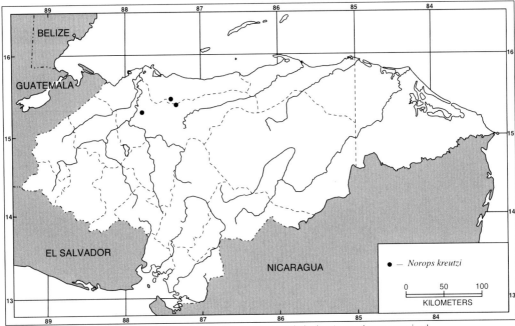

Map 13. Localities for *Norops kreutzi*. Solid symbols denote specimens examined.

second canthals; 6–10 (7.7 ± 1.5) scales between posterior canthals; loreal region slightly concave, 17–30 (22.4 ± 3.5) mostly keeled (some smooth or rugose) loreal scales in maximum of 4–5 (4.4 ± 0.5) horizontal rows; 6–8 (6.7 ± 0.7) supralabials and 6–7 (6.3 ± 0.4) infralabials to level below center of eye; suboculars weakly to strongly keeled, in broad contact with supralabial scales; ear opening vertically oval; scales anterior to ear opening not granular, slightly larger than those posterior to ear opening; 5–6 (5.7 ± 0.5) postmentals, outer pair usually largest; gular scales faintly keeled; male dewlap relatively small, extending to level of axilla; male dewlap with about 6–7 horizontal gorgetal-sternal scale rows, about 6–17 scales per row (*n* = 4); 2–3 (modal number) anterior marginal pairs in male dewlap; female dewlap absent; no nuchal crest or dorsal ridge; about 6–11 middorsal scale rows slightly enlarged, keeled, dorsal scales lateral to middorsal series grading into slightly enlarged lateral scales; a few or no enlarged scales scattered among laterals; 46–87 (65.5 ± 11.1) dorsal

scales along vertebral midline between levels of axilla and groin in males, 48–79 (61.8 ± 8.8) in females; 34–52 (42.5 ± 6.1) dorsal scales along vertebral midline contained in 1 head length in males, 28–47 (36.5 ± 5.3) in females; ventral scales on midsection larger than largest dorsal scales; ventral body scales keeled, imbricate; 43–59 (51.1 ± 5.7) ventral scales along midventral line between levels of axilla and groin in males, 37–56 (45.3 ± 5.2) in females; 29–41 (33.8 ± 4.2) ventral scales contained in 1 head length in males, 23–34 (28.1 ± 2.8) in females; 96–140 (121.5 ± 12.3) scales around midbody in males, 117–144 (126.6 ± 7.3) in females; tubelike axillary pocket absent; precloacal scales not keeled; enlarged postcloacal scales almost always present in males; tail slightly to distinctly compressed, TH/TW 1.07–1.58 in 21; basal subcaudal scales smooth to keeled; lateral caudal scales keeled, homogeneous; dorsal medial caudal scale row slightly enlarged, keeled, not forming crest; most scales on anterior surface of antebrachium unicarinate; 18–26 (23.1 ± 1.9) subdigital lamellae on Phalanges II–IV of Toe IV of hind limbs; 5–9 (6.2 ± 1.0) subdigital scales on Phalanx I of Toe IV of hind limbs; SVL 28.8–42.1 (37.2 ± 3.5) mm in males, 36.0–45.0 (39.9 ± 2.5) mm in females; TAL/SVL 1.71–1.98 in seven males, 1.51–1.76 in 10 females; HL/SVL 0.24–0.29 in males, 0.24–0.27 in females; SHL/SVL 0.19–0.23 in males, 0.18–0.23 in females; SHL/HL 0.68–0.82 in males, 0.68–0.92 in females; longest toe of adpressed hind limb usually reaching ear opening, or just anterior to ear opening.

Color in life of an adult male (SMF 79174): dorsal surface of head pale grayish tan with dark grayish brown markings in occipital region; dorsal surface of body grayish tan, middorsal region dark pale grayish brown; dorsal surfaces of forelimbs pale grayish tan with dark brown spot at elbow; dorsal surfaces of hind limbs pale grayish tan with gray-brown markings; dorsal surface of tail pale grayish tan with some gray-brown spotting; belly pale cream; dewlap cream; iris pale bronze.

Color in alcohol: dorsal surfaces of head and body grayish brown with small dark brown linear shaped marks, or with brown, dark brown–outlined broad, middorsal stripe (latter present in females only); dorsal surfaces of limbs grayish brown to brown, with indistinct, narrow, pale brown crossbars; dorsal surface of tail brown with dark brown, narrow crossbands; ventral surfaces of head and body white or grayish brown, with some brown flecking; subcaudal surface mostly brown.

Hemipenis: the fully everted hemipenis of SMF 78843 is a stout and distinctly bilobed organ; sulcus spermaticus bordered by well-developed sulcal lips, bifurcating at base of apex and continuing almost to tip of lobes; small asulcate-side processus present; no particular surface structure discernible on most of truncus; apex extensively calyculate.

*Diagnosis/Similar Species.* Norops laeviventris is distinguished from all other Honduran *Norops*, except *N. cusuco* and *N. kreutzi*, by the combination of having several slightly enlarged middorsal scale rows, heterogeneous lateral scales in some, keeled ventral scales, a dirty white to cream male dewlap in life, and usually a pair of enlarged postcloacal scales in males. *Norops laeviventris* differs from *N. cusuco* and *N. kreutzi* by having a smaller male dewlap (extending to about level of axilla in *N. laeviventris* versus extending well onto chest posterior to level of axilla in *N cusuco* and *N. kreutzi*). Also, *N. laeviventris* differs from *N. kreutzi* by having a dirty white to cream male dewlap in life with gorgetal scales of same color (male dewlap pale yellow in life with purple gorgetal scales in *N. kreutzi*).

*Illustrations* (Figs. 27, 28). Duméril et al., 1870–1909 (head scales; as *Anolis*); Köhler, 2000 (head and dewlap; as *N.* cf. *laeviventris*), 2003 (adult, head and dewlap), 2008 (adult, head and dewlap).

*Remarks.* Köhler (2000) and McCranie et al. (2000) considered *N. laeviventris* to range from Mexico to western Panama. Savage (2002) applied the name *N. intermedius* (Peters) to the Costa Rican and Panamanian *N. laeviventris* complex populations. Savage (2002) considered the Nicaraguan populations to possibly represent *N. intermedius* as well. We follow Savage (2002) in considering *N. intermedius* a species distinct from those of Honduras northward. *Norops laeviventris* is a member of the *N. laeviventris* species subgroup (see Remarks for *N. cusuco*) of the *N. auratus* species group of Nicholson et al. (2012). Köhler and Obermeier (1998) and Köhler et al. (1999) considered *N. laeviventris* a member of the *N. crassulus* species group. However, the analyses by Nicholson (2002), Poe (2004), and Nicholson et al. (2005, 2012) do not support those relationships (also see Remarks for *N. crassulus*).

*Natural History Comments.* Norops laeviventris is known from 1,000 to 2,000 m elevation in the Premontane Wet Forest, Premontane Moist Forest, Lower Montane Wet Forest, and Lower Montane Moist Forest formations. This species is usually found in low vegetation, low on tree trunks, on rock walls, on old buildings, and in roofing tile and other debris on the ground,

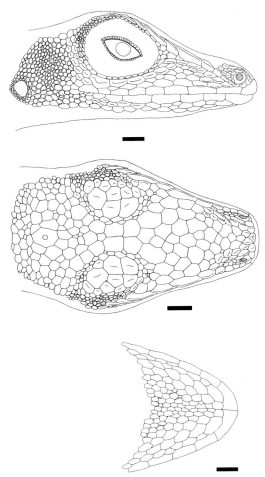

Figure 27. *Norops laeviventris* head in lateral, dorsal, and ventral views. UTA R-29481, adult female from Los Amates, Sierra del Espiritu Santo, Izabal, Guatemala. Scale bar = 1.0 mm. Drawing by Gunther Köhler.

frequently in forest edge situations (including along streams). It was found during every month of the year except March, October, and December; thus, it is likely active throughout the year under suitable conditions. Stuart (1948: 50, as *Anolis nannodes* [Cope]) reported females collected in Guatemala during May contained "well-developed eggs." Nothing else has been reported on reproduction in this species, nor has anything been reported on its diet.

*Etymology.* The name *laeviventris* is formed from the Latin *laevis* (smooth,

polished, bald) and *ventris* (belly). The name is a misnomer as the species has keeled ventral scales (stated "*abdominalibus sublaevibus*" by Wiegmann, 1834: 47, for the holotype).

*Specimens Examined* (58 [3]; Map 14). COMAYAGUA: Cerro El Volcán, UF 166265, 166278; near Río Negro, UF 166289–97; Río Varsovia, UF 166179. EL PARAÍSO: Montaña del Volcán, MCZ R-49951. FRANCISCO MORAZÁN: Cerro Uyuca, KU 103239; Los Planes, USNM 581193; Montaña de la Sierra, UF 166214; Parque Nacional La Tigra Centro de Visitantes El Rosario, UF 150154; 7.2 km SW of San Juancito, KU 193099. INTIBUCÁ: 2.4 km NE of La Esperanza, LACM 47291; La Esperanza, LACM 47292; San Pedro La Loma, SMF 79181, UF 166282. LA PAZ: Opatoro, SMF 78118–19. LEMPIRA: Erandique, CM 64614; Río Arcáqual near Parque Nacional Celaque Centro de Visitantes, SMF 78843, 79139, 79143, 79174–78, USNM 532570. OLANCHO: Babilonia, SMF 91751; near Cuaca, UTA R-53694; El Díctamo, USNM 565444–45; Monte Escondido, MCZ R-191106–12, SMF 79180, USNM 565446–48; Parque Nacional La Muralla Centro de Visitantes, USNM 565449; Quebrada El Pinol, USNM 344797–801; between Ríos Catacamas and Seco, FMNH 236389. SANTA BÁRBARA: El Ocotillo, USNM 532573. YORO: Cerro de Pajarillos, USNM 532572.

*Norops lemurinus* (Cope)

*Anolis* (*Gastrotropis*) *lemurinus* Cope, 1861: 213 (three syntypes originally in ANSP, now lost [see Stuart, 1963: 63; also not listed by Malnate, 1971]; type locality: "Veragua, New Grenada" [= Panama]).

*Anolis palpebrosus*: Dunn and Emlen, 1932: 27.

*Anolis lemurinus*: Barbour, 1934: 137; Meyer, 1969: 222 (in part); Williams, 1969: 362; Meyer and Wilson, 1973: 17 (in part); Cruz Díaz, 1978: 26; O'Shea, 1986: 36; Wilson et al., 1991: 70; Townsend, 2006:

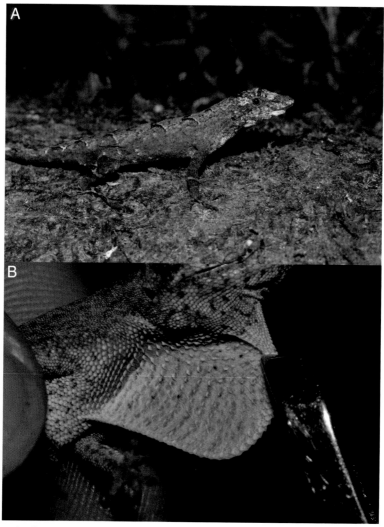

Figure 28.   *Norops laeviventris*. (A) Adult male (MCZ R-191106) from Monte Escondido Campground, Olancho; (B) adult male dewlap (MCZ R-191106). Photographs by Leonardo Valdés Orellana.

35; Townsend and Wilson, 2010b: 697; Logan et al., 2012: 215; Köhler, 2014: 210.
*Anolis lemurinus lemurinus*: Meyer, 1966: 175.
*Anolis tropidonotus*: Meyer and Wilson, 1973: 20 (in part).
*Norops lemurinus*: Savage, 1973: 11; Wilson and Cruz Díaz, 1993: 17 (in part); Köhler, 1996b: 28, 1999a: 50; Cruz D., 1998: 29 (in part), *in* Bermingham et al., 1998; Monzel, 1998: 160; Wilson and McCranie, 1998: 16; Köhler et al., 2000: 425; Nicholson et al., 2000: 30; Köhler and McCranie, 2001: 244; Wilson et al., 2001: 136 (in part); Lundberg, 2002a: 5; McCranie and Castañeda, 2005: 14; McCranie et al., 2005: 104, 2006: 121; Wilson and Townsend, 2006: 105; Klütsch et al., 2007: 1125; Diener, 2008: 13; McCranie and Solís, 2013:

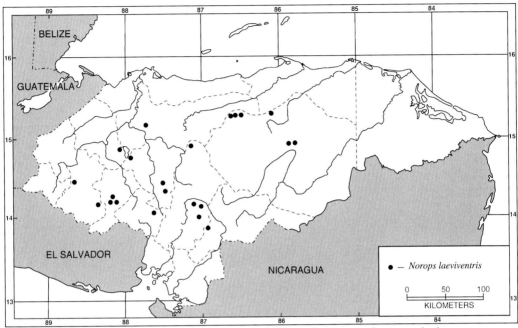

Map 14.   Localities for *Norops laeviventris*. Solid symbols denote specimens examined.

242; McCranie and Valdés Orellana, 2014: 45.

*Anolis bourgeaei*: Fitch and Hillis, 1984: 318.

*Norops unilobatus*: Köhler and Vesely, 2010: 225 (in part).

*Geographic Distribution. Norops lemurinus* occurs at low and moderate elevations on the Atlantic versant from central Veracruz, Mexico, to central Panama and disjunctly from northwestern Costa Rica to central Panama on the Pacific versant. In Honduras, this species occurs across the northern portion of the country, in southeastern Honduras, and on the Cayos Cochinos.

*Description.* The following is based on 24 males (KU 101392, 194839–41, 194851–52, 194874, 194877, 194882–85; SMF 79196, 79199–200, 79231, 79233, 80816, 81129, 86985; UMMZ 70320, 71290 [1–3]) and 19 females (SMF 79195, 79197–98, 79201, 79204, 79216–20, 79225–30, 79232, 80815, 81128). *Norops lemurinus* is a moderately large anole (SVL 64 mm in largest Hon-

duran male examined [KU 194882], 73 mm in largest Honduran female examined [SMF 79220]; maximum reported SVL 77 mm [Köhler, 1996b]); dorsal head scales keeled in internasal region, smooth or tuberculate in prefrontal, frontal, and parietal areas; weak to moderate frontal and parietal depressions present; 5–8 (5.6 ± 1.0) postrostrals; anterior nasal divided, lower section contacting rostral and first supralabial; 6–11 (8.3 ± 1.2) internasals; canthal ridge sharply defined; scales comprising supraorbital semicircles keeled, largest scale in semicircles about same size as, or larger than, largest supraocular scale; supraorbital semicircles well defined; 0–2 (1.2 ± 0.5) scales separating supraorbital semicircles at narrowest point; 1–5 (3.0 ± 0.8) scales separating supraorbital semicircles and interparietal at narrowest point; interparietal well defined, greatly enlarged relative to adjacent scales, surrounded by scales of moderate size, longer than wide, about same size as, or slightly larger than, ear opening; 2–3 rows of about 4–8 (total number) enlarged, keeled supraocular scales; 1–2 enlarged supraoculars in broad contact with supraorbital semicircles, or all enlarged supraoculars completely separated from supraorbital semicircles by 1 row of small scales; 2 elongate, overlapping superciliaries, posterior much shorter than anterior; usually 3 enlarged canthals; 7–11 (8.9 ± 1.0) scales between second canthals; 8–14 (10.3 ± 1.5) scales between posterior canthals; loreal region slightly concave, 31–71 (47.0 ± 10.7) mostly strongly keeled (some smooth or rugose) loreal scales in maximum of 5–9 (6.9 ± 0.8) horizontal rows; 6–9 (7.2 ± 0.7) supralabials and 6–9 (6.9 ± 0.7) infralabials to level below center of eye; suboculars weakly to strongly keeled, in contact with supralabials, or separated (most often) from supralabials, by 1 row of scales; ear opening vertically oval; scales anterior to ear opening granular, similar in size to those posterior to ear opening; 4–8 (6.1 ± 0.7) postmentals, outer pair largest; keeled granular scales present on chin and throat; male dewlap moderately large, extending to level of axilla, or slightly beyond; male dewlap with 6–7 horizontal gorgetal-sternal scale rows, about 4–8 scales per row (*n* = 4); 2 (modal number) anterior marginal pairs in male dewlap; female dewlap small; indistinct nuchal crest and dorsal ridge present in males; about 2 middorsal scale rows slightly enlarged, strongly keeled, dorsal scales grading into smaller granular lateral scales; no enlarged scales scattered among laterals; 51–83 (68.2 ± 7.8) dorsal scales along vertebral midline between levels of axilla and groin in 13 males, 62–85 (73.5 ± 6.5) in females; 37–54 (44.3 ± 3.5) dorsal scales along vertebral midline contained in 1 head length in males, 36–52 (43.2 ± 4.7) in females; ventral scales on midsection slightly larger than largest dorsal scales; ventral body scales keeled, mucronate, imbricate; 38–60 (48.2 ± 6.3) ventral scales along midventral line between levels of axilla and groin in 13 males, 37–52 (43.1 ± 4.5) in females; 28–40 (33.8 ± 3.6) ventral scales contained in 1 head length in males, 22–42 (28.1 ± 5.3) in females; 108–154 (135.0 ±

11.4) scales around midbody in 13 males, 108–146 (132.2 ± 10.6) in females; tubelike axillary pocket absent, but shallow, scaled axillary pocket present; precloacal scales weakly keeled; no enlarged postcloacal scales in males; tail rounded to slightly compressed, TH/TW 1.00–1.56 in 28; all subcaudal scales keeled; lateral caudal scales keeled, homogeneous, although indistinct division in segments discernable; dorsal medial caudal scale row enlarged, keeled, not forming crest; scales on anterior surface of antebrachium distinctly keeled, unicarinate; 24–29 (25.9 ± 1.4) subdigital lamellae on Phalanges II–IV of Toe IV of hind limbs; 6–10 (7.5 ± 1.1) subdigital scales on Phalanx I of Toe IV of hind limbs; SVL 46.0–64.0 (56.1 ± 3.8) mm in males, 55.0–73.0 (61.3 ± 4.6) mm in females; TAL/SVL 1.84–2.30 in 15 males, 1.91–2.27 in 10 females; HL/SVL 0.25–0.28 in males, 0.24–0.28 in females; SHL/SVL 0.25–0.31 in males, 0.23–0.32 in females; SHL/HL 0.99–1.20 in males, 0.97–1.22 in females; longest toe of adpressed hind limb usually reaching between posterior and anterior borders of eye.

Color in life of an adult male (USNM 565454): middorsum of body Drab-Gray (119D) with dark brown transverse markings, middorsum bordered below by longitudinal dirty white stripe; Fuscous (21) edged brown ventrolateral stripe present below longitudinal dirty white stripe; another dirty white stripe borders lower edge of Fuscous edged stripe; also narrow Fuscous stripe borders second dirty white stripe; dorsal surface of head pale brown with Fuscous snout, Fuscous interorbital bar, Fuscous mottling posterior to interorbital bar, and Fuscous U-shaped line that extends onto neck; Fuscous lines also radiate outward from eye; dorsal surface of tail with alternating pale and dark brown crossbars; dorsal surfaces of limbs golden brown with heavy dark brown mottling, some mottling suggesting crossbars; belly and anterior subcaudal surface brown, remainder of subcaudal surface with alternating pale brown and dark brown crossbars; dewlap Kingfisher Rufous (240) with white gorgetal scales, some gorgetal scales dark brown edged. Color in life of another adult male (KU 220105): dorsal surface of head tan with brown markings including lyriform mark on nuchal area; dorsal surface of body pale brown with dorsolateral and lateral tan stripes, stripes suffused with pale brown; dorsal surfaces of limbs mottled tan and brown; tail tan with brown crossbars; venter cream with brown punctations; dewlap pinkish orange with white scales. Color in life of another adult male (KU 220102) was described by Wilson and Cruz Díaz (1993: 17): "dorsum pale brown, dorsolateral and lateral stripes tan, suffused with pale brown; limbs mottled tan and brown; head tan with brown markings; dewlap pinkish orange with white spots [= scales]; tail tan with brown crossbars; venter cream with brown punctations." Color in life of an adult female (USNM 570004): dorsal surface of head golden yellow with dark brown–outlined bronze interocular bar, dark brown U-shaped mark on parietal at anterior edge of middorsal stripe; dorsal surface of body golden yellow with middorsal stripe beginning on nape and extending onto tail, stripe orange anteriorly, grading to red-orange posteriorly; brown stripe outlined with golden yellow scalloping present along either side of middorsal stripe; broad dark brown lateral band extending from axilla to groin; dorsal surfaces of forelimbs with paler brown spot at elbow; dorsal surfaces of hind limbs brown on thighs, golden yellow with brown crossbars on shanks; lateral surface of head golden yellow with dark brown marks radiating outward from eye; chin and belly pale yellow; dewlap orange with pale yellow scales; iris rust red. Logan et al. (2012) reported Cayos Cochinos males have white spots (= gorgetal scales) on their dewlaps (see quote above).

Color in alcohol: dorsal surfaces of head, body, and tail some shade of brown with a variety of possible dorsal markings (e.g., broad dark brown transverse bands, diamonds, a pale vertebral stripe) or patternless; dorsal surfaces of hind limbs with oblique dark brown bands; tail with faint to distinct dark brown bands; distinct dark brown lines radiate outward from eye; distinct dark brown interorbital bar present; dark brown lyriform mark usually present in nuchal area.

Hemipenis: the completely everted hemipenis of SMF 77169 is a medium-sized bilobed organ; sulcus spermaticus bifurcating at base of apex, branches continuing to tips of lobes; asulcate processus absent; lobes strongly calyculate; truncus with transverse folds.

*Diagnosis/Similar Species.* Norops lemurinus is distinguished from all other Honduran *Norops*, except *N. bicaorum* and *N. roatanensis*, by the combination of having a dark brown lyriform mark in the nuchal region (usually present, but obscure or absent in some specimens), long hind legs (longest toe of adpressed hind limb usually reaching to between posterior and anterior borders of eye), keeled, mucronate, imbricate ventral scales, a shallow and scaled axillary pocket, about two middorsal scale rows slightly enlarged, and no enlarged postcloacals in males. *Norops lemurinus* differs from *N. bicaorum* and *N. roatanensis* by having a red to red-orange male dewlap in life without suffusion of black pigment, frequently with black or dark brown-edged gorgetal scales (pink-red or orange-red in life, with suffusion of black pigment centrally, and white gorgetal scales in *N. bicaorum* and *N. roatanensis*). *Norops lemurinus* differs further from *N. bicaorum* in having hemipenial sulcal branches that continue to the tips of each lobe, an average SVL of about 56 mm in males and 61 mm in females, and in lacking an asulcate processus (sulcal branches opening into broad concave area distal to point of bifurcation on each lobe, an average SVL of about 64 mm in males and about 66 mm in females, and a low asulcate processus present in *N. bicaorum*).

*Illustrations* (Figs. 29, 30, 87, 105, 114, 115, 117). Stafford, 1994 (adult); Fläschendräger and Wijffels, 1996 (adult; as *Anolis l. lemurinus*); Lee, 1996 (adult; as *Anolis*), 2000 (adult); Campbell, 1998 (adult, head

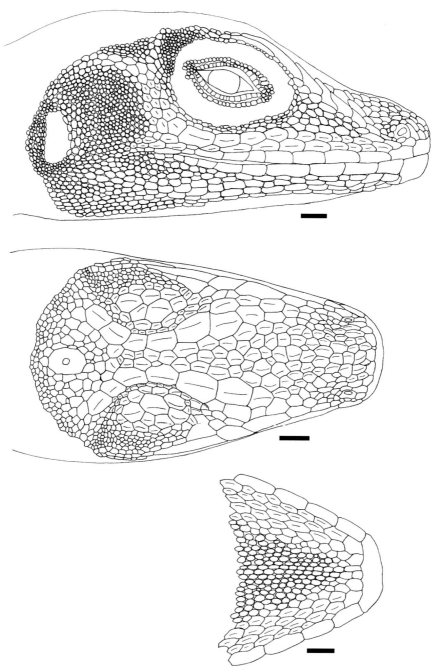

Figure 29.    *Norops lemurinus* head in lateral, dorsal, and ventral views. SMF 85887, adult female from Raudal Kiplatara, Rio Plátano Biosphere Reserve, Gracias a Dios. Scale bar = 1.0 mm. Drawing by Linda Acker.

Figure 30. *Norops lemurinus*. (A) Adult male (in UNAH collection); (B) adult male dewlap (in UNAH collection), both from Los Pinos, Cortés. Photographs by James R. McCranie (A) and Leonardo Valdés Orellana (B).

and dewlap; as *N. bourgeaei*); Köhler, 1999a (head and dewlap, hemipenis), 2000 (adult, head and dewlap), 2001b (head and dewlap), 2003 (adult, head and dewlap), 2008 (adult, head and dewlap), 2014 (dorsal, ventral, postmental, and supraocular scales, scales surrounding ear opening; as *Anolis*); Stafford and Meyer, 1999 (adult); Lundberg, 2002a (adult); Savage, 2002 (adult); D'Cruze, 2005 (adult, head and dewlap); Guyer and Donnelly, 2005 (adult, female dewlap); McCranie et al., 2005 (adult, head and dewlap), 2006 (adult); Calderón-Mandujano et al., 2008 (adult; as *Anolis*); Diener, 2008 (adult).

*Remarks.* Köhler (2001a) demonstrated that *A. bourgeaei* Bocourt is a junior synonym of *Dactyloa laeviventris* Wiegmann (= *Norops laeviventris*). Numerous authors have included the nominal form *bourgeaei* as a subspecies of *N. lemurinus* (see Köhler, 2001a).

*Norops lemurinus* is a member of the *N. lemurinus* species subgroup of the *N. auratus* species group of Nicholson et al. (2012). Williams (1976b) also recognized a *N. lemurinus* species group. The three Honduran species (*N. bicaorum*, *N. lemurinus*, and *N. roatanensis*) form a closely related group based on their similar external morphology. Members of this subgroup are characterized by having distinct dark lines radiating outward from the eye, a red male dewlap in life, strongly keeled midventral scales, a shallow and scaled axillary pocket, and usually a distinct dark-colored lyriform nuchal mark. The recovered phylogenetic analyses in Nicholson et al. (2005, 2012) showed *N. bicaorum* and *N. lemurinus* to be sister to a clade containing *N. limifrons* and *N. zeus*. The recent recognition of *N. bicaorum* and *N. roatanensis* has likely rendered the mainland populations of *N. lemurinus* paraphyletic. Thus, more works appears to be needed regarding *N. lemurinus* from throughout its geographic distribution.

Fitch and Hillis (1984) examined 18 specimens of *N. lemurinus* (as *Anolis bourgeaei*) from "Santa Barbara, Honduras." However, we were unable to locate such a large series from that town or department.

*Natural History Comments. Norops lemurinus* is known from near sea level to 960 m elevation in the Lowland Moist Forest, Lowland Dry Forest, Lowland Arid Forest, and Premontane Wet Forest formations. Most active specimens of this shade-tolerant species are encountered in low vegetation or low on tree trunks in forest edge situations. It sleeps at night on leaves, stems, and branches of low vegetation. The species is sometimes active on the ground near tree trunks, which they climb in an effort to escape. It has been collected during every month of the year except September. Costa Rican individuals of this species feed on a variety of arthropods. Savage (2002: 464) stated "adults prey extensively on beetles, orthopterans, and caterpillars, but juveniles favor smaller and softer food items (e.g., flies and bark lice)."

Its principal reproductive activity in Costa Rica is during the rainy season (May through December) with a definite decline during the dry season (Savage, 2002).

*Etymology.* The name *lemurinus* is derived from the Latin *lemur* (shade, ghost of the departed) and *-inus* (pertaining to). The name probably refers to the lemurlike "immaculate" belly (Cope, 1861: 213). Individuals also appear a ghost white when sleeping at night, but Cope would not have had any field experience with the species.

*Specimens Examined* (435 [137] +10 skeletons; Map 15). ATLÁNTIDA: mountains S of Corozal, LACM 47696–97, 47699; Corozal, LACM 47698, 47700; Estación Forestal CURLA, SMF 77169–70, 77172–74, 78121–22, USNM 539841, 570001–02; 8 km SE of La Ceiba, KU 101392; 0.8 km S of La Ceiba, USNM 53359; near La Ceiba, SMF 79204, UTA R-41943; La Ceiba, MCZ R-32210, USNM 55242, 55244, 63206; near Lancetilla, SMF 79233; Lancetilla, AMNH 70407–21, 70433–34, MCZ R-29812–15, 29824–26, 29827–29 (skeletons), 29831–32, 31613–15, 32206 (skeleton), 38822, 175281, TCWC 30122, UMMZ 58390, 70320 (6), 71290 (7), 74039 (4), 172051, USNM 578750–53; Las Mangas, SMF 79996–80000; near Los Planes, UTA R-41942; Los Planes, UTA R-41940–41; Mezapita, USNM 580301–03; Parque Nacional Cuero y Salado, SMF 77171; mountains S of Piedra Pintada, LACM 47716–17; Punta Sal, SMF 79195–201; Quebrada de Oro, SMF 79220–21, USNM 539840, 570003–07; Río Cangrejal, USNM 578816; Salado Barra, MCZ R-191086–87; San Alejo, MCZ R-34386; San Marcos, USNM 570008–13; Santa Ana, SMF 91335; 32.2 km ESE of Tela, LACM 47715; about 80 km ESE of Tela, FMNH 13010–12; Tela, MCZ R-27578–81, 27582 (skeleton), 27583–91, 27592 (skeleton), 27593–95, 27596–97 (both skeletons), 27598–600, 175278–80, 177866 (skeleton), UMMZ 67695 (10), 118061. COLÓN: Amarillo, BMNH 1985.1110; Balfate, AMNH 58611; Cerro Calentura, CM 64594, LSUMZ 22459–60, 22489, 33674, 33677–78; Laguna Guaimoreto, LSUMZ

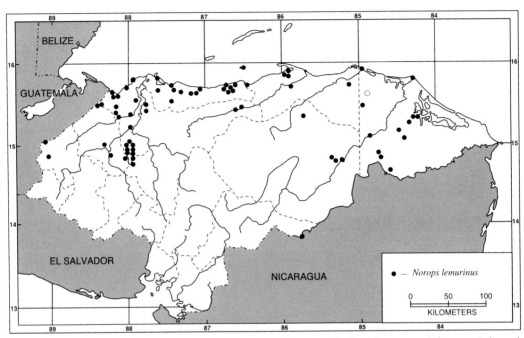

Map 15. Localities for *Norops lemurinus*. Solid symbols denote specimens examined and the open symbol an accepted record.

33676; Salamá, USNM 242034–55, 243279; 1–3 km W of Trujillo, KU 101393; 0.5 km S of Trujillo, LACM 47701; 2 km E of Trujillo, LACM 47702–04; about 15 km E of Trujillo, LSUMZ 27737–39; Trujillo, CM 64617, 64622, LACM 47705. COMAYA-GUA: Pito Solo, FMNH 5273–74. COPÁN: El Gobiado, USNM 573132–33; La Playona, FMNH 236390–91, SMF 79216–19, USNM 570014–15. CORTÉS: 8.0 km N of Agua Azul, USNM 243277; 6 km N of Agua Azul, AMNH 149607–10; 4.8 km S of Agua Azul, LACM 47707; Agua Azul, AMNH 70402, 70501, 70504, SMF 78126; Amapa, AMNH 70502; Cañon Santa Ana, SMF 79232; 4.9 km S of Chamelecón, KU 194868–74; between Cofradía and Buenos Aires, SMF 77742; near Cofradía, MCZ R-32523–24; 1.6 km W of El Jaral, LACM 47706, 47708–09, LSUMZ 88074–75; El Paraíso, UF 144753–55, 149661–62; Finca Fé, LSUMZ 12004, SMF 79225–27; Hacienda Santa Ana, FMNH 5260–62; NE shore of Lago de Yojoa, KU 194866, 194875–89; Lago de Yojoa, SMF 77175, 78123–25, UMMZ 70321 (3); Laguna Ticamaya, FMNH 5265–72; La Lima, BYU 22559–66; Los Pinos, SMF 92848, UF 150153, 166184, 166215–19, 166222–24, 166226–30, USNM 565452–53, 573134; near Peña Blanca, UF 166220–21, 166225; 16.1 km S of Potrer-illos, USNM 128102; 27.2 km W of Puerto Cortés, SMF 81128–29; Puerto Cortés, FMNH 5275; 10.9 km S of turnoff to Pulhapanzak on highway 1, KU 194836–65; 3.2 km W of San Pedro Sula, LACM 47718–19, TCWC 19188; W of San Pedro Sula, FMNH 5254–56, 5263–64; near San Pedro Sula, FMNH 5257–59; San Pedro Sula, UF 147667, USNM 565454; Santa Teresa, SMF 91275; about 0.5 km SSE of Tegucigalpita, SMF 79141; about 1 km SSE of Tegucigalpita, SMF 79105, 79107, 79230–31; 4.7 km N of Tegucigalpita, SMF 79228–29; 7.2 km ENE of Villanueva, LACM 47710–14. EL PARAÍSO: Boca Español, SMF 86985. GRACIAS A DIOS: Bachi Tingni, FMNH 282556; Barra Patuca, USNM 20304; Krausirpe, LSUMZ 28485–87, 52490; Leimus (Río Warunta), SMF

91338; Mocorón, UTA R-43563; Palacios, BMNH 1985.1108–09; Raudal Kiplatara, SMF 85885–87; Río Plátano, CM 59120; Rus Rus, UF 150336; Samil, SMF 91276; Urus Tingni Kiamp, USNM 570016; near Warunta, SMF 91337; Warunta Tingni Kiamp, USNM 565451. ISLAS DE LA BAHÍA: Cayo Cochino Grande near La Ensenada, KU 220102–10; Cayo Cochino Grande, La Ensenada, KU 220111; Cayo Cochino Grande, SMF 78120, 80800–04; Cayo Cochino Pequeño, KU 220112–17; Cayo Cochinos, UTA R-53993. OLANCHO: Caobita, SMF 80816; Matamoros, SMF 86985; Quebrada de la Chilantro, SMF 91336; Quebrada El Mono, SMF 80815. SANTA BÁRBARA: mountains W of Lago de Yojoa, KU 66937; tributary of Río Listón, SMF 91312; 1 km S of San José de Colinas, SMF 91277; Santa Bárbara, TCWC 23626. YORO: 5 km E of Coyoles, LACM 47692; Coyoles, LACM 47693–95, UTA R-38526; 24.1 km NNE of El Progreso, UF 20441; Los Indios, MCZ R-38823. "HONDURAS": AMNH 70500, CM 29375, UF 84760–61, 99315 (skeleton), ZSM 73/1998.

*Other Records* (Map 15). GRACIAS A DIOS: Baltiltuk, UNAH 5457, 5462, 5464, 5479, 5481, 5514 (Cruz Díaz, 1978; specimens now lost).

*Norops limifrons* (Cope)

*Anolis* (*Dracontura*) *limifrons* Cope, 1862: 178 (two syntypes, ANSP 7900–01 [see Malnate, 1971: 358]; type locality: "Veragua" [= Panama]).

*Anolis limifrons*: Meyer, 1969: 225 (in part); Meyer and Wilson, 1973: 18 (in part); Cruz Díaz, 1978: 26; O'Shea, 1986: 36; Poe, 2004: 63; Köhler and Sunyer, 2008: 97.

*Norops limifrons*: Savage, 1973: 11; Wilson and McCranie, 1994: 418; Köhler, 2000: 57; Köhler et al., 2000: 425; Nicholson et al., 2000: 30; Köhler and McCranie, 2001: 244; McCranie and Köhler, 2001: 232; Wilson et al., 2001: 136 (in part); Castañeda, 2002: 15; McCranie et al., 2002a: 27, 2006: 122; Wilson and Townsend, 2006: 105; Diener, 2008: 13; McCranie, 2011: 175.

*Geographic Distribution. Norops limifrons* occurs at low and moderate elevations on the Atlantic versant from northeastern Honduras to west-central Panama and from west-central Costa Rica to central Panama on the Pacific versant. In Honduras, this species occurs in the northeastern portion of the country southeast and east of the Cordillera Nombre de Dios.

*Description.* The following is based on 10 males (SMF 80706, 80708–09, 80712, 80715–16, 86172, 86184, 86198, 86223) and 10 females (SMF 80704–05, 80710–11, 80713–14, 86176, 86197, 86212, 86220). *Norops limifrons* is a small anole (SVL 41 mm in largest Honduran males examined [SMF 86172, 86198], 44 mm in largest Honduran females examined [SMF 86197, 86212, 86220]); dorsal head scales weakly keeled in internasal region, rugose to tuberculate in prefrontal, frontal, and parietal areas; deep frontal depression present; parietal depression absent; 5–8 (6.6 ± 1.0) postrostrals; anterior nasal entire, contacting rostral and first supralabial, occasionally contacting only rostral; 6–11 (8.8 ± 1.4) internasals; canthal ridge sharply defined; scales comprising supraorbital semicircles weakly keeled, largest scale in semicircles about same size as largest supraocular scale; supraorbital semicircles well defined; 1–3 (1.6 ± 0.7) scales separating supraorbital semicircles at narrowest point; 1–3 (2.2 ± 0.6) scales separating supraorbital semicircles and interparietal at narrowest point; interparietal well defined, greatly enlarged relative to adjacent scales, surrounded by scales of moderate size, longer than wide, usually larger than ear opening; 2–3 rows of about 6–9 (total number) enlarged, keeled supraocular scales; 1–2 enlarged supraoculars in broad contact with supraorbital semicircles; 2 elongate superciliaries, posterior much shorter than anterior; usually 3–4 enlarged canthals; 8–16 (11.3 ± 1.8) scales between second canthals; 12–18 (14.4 ± 1.9) scales between posterior canthals;

loreal region slightly concave, 36–69 (49.2 ± 9.4) mostly keeled (some smooth or rugose) loreal scales in maximum of 5–8 (6.3 ± 0.8) horizontal rows in 16; 6–8 (6.7 ± 0.7) supralabials and 5–8 (6.5 ± 0.8) infralabials to level below center of eye; suboculars weakly keeled, 2 suboculars usually in broad contact with supralabials; ear opening vertically oval; scales anterior to ear opening granular, similar in size to those posterior to ear opening; 6–8 (7.1 ± 0.9) postmentals, outer pair usually largest; keeled granular scales present on chin and throat; male dewlap small, extending to level of axilla; male dewlap with 10–14 horizontal gorgetal-sternal scale rows, about 16–19 scales per row ($n$ = 2); 2 (modal number) anterior marginal pairs in male dewlap; female dewlap absent; no nuchal crest or dorsal ridge; 2 middorsal scale rows slightly enlarged, smooth to rugose, dorsal scales lateral to middorsal series gradually larger than granular lateral scales; no enlarged scales scattered among laterals; 79–105 (89.4 ± 9.4) dorsal scales along vertebral midline between levels of axilla and groin in nine males, 90–114 (100.3 ± 12.0) in females; 50–72 (57.6 ± 6.2) dorsal scales along vertebral midline contained in 1 head length in males, 48–82 (57.8 ± 10.7) in nine females; ventral scales on midsection about same size as largest dorsal scales; ventral body scales smooth, subimbricate; 57–67 (61.9 ± 3.2) ventral scales along midventral line between levels of axilla and groin in nine males, 52–78 (64.0 ± 13.0) in females; 26–48 (40.4 ± 7.0) ventral scales contained in 1 head length in males, 32–48 (37.4 ± 8.1) in females; 122–155 (139.0 ± 12.0) scales around midbody in nine males, 119–188 (146.3 ± 34.7) in females; tubelike axillary pocket absent; precloacal scales not keeled; usually pair of slightly enlarged postcloacal scales in males; tail rounded to slightly compressed, TH/TW 1.00–1.31; basal subcaudal scales smooth; lateral caudal scales faintly keeled, homogeneous, although indistinct division in segments discernable; dorsal medial caudal scale row not enlarged, keeled, not forming crest;

most scales on anterior surface of antebrachium weakly keeled, unicarinate; 20–29 (23.7 ± 2.2) subdigital lamellae on Phalanges II–IV of Toe IV of hind limbs; 6–10 (8.1 ± 1.0) subdigital scales on Phalanx I of Toe IV of hind limbs; SVL 35.0–41.0 (38.3 ± 2.0) mm in males, 33.0–44.0 (41.2 ± 5.7) mm in females; TAL/SVL 1.53–2.22 in eight males, 1.39–2.50 in females; HL/SVL 0.24–0.27 in males, 0.24–0.29 in females; SHL/SVL 0.25–0.32 in males, 0.25–0.34 in females; SHL/HL 0.98–1.22 in males, 0.97–1.17 in females; longest toe of adpressed hind limb usually reaching between anterior border of eye and tip of snout.

Color in life of an adult male (USNM 570038): dorsum of body Brownish Olive (29); dorsum of head Grayish Horn Color (91) with radiating darker bars on supraocular scales; dorsal surfaces of limbs Citrine (51) with paler banding; chin white with tan bars along periphery; dorsal surface of tail Citrine with darker bands; venter Straw Yellow (56); dewlap white with basal Orange Yellow (18) spot. Dewlap color of another adult male (SMF 86897) was white with a basal yellow spot; the belly was also yellow.

Color in alcohol: dorsal surfaces of head and body brown to grayish brown; middorsal pattern of small dark brown spots or blotches present in some, some females have a pale brown middorsal stripe that is bounded on both sides by thin, dark brown border, pale middorsal stripe extending well onto tail; lateral surface of head pale brown; dorsal surfaces of limbs brown with paler brown crossbands; dorsal surface of tail brown with indistinct darker brown crossbars in those specimens in which pale middorsal stripe not extending onto tail; ventral surface of head white with brown flecks on many scales; ventral surface of body white with sparse to numerous brown flecks on scales of chest region, brown flecking on scales becoming more prominent posteriorly; subcaudal surface with sparse to numerous brown flecks on scales proximally, flecking becoming more prominent distally until subcaudal surface mostly brown for distal half.

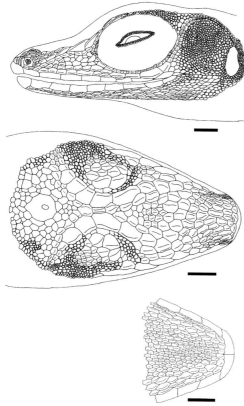

Figure 31. *Norops limifrons* head in lateral, dorsal, and ventral views. SMF 86900, adult female from confluence of Ríos Sausa and Wampú, Olancho. Scale bar = 1.0 mm. Drawing by Gunther Köhler.

Hemipenis: the completely everted hemipenis of SMF 86913 is a small unilobed organ; sulcus spermaticus bordered by well-developed sulcal lips, opening at base of apex; no discernable surface structure on truncus and lobe; asulcate processus absent.

*Diagnosis/Similar Species. Norops limifrons* is distinguished from all other Honduran *Norops*, except *N. carpenteri, N. ocelloscapularis, N. rodriguezii, N. yoroensis,* and *N. zeus,* by having a single elongated prenasal scale, smooth, subimbricate ventral scales, and slender habitus. *Norops limifrons* differs from *N. carpenteri, N. ocelloscapularis, N. rodriguezii,* and *N. yoroensis* by having long hind legs (longest toe of adpressed hind limb usually reaching between anterior border of eye and tip of snout in *N. limifrons* versus not reaching anterior border of eye in those four species). *Norops limifrons* also differs from those four species by having a dirty white dewlap with a basal orange-yellow spot in life (male dewlap orange in *N. carpenteri, N. ocelloscapularis, N. rodriguezii,* and *N. yoroensis*). *Norops limifrons* differs further from *N. carpenteri* in having a brown dorsum without paler spots in life (dorsum greenish brown with paler spots in life in *N. carpenteri*). *Norops limifrons* differs from *N. zeus* by having a dirty white dewlap with a basal orange-yellow spot in life (uniformly dirty white, usually without basal orange-yellow spot in life in *N. zeus*).

*Illustrations* (Figs. 31, 32). Fläschendräger and Wijffels, 1996 (adult; as *Anolis*); Köhler, 1999d (adult), 2000 (adult, ventral scales, head and dewlap), 2001b (adult, ventral scales, head and dewlap), 2003 (adult, ventral scales, head and dewlap), 2008 (adult, ventral scales, head and dewlap; Nicaragua specimen only); Savage, 2002 (adult); Guyer and Donnelly, 2005 (adult, dewlap); McCranie et al., 2006 (adult); Diener, 2008 (adult); Köhler and Sunyer, 2008 (head scales, hemipenis, head and dewlap; as *Anolis*).

*Remarks.* Köhler and Sunyer (2008) described two new species for populations from central and eastern Panama and from northwestern Panama and extreme eastern Costa Rica that had been previously identified as *N. limifrons*. Köhler and McCranie (2001) had earlier described the Honduran Cordillera Nombre de Dios populations of this complex as a new species (*N. zeus*). These nomenclatural changes were based largely on hemipenial morphology and/or differences in dewlap color.

The tissues identified as those of *N. limifrons* used in the phylogenetic analyses in Nicholson (2002) and Nicholson et al. (2005, 2012) actually represent those of the recently described and closely related *N. apletophallus* (Köhler and Sunyer). *Norops limifrons* is usually placed in the *N. fuscoauratus* species group (Nicholson, 2002). However, phylogenetic analyses

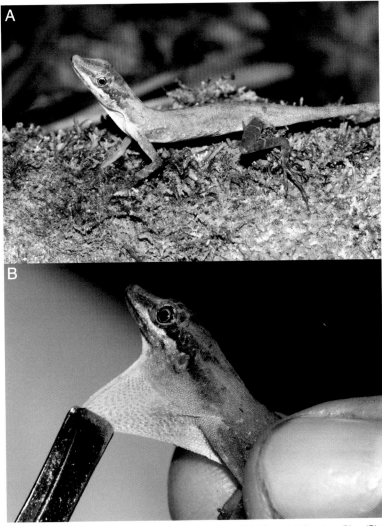

Figure 32.  *Norops limifrons*. (A) Adult male (FMNH 282550) from Leimus (Río Warunta), Gracias a Dios; (B) adult male dewlap (FMNH 282550). Photographs by James R. McCranie.

based on molecular data do not support the monophyly of the *N. fuscoauratus* species group (Nicholson, 2002; Nicholson et al., 2005, 2012; also see Poe, 2004, based largely on morphology) as recognized by Savage and Guyer (1989). There seems little doubt that the two Honduran species *N. limifrons* and *N. zeus* are closely related based on their similar external morphology (also see Remarks for *N. ocelloscapularis*). However, we take a conservative approach herein and retain these two species in the *N. fuscoauratus* species subgroup of the *N. auratus* species group of Nicholson et al. (2012).

*Natural History Comments. Norops limifrons* is known from near sea level to 900 m elevation in the Lowland Moist Forest and Premontane Wet Forest formations. It can be extremely common in primary to secondary broadleaf forest and is active on the ground, on logs, low on tree trunks, and on leafy vegetation. *Norops limifrons* also

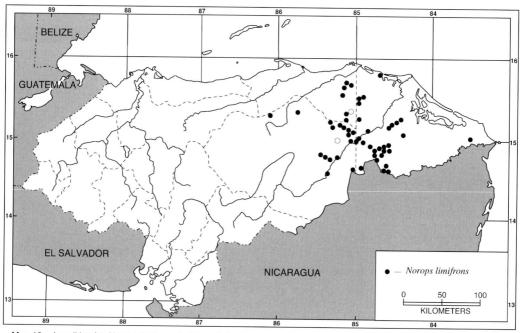

Map 16.  Localities for *Norops limifrons*. Solid symbols denote specimens examined and open symbols accepted records.

sleeps on leaves at night. It appears equally common in forest and in gap or forest edge situations such as along streams and trails. The species has been found during every month of the year. Its ecology has been extensively studied in Costa Rica and Panama (studies cited by Savage, 2002, from Atlantic central, west-central, and eastern Panama represent newly described species in the *N. limifrons* complex). Savage (2002) summarized those studies and the following is from that summary (see the literature cited therein for more information). Those populations of *N. limifrons* were found to be a gap inhabitant with their highest populations in relatively open or disturbed places in humid forests. They forage on the ground and head downward from slightly elevated perch sites. The species feeds on a variety of arthropods, especially araneid spiders and beetles, but avoids ants. Females lay a single egg every 7–10 days during the rainy season, but only about every three weeks during the drier season. One Honduran female (SMF 91283)

collected during July deposited an egg in the collecting bag.

*Etymology.* The name *limifrons* is formed from the Latin *lima* (file) and *frons* (brow, forehead), and apparently refers to the "facial rugae [being] moderately developed" (Cope, 1862: 179).

*Specimens Examined* (230 [2] + 2 skeletons, 1 egg; Map 16). COLÓN: El Ocotillal, SMF 86215–18; Quebrada Machín, USNM 526490–91 (skeletons), 541026–29, 570017–21; Río Guaraska, BMNH 1985.1112–13; Río Kinikisne, BMNH 1985.1118; Río Paulaya, BMNH 1985.1114–17; Tulito, BMNH 1985.1119–20. GRACIAS A DIOS: Awasbila, SMF 88689, USNM 570022–26; Bachi Kiamp, MVZ 267188–90, SMF 91278, 91280–81, 91342–43, 92844–45, USNM 565458; Bachi Tingni, SMF 91282; Bodega de Río Tapalwás, UF 150339, USNM 570027–31; Cabeceras de Río Rus Rus, SMF 86888, USNM 570032–34; Caño Awalwás, UF 150333, 150340, USNM 570035–40; Cerro Wahatingni, UF 150332,

150337; Crique Curamaira, USNM 570041–42; Crique Unawás, SMF 86220–29; near Cueva de Leimus, SMF 91341; Dos Bocas, MVZ 267191; Hiltara Kiamp, USNM 570043, 573938; Kakamuklaya, USNM 573135–36; Kalila Plapan Tingni, USNM 570044; Karasangkan, USNM 570045; Kaska Tingni, SMF 88709, USNM 570046–48; Kipla Tingni Kiamp, USNM 570049, 573137–39; Krahkra, USNM 570050; Krausirpe, LSUMZ 28497; Kyras, LACM 16857; Laguna Baraya, ROM 19269; Leimus (Río Warunta), FMNH 282550, USNM 565456; Mocorón, UTA R-46171–72; Pomokir, SMF 86207–14; confluence of Quebrada Waskista and Río Wampú, SMF 80708–09, 86887, USNM 330183–84; Raudal Kiplatara, SMF 86172–87; Rawa Kiamp, SMF 91283, 91284 (egg from SMF 91283); Río Cuyamel, SMF 86188–99, 86203–05; Río Sutawala, USNM 570051–53; confluence of Ríos Wampú and Patuca, USNM 330181–82; Rus Rus, USNM 570054–57; Sadyk Kiamp, SMF 91279, 92843, USNM 565457; San San Hil Kiamp, USNM 570058–59; Urus Tingni Kiamp, USNM 570060–65; Warunta Tingni Kiamp, USNM 570066–67. OLANCHO: Caobita, SMF 80711; Cuaca, UTA R-53901–02, 53917; Matamoros, SMF 80710, 80712, 86910–14, USNM 570068–70; Nueva Esperanza, USNM 570071; Quebrada de la Chilantro, SMF 91340; Quebrada de Las Marías, SMF 86909, USNM 570074–78; Quebrada El Guásimo, SMF 80713–16, 86916, USNM 570072; Quebrada El Mono, SMF 86915; Quebrada El Robalo, USNM 570079; Quebrada Kuilma, USNM 570073; confluence of Quebrada Siksatara and Río Wampú, SMF 86889–90, USNM 330180; Qururia, USNM 570080; confluence of Ríos Aner and Wampú, SMF 80704, 86894–96, USNM 330176–77; Río Cuaca, SMF 91339; Río Kosmako, USNM 570081–84; confluence of Ríos Sausa and Wampú, SMF 80705–07, 86897–903, USNM 330178–79; along Río Wampú between Ríos Aner and Sausa, SMF 86907–08; confluence of Ríos Yanguay and Wampú, SMF 86891–93, USNM 330175; Yapuwás, SMF 86904–06.

*Other Records* (Map 16). COLÓN: Empalme Río Chilmeca, UNAH 5447 (Cruz Díaz, 1978; specimen now lost). OLANCHO: about 40 km E of Catacamas, TCWC 23627 (Meyer and Wilson, 1973; specimen not found during February 2010).

### Norops loveridgei (Schmidt)

*Anolis loveridgei* Schmidt, 1936: 47 (holotype, MCZ R-38700; type locality: "Portillo Grande, 4100 feet altitude, Yoro, Honduras"); Smith, 1937: 42; Barbour and Loveridge, 1946: 72; Peters, 1952: 28; Marx, 1958: 453; Etheridge, 1959: 215; Williams, 1963: 479; Meyer, 1969: 228; Peters and Donoso-Barros, 1970: 59; Meyer and Wilson, 1973: 18; Kluge, 1984: 11; Poe, 2004: 63; Nicholson et al., 2005: 933, 2006: 452; Köhler et al., 2007: 391; D'Angiolella et al., 2011: 38; Townsend et al., 2012: 100; Pyron et al., 2013: fig. 19; Köhler, 2014: 210.

*Norops loveridgei*: Savage and Talbot, 1978: 480; McCranie and Cruz, 1992: 233; Köhler et al., 2001: 254; Wilson et al., 2001: 136; Köhler, 2003: 102, 2008: 108; Wilson and McCranie, 2003: 59, 2004b: 43; McCranie and Castañeda, 2005: 14; McCranie et al., 2006: 217; Wilson and Townsend, 2006: 105; Nicholson et al., 2012: 12; McCranie and Solís, 2013: 242.

*Geographic Distribution. Norops loveridgei* occurs at moderate and intermediate elevations and the upper limits of low elevations in the Cordillera Nombre de Dios of Atlántida and Yoro and in the mountains of central Yoro in northwestern and north-central Honduras.

*Description.* The following is based on nine males (FMNH 21776; KU 219982; MCZ R-38700, 38831; SMF 91730; UNAH 1896; USNM 344803, 578754, 578758) and seven females (FMNH 21870; SMF 78793, 86951; UMMZ 90665; USNM 344802, 344804, 578756). *Norops loveridgei* is a large anole (SVL 118 mm in largest males [UNAH 1896; USNM 578758], 116 mm in largest female [USNM 578756]); dorsal head scales keeled in internasal region, rugose or tuberculate in prefrontal area, rugose to weakly keeled in frontal and parietal areas; moderate frontal and parietal depressions present; 5–8 (5.8 ± 0.9) post-rostrals; anterior nasal divided, lower section usually contacting rostral and first supralabial, or only first supralabial, occasionally lower anterior nasal separated from rostral and first supralabial by 1 scale row; 6–10 (8.0 ± 1.2) internasals; canthal ridge sharply defined; scales comprising supraorbital semicircles rugose to weakly keeled, largest scale in semicircles larger than largest supraocular scale; supraorbital semicircles well defined; 3–7 (4.6 ± 1.2) scales separating supraorbital semicircles at narrowest point; 5–10 (6.4 ± 1.3) scales separating supraorbital semicircles and interparietal at narrowest point; interparietal well defined, only slightly enlarged relative to adjacent scales, surrounded by scales of moderate size, longer than wide, smaller than ear opening; 5–6 rows of about 5–8 (total number) only slightly enlarged, weakly keeled supraocular scales; enlarged supraoculars in broad contact with supraorbital semicircles; 3 elongate superciliaries, anteriormost longest; 3–6 enlarged canthals; 8–15 (10.4 ± 1.8) scales between second canthals; 10–16 (12.0 ± 1.9) scales between posterior canthals; loreal region slightly concave, 61–100 (78.4 ± 10.5) mostly weakly keeled loreal scales in maximum of 8–12 (9.4 ± 0.9) horizontal rows; 9–12 (10.5 ± 0.9) supralabials and 5–12 (10.1 ± 1.8) infralabials to level below center of eye; suboculars weakly keeled, usually separated from supralabials by 1 row of scales (occasionally 1 subocular contacts 1 supralabial scale); ear opening vertically oval; scales anterior to ear opening not granular, slightly larger than those posterior to ear opening; 4–6 (5.6 ± 0.6) postmentals, outer pair usually largest; keeled granular scales present on chin and throat; male dewlap huge, extending well onto chest; male dewlap with 7–11 horizontal gorgetal-sternal scale rows, about 21–27 scales per row

($n$ = 4); 4–5 (usually 4; modal number) anterior marginal pairs in male dewlap; female dewlap well developed, much smaller than male dewlap; low nuchal crest present but no dorsal ridge; about 2 middorsal scale rows slightly enlarged, weakly keeled, nonimbricate, dorsal scales lateral to middorsal series abruptly larger than granular lateral scales; no enlarged scales scattered among laterals; 59–92 (73.6 ± 12.7) dorsal scales along vertebral midline between levels of axilla and groin in males, 49–95 (67.8 ± 16.7) in females; 40–59 (46.1 ± 6.9) dorsal scales along vertebral midline contained in 1 head length in males, 36–52 (41.0 ± 5.7) in females; ventral scales on midsection larger than largest dorsal scales; ventral body scales smooth, bulging, subimbricate; 66–81 (70.1 ± 4.8) ventral scales along midventral line between levels of axilla and groin in males, 35–76 (55.0 ± 14.6) in females; 38–62 (49.7 ± 7.0) ventral scales contained in 1 head length in males, 34–48 (41.9 ± 4.6) in females; 126–179 (162.9 ± 17.0) scales around midbody in males, 130–174 (155.7 ± 18.0) in females; tubelike axillary pocket absent; precloacal scales rugose; pair of enlarged postcloacal scales present in males; tail slightly to distinctly compressed, TH/TW 1.21–2.31 in 14; all subcaudal scales keeled; lateral caudal scales smooth to weakly keeled, homogeneous, although indistinct division in segments discernable; dorsal medial caudal scale row slightly enlarged, weakly keeled, not forming crest; most scales on anterior surface of antebrachium weakly keeled, unicarinate; 33–43 (37.7 ± 3.2) subdigital lamellae on Phalanges II–IV of Toe IV of hind limbs; 11–15 (13.0 ± 1.4) subdigital scales on Phalanx I of Toe IV of hind limbs; SVL 69–118 (101.8 ± 18.5) mm in males, 60–116 (92.3 ± 20.6) mm in females; TAL/SVL 1.68–1.97 in five males, 1.80–1.92 in five females; HL/SVL 0.26–0.29 in males, 0.25–0.30 in females; SHL/SVL 0.25–0.29 in males, 0.25–0.28 in females; SHL/HL 0.88–1.06 in males, 0.88–1.00 in females; longest toe of adpressed hind limb usually reaching between posterior and anterior borders of eye.

Color in life of an adult female (USNM 344802): dorsum brown with darker brown bands on body; middorsal stripe on body and length of tail rust brown; belly pale brown; dewlap dark dirty red with pale brown scales; iris rust brown. Color in life of a subadult male (KU 219982) was described by McCranie and Cruz (1992: 234): "dorsum of body Smoke Gray (color 44) with four Olive-Gray (color 42) posteriorly directed crossbands; scales of vertebral ridge slightly paler with about 12 brownish spots; dorsal surfaces of limbs Smoke Gray with obscure, slightly darker crossbands or blotches; dorsal and lateral surfaces of head Smoke Gray with darker smudging laterally posterior to eye; a small, pale yellowish spot below eye; dorsal surface of tail Smoke Gray with dark brownish spots proximally on caudal ridge, with Olive-Gray crossbands medially, tail dark brownish distally; chin cream with dark brownish smudging; venter of body similar to Smoke Gray dorsal surfaces, but slightly paler; dewlap Spectrum Orange (color 17) with two crescent-shaped Magenta (color 2) streaks; dewlap scales pale yellowish." Color in life of another subadult male (USNM 344803): dorsal surface of body Smoke Gray (45) with dark brown crossbars; dewlap Spectrum Orange (17) with two Scarlet (14) streaks and basal spot; dewlap scales brown, except those of basal two rows dirty white. Color in life of a subadult female (USNM 344804) was similar to that of USNM 344803, except that the dorsal crossbars were bordered anteriorly by dirty white and the dewlap was Scarlet (14).

Color in alcohol: dorsal and lateral surfaces of head and body grayish brown, patternless or with posteriorly directed dark brown crossbands that are expanded middorsally and becoming narrower laterally; some females with cream-colored, brown-bordered vertebral stripe; usually no dark brown interorbital bar; dark brown lines radiating outward from eye present; one female (SMF 78793) has a pair of small pale brown blotches on dorsal surface of base of tail just above cloaca; dorsal surfaces of limbs and tail with dark brown crossbands; ventral surfaces of head and body grayish brown, darker brown in throat region; supraocular region grayish turquoise in some.

Hemipenis: unknown.

*Diagnosis/Similar Species. Norops loveridgei* is distinguished from all other Honduran *Norops*, except *N. biporcatus* and *N. petersii*, by the combination of its large size (adult males average about 102 mm SVL, females about 92 mm SVL) and 9–12 supralabials to level below center of eye. *Norops loveridgei* differs from *N. biporcatus* and *N. petersii* by having long hind legs (longest toe of adpressed hind limb usually reaching between posterior and anterior borders of eye in *N. loveridgei* versus usually reaching about ear opening in those two species). *Norops loveridgei* further differs from *N. biporcatus* by having a brown coloration in life (green coloration in life in *N. biporcatus*), more than 60 loreal scales (usually fewer than 60 in *N. biporcatus*), and usually more 130 scales around midbody (fewer than 130 in *N. biporcatus*). *Norops loveridgei* differs further from *N. petersii* by having an orange male dewlap with purple streaks in life, 3–7 scales separating supraorbital semicircles at narrowest point, and 61–100 loreals in a maximum of 8–12 horizontal rows (male dewlap pinkish brown with pale yellow margin in life, 1–2 scales separating semicircles, and 40–67 loreals in maximum of 6–9 horizontal rows in *N. petersii*).

*Illustrations* (Figs. 33, 34, 88). McCranie and Cruz, 1992 (subadult, head and dewlap); Köhler, 2003 (adult), 2008 (adult), 2014 (cloacal region; as *Anolis*); Townsend et al., 2012 (adult).

*Remarks. Norops loveridgei* is usually placed in the *N. biporcatus* species group in the earlier literature. However, phylogenetic analyses based on molecular data do not support a close relationship between *N. biporcatus* and *N. loveridgei* (Nicholson, 2002; Nicholson et al., 2005, 2012; also see Poe, 2004, for an analysis based largely on

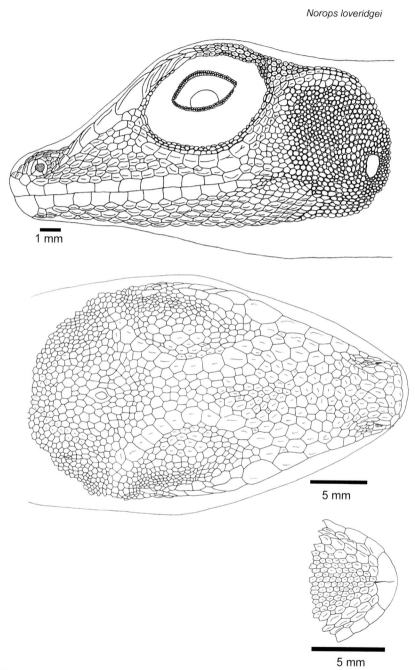

Figure 33.   *Norops loveridgei* head in lateral, dorsal, and ventral views. SMF 86951, adult female from Quebrada de Oro, Atlántida. Drawing by Lara Czupalla.

Figure 34. *Norops loveridgei.* (A) Adult male (SMF 91730) from La Liberación, Atlántida; (B) adult male dewlap (SMF 91730); (C) adult female dewlap (USNM 344802) from ridge above Quebrada de Oro, Atlántida. Photographs by James R. McCranie.

morphology). In both most recent phylogenetic analyses based on molecular data (Nicholson et al., 2005, 2012) *N. loveridgei* was recovered in a clade with *N. purpurgularis*. *Norops purpurgularis*, along with *N. johnmeyeri* and *N. pijolense*, are usually placed in the *N. schiedii* species group (see Remarks for *N. johnmeyeri*). *Norops loveridgei* also resembles *N. johnmeyeri*, *N.*

*pijolense*, and *N. purpurgularis* in having a well-developed female dewlap, long hind limbs, smooth ventral scales, and deliberant movements. Therefore, we place *N. loveridgei* in the *N. schiedii* species subgroup of the *N. auratus* species group of Nicholson et al. (2012).

*Natural History Comments.* *Norops loveridgei* is known from about 550 to 1,600 m elevation in the Premontane Wet Forest and Lower Montane Wet Forest formations and peripherally in the Lowland Moist Forest formation. This is a highly arboreal species that is active during the day low on tree trunks, in low vegetation, or on boulders along stream edges. It sleeps at night on tree trunks and branches, and on leaves and stems of vegetation; a juvenile was sleeping at night on top of a log. An adult female that laid an egg in the collecting bag was taken on a tree trunk about 2 m above the ground during August. That lizard was slowly descending the tree trunk when first seen and was likely looking to lay its egg in leaf litter. Individuals of *N. loveridgei* are deliberate in their movements and usually do not attempt to escape before capture. Townsend et al. (2012: 101) reported finding this species sleeping at night on vegetation ("including palm fronds and woody lianas, 1.5 to 4 m above the ground"). *Norops loveridgei* has been found during February, April to June, and during August and is probably active throughout the year during suitable weather. Nothing has been published on diet in this species.

*Etymology.* The name *loveridgei* is a patronym honoring Arthur Loveridge, who at the time of the description of this species was the curator of the herpetological collections at the Museum of Comparative Zoology, Harvard University.

*Specimens Examined* (20 [6]; Map 17). ATLÁNTIDA: S slope of Cerro Búfalo, USNM 344802–04, 565459; about 1 km N of Cerro Cabeza de Negro, UNAH 1896; La Liberación, SMF 91730, USNM 578754–58; Quebrada de Oro, KU 219982, SMF 78793, 86951. YORO: Mataderos Mountains, FMNH 21776, MCZ R-38831,

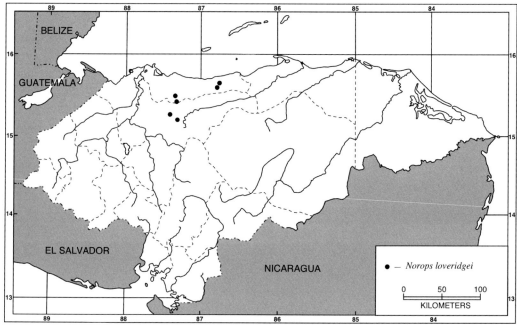

Map 17.   Localities for *Norops loveridgei*. Solid symbols denote specimens examined.

UMMZ 90665 (was MCZ R-38833); Portillo Grande, FMNH 21870, MCZ R-38700, 38834.

### *Norops morazani* (Townsend and Wilson)

*Anolis* sp.: Townsend et al., 2007: 10.
*Anolis morazani* Townsend and Wilson, 2009: 63 (holotype, SMF 87153; type locality: "Honduras, Departamento de Francisco Morazán, Municipio de Marale, Parque Nacional Montaña de Yoro, Cataguana, 15°01′N, 87°06′W, 1910 m elevation"); Wilson et al., 2013: 66.
*Norops morazani*: McCranie, 2009: 10.

*Geographic Distribution. Norops morazani* is known to occur at moderate and intermediate elevations in Parque Nacional Montaña de Yoro and environs in northern Francisco Morazán, Honduras.

*Description.* The following is based on 10 males (UF 150000, 151759, 151771–72, 151774, 151776, 151778, 151782, 151788, 151795) and 10 females (UF 151756, 151761, 151766–67, 151780, 151783–84,

151787, 151789, 151792). *Norops morazani* is a medium-sized anole (SVL 52 mm in largest male examined [UF 150000], 54 mm in largest female examined [UF 151756]; maximum reported SVL 59 mm [Townsend and Wilson, 2009]); dorsal head scales rugose to strongly keeled in internasal, prefrontal, frontal, and parietal areas; deep frontal depression present; parietal depression shallow; 3–6 (4.0 ± 1.1) postrostrals; anterior nasal divided, contacting rostral and first supralabial; 3–6 (4.6 ± 0.8) internasals; canthal ridge sharply defined; scales comprising supraorbital semicircles rugose, largest scale in semicircles larger than largest supraocular scale; supraorbital semicircles well defined; 0–2 (1.2 ± 0.7) scales separating supraorbital semicircles at narrowest point; 2–4 (2.7 ± 0.4) scales separating supraorbital semicircles and interparietal at narrowest point; interparietal well defined, slightly enlarged relative to adjacent scales, surrounded by scales of moderate size, longer than wide, about same size as ear opening; 2–3 rows of about 5–7 (total number) enlarged, faintly to strongly keeled supraocular scales; usually 1 enlarged supraocular in broad contact with supraorbital semicircles; 2–3 elongate superciliaries, posterior one or two only slightly shorter than anteriormost; usually 3 enlarged canthals; 4–8 (5.6 ± 0.9) scales between second canthals; 6–10 (7.6 ± 1.4) scales between posterior canthals; loreal region slightly concave, 18–35 (24.1 ± 4.8) mostly strongly keeled (some smooth or rugose) loreal scales in maximum of 4–6 (4.5 ± 0.6) horizontal rows; 5–7 (6.5 ± 0.6) supralabials and 5–7 (6.0 ± 0.7) infralabials to level below center of eye; suboculars weakly to strongly keeled, usually 2 suboculars in broad contact with supralabials; ear opening vertically oval; scales anterior to ear opening not granular, slightly larger than those posterior to ear opening; 4–6 (4.3 ± 0.6) postmentals, outer pair usually largest; keeled granular scales present on chin and throat; male dewlap small, extending to level of axilla; male dewlap with 5–7 horizontal gorgetal-sternal scale rows, about 5–7 scales per row (*n* = 7); 2 (modal number) anterior marginal pairs in male dewlap; female dewlap absent; low nuchal crest and low dorsal ridge present in males; about 6–10 middorsal scale rows distinctly enlarged, keeled; dorsal scales lateral to middorsal series grading into keeled, nongranular lateral scales; flank scales slightly heterogeneous, solitary, slightly enlarged keeled scales scattered among laterals; 34–48 (41.8 ± 4.5) dorsal scales along vertebral midline between levels of axilla and groin in males, 37–49 (43.7 ± 3.5) in females; 23–30 (26.2 ± 2.4) dorsal scales along vertebral midline contained in 1 head length in males, 23–43 (28.4 ± 6.1) in females; ventral scales on midsection slightly larger than largest dorsal scales; ventral body scales strongly keeled, mucronate, imbricate, 33–44 (38.7 ± 3.3) ventral scales along midventral line between levels of axilla and groin in males, 34–43 (38.4 ± 2.5) in females; 21–26 (23.3 ± 1.6) ventral scales contained in 1 head length in males, 21–24 (22.0 ± 1.2) in females; 97–116 (103.3 ± 5.6) scales around midbody in males, 96–126 (109.9 ± 9.5) in females; tubelike axillary pocket absent; precloacal scales strongly keeled; pair of greatly enlarged postcloacal scales in males; tail slightly compressed, TH/TW 1.07–1.40; all subcaudal scales keeled; lateral caudal scales keeled, homogeneous, although indistinct division in segments discernable; dorsal medial caudal scale row slightly enlarged, keeled, not forming crest; scales on anterior surface of antebrachium distinctly keeled, unicarinate; 20–26 (23.0 ± 1.6) subdigital lamellae on Phalanges II–IV of Toe IV of hind limbs; 5–8 (6.8 ± 0.7) subdigital scales on Phalanx I of Toe IV of hind limbs; SVL 37.8–51.7 (45.2 ± 3.8) mm in males, 48.0–53.9 (50.2 ± 2.2) mm in females; TAL/SVL 1.61–2.73 in seven males, 1.93–2.62 in nine females; HL/SVL 0.26–0.28 in males and females; SHL/SVL 0.24–0.29 in males, 0.24–0.28 in females; SHL/HL 0.87–1.06 in males, 0.87–1.02 in females; longest toe of adpressed hind limb usually reaching between posterior margin of eye and mideye.

Color in life of an adult male (USNM 581187): dorsal and lateral surfaces of body and tail Natal Brown (49 in Köhler, 2012) with Clay Color (18) mottling; top of head mottled Clay Color and Natal Brown, with Natal Brown interorbital crossbar; top of forelimbs mottled Clay Color and Natal Brown, hind limbs Clay Color with Natal Brown crossbars; ventral surfaces of head and body Beige (254), that of limbs Pale Cinnamon (55); subcaudal surface Pale Cinnamon basally, mottled Pale Cinnamon and Natal Brown medially, becoming Natal Brown on posterior half; dewlap Pratt's Ruby (68). Color in life of an adult male (UF 150000) was described by Townsend and Wilson (2009: 67): "middorsum pale rust brown with a series of rust brown, cream-outlined scallops bounded below by a narrow cream lateral line that originates above ear opening; lateral surfaces of body pale rust brown; dorsal surface of head pale rust brown with slightly darker interorbital bar; dorsal surface of limbs pale rust brown with tan-outlined rust brown crossbars; tail pale rust brown; venter uniform cream; dewlap red with cream gorgetal scales." Color in life of an adult female (MCZ R-185612) was described by those same authors (p. 67): "the middorsum was rust brown with a broad, boldly defined, pale tan middorsal stripe bordered by rust brown stripes, fading to yellowish brown on lateral surfaces. The dorsal surface of head rust brown, lateral surfaces were yellowish brown with a dark brown edge, pale yellow spot below the eye. The ventral surface was pale yellow with pale orange smudging on posterior portion of venter. The small dewlap was orange with yellow and the ventral side of the hind limbs was yellow with an orange wash. The dorsal surface of the tail was pale rust brown grading to dark brown distally, and the ventral surface had an orange wash."

Color in alcohol: dorsal and lateral surfaces of head and body with brown to grayish brown ground color; dark brown interorbital bar usually present; sides of body usually with pale brown longitudinal lateral stripe; ventral surfaces of head and body pale grayish brown or dirty white; dorsum uniform in males, or with brown, cream-outlined scallops bounded below by narrow cream lateral stripe that originates above ear opening; some males with irregular dark brown middorsal stripe; females show wide degree of variation in dorsal pattern including narrow pale brown middorsal stripe, middorsal stripe bordered by dark, scalloped or triangular blotches in some; all ventral surfaces pale brown, patternless, or with numerous dark brown punctations.

Hemipenis: the completely everted hemipenis of SMF 87153 is a stout, bulbous, bilobed organ; sulcus spermaticus bordered by well-developed sulcal lips that open into broad concave area at base of apex; lobes calyculate; truncus with transverse folds; asulcate processus divided, with medial flap.

*Diagnosis/Similar Species. Norops morazani* is distinguished from all other Honduran *Norops*, except *N. crassulus* and *N. rubribarbaris*, by the combination of having about 6–10 distinctly enlarged middorsal

scale rows, slightly heterogeneous lateral scales, strongly keeled, mucronate, imbricate ventral scales, and a pair of greatly enlarged postcloacal scales in males. *Norops morazani* differs from *N. crassulus* and *N. rubribarbaris* by having a hemipenis with a divided asulcate processus (undivided in *N. rubribarbaris* and *N. crassulus*).

*Illustrations* (Figs. 35, 36, 97). Townsend and Wilson, 2009 (adult, head scales, hemipenis).

*Remarks.* Townsend and Wilson (2009) in their description of the species listed SMF 87513 as both the holotype and a paratype. This species was placed in the *N. crassulus* species group by its describers (see Remarks for *N. amplisquamosus*). *Norops morazani* tissues were not utilized in the phylogenetic analysis of Nicholson et al. (2012).

*Natural History Comments. Norops morazani* is known from 1,275 to 2,150 m elevation in the Premontane Wet Forest and Lower Montane Wet Forest formations. Townsend and Wilson (2009: 68) stated, "adults and juveniles were commonly found along the edges of a fallow field overgrown with blackberry (*Rubus* spp.) and bracken fern (*Pteridium* sp.); individuals were also common along streams and trails in heavily to lightly disturbed cloud forest." McCranie (June 2013), in addition to finding it in the above-described situations, also found it sleeping at night in low vegetation, including that growing on a fallen log, along the Quebrada Cataguana and a small river near Los Planes (most seen were not collected). All adult females collected during June were "gravid" (Townsend and Wilson, 2009: 68). Other specimens were collected during March and September. Nothing has been published on its diet.

*Etymology.* The name *morazani* is a patronym honoring José Francisco Morazán Quesada, a postcolonial Honduran statesman and national hero, and for whom the Honduran department Francisco Morazán is named.

*Specimens Examined* (76 [0]; Map 18). FRANCISCO MORAZÁN: Cataguana, MVZ 257260–64, 257266–68, UF 150000,

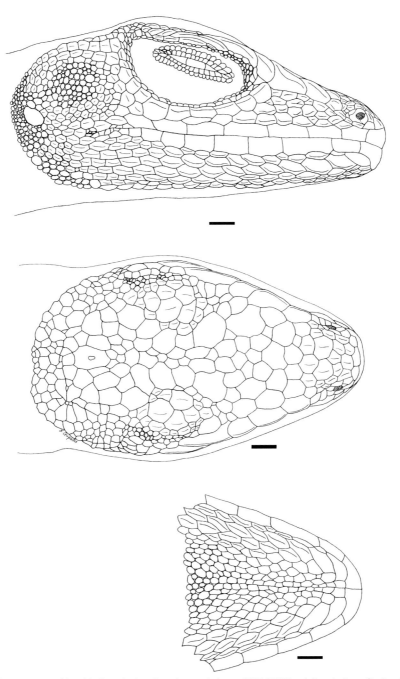

Figure 35.  *Norops morazani* head in lateral, dorsal, and ventral views. SMF 87153, adult male from Quebrada Cataguana, Parque Nacional Montaña de Yoro, Francisco Morazán. Scale bar = 1.0 mm. Drawing by Gunther Köhler.

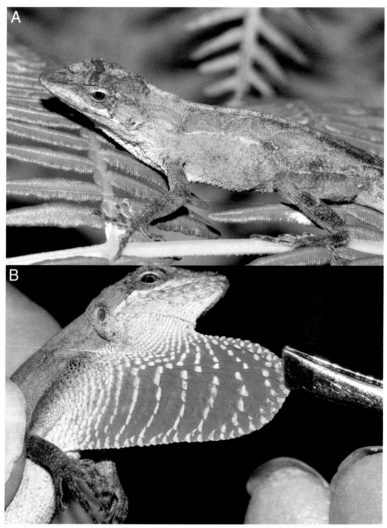

Figure 36.   *Norops morazani.* (A) Adult male (USNM 581184) from Cataguana, Parque Nacional Montaña de Yoro, Francisco Morazán; (B) adult male dewlap (USNM 581184). Photographs by James R. McCranie.

151756, 151758–78, 151780–85, 151787–89, 151791–96, USNM 565032–33, 565040, 581184–85, 581187–89; Los Planes, USNM 581186; Montaña de La Sierra, UF 166239–45; Quebrada Cataguana, MVZ 257265, SMF 87152–55, UF 151757, 151779, 151790, USNM 565034–39.

*Norops muralla* Köhler, McCranie, and Wilson

*Norops sminthus*: Espinal, 1993: table 3; Wilson et al., 2001: 136 (in part).

*Norops* cf. *sminthus*: Köhler and Obermeier, 1998: 136 (in part).

*Norops muralla* Köhler, McCranie, and Wilson, 1999: 285 (holotype, SMF 78093; type locality: "Honduras, Departamento de Olancho, Parque Nacional La Muralla, along trail to Cerro de Enmedio, 1500 m"); Köhler, 2000: 63, 2003: 102,

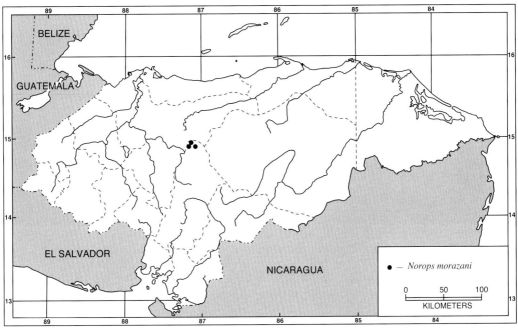

Map 18.   Localities for *Norops morazani*. Solid symbols denote specimens examined.

2008: 108; Köhler et al., 2001: 254; Wilson and McCranie, 2003: 59, 2004b: 43.

*Norops "sminthus"*: Espinal et al., 2001: 106.

*Anolis muralla*: Köhler et al., 2007: 391; Townsend and Wilson, 2009: 63; Wilson et al., 2013: 66; Köhler, 2014: 210.

*Geographic Distribution. Norops muralla* occurs at moderate (upper level) and intermediate elevations in Refugio de Vida Silvestre Muralla in northwestern Olancho, Honduras.

*Description.* The following is based on nine males (SMF 78093, 78375–76, 87000–01; USNM 521912, 521916–18) and 19 females (SMF 78372–74, 78377–83; USNM 521910–11, 521913, 521915–16, 521920–23). *Norops muralla* is a medium-sized anole (SVL 48 mm in largest male examined [USNM 521912], 56 mm in largest female examined [USNM 521911]; maximum reported SVL 57 mm [Townsend and Wilson, 2009]); dorsal head scales smooth or rugose in internasal, prefrontal, and frontal regions, most scales weakly keeled or conical in parietal area; frontal depression present; parietal depression absent; 2–5 (4.1 ± 1.0) postrostrals; anterior nasal divided, lower section contacting rostral and first supralabial; 4–7 (4.6 ± 0.8) internasals; canthal ridge sharply defined; scales comprising supraorbital semicircles keeled, largest scale in semicircles larger than largest supraocular scale; supraorbital semicircles well defined; 0–2 (1.0 ± 0.4) scales separating supraorbital semicircles at narrowest point; 2–4 (3.0 ± 0.6) scales separating supraorbital semicircles and interparietal at narrowest point; interparietal well defined, slightly to distinctly enlarged relative to adjacent scales, surrounded by scales of moderate size, longer than wide, smaller than ear opening; 2–3 rows of about 5–8 (total number) enlarged, smooth to faintly keeled supraocular scales; 1–2 enlarged supraoculars in broad contact with supraorbital semicircles; 2 elongate superciliaries, posterior only slightly shorter than anterior; usually 3 enlarged canthals; 5–7 (5.7 ± 0.8) scales between second canthals; 6–9 (7.8 ± 1.0) scales between posterior canthals;

loreal region slightly concave, 16–28 (22.4 ± 3.2) mostly strongly keeled loreal scales in maximum of 4–6 (5.0 ± 0.8) horizontal rows; 5–8 (6.6 ± 0.8) supralabials and 6–8 (7.2 ± 0.6) infralabials to level below center of eye; suboculars weakly to strongly keeled, usually 2 suboculars in broad contact with supralabials; ear opening vertically oval; scales anterior to ear opening not granular, slightly larger than those posterior to ear opening; 4–7 (5.1 ± 1.0) postmentals, outer pair usually largest; keeled granular scales present on chin and throat; male dewlap small, extending to level of axilla; male dewlap with 6–7 horizontal gorgetal-sternal scale rows, about 3–4 scales per row ($n = 2$); 2 (modal number) anterior marginal pairs in male dewlap; female dewlap absent; low nuchal crest present in males, no dorsal ridge; about 6–8 middorsal scale rows distinctly enlarged, keeled, small scales irregularly interspersed among enlarged dorsal scales; dorsal scales lateral to middorsal series abruptly larger than granular lateral scales; flank scales heterogeneous, solitary enlarged keeled or elevated scales scattered among laterals; 36–44 (38.6 ± 2.4) dorsal scales along vertebral midline between levels of axilla and groin in males, 37–52 (43.9 ± 4.9) in females; 23–32 (27.6 ± 3.2) dorsal scales along vertebral midline contained in 1 head length in males, 21–40 (26.8 ± 4.6) in females; ventral scales on midsection about same size as largest dorsal scales; ventral body scales smooth, flat, imbricate; 31–37 (33.6 ± 2.0) ventral scales along midventral line between levels of axilla and groin in males, 29–38 (33.7 ± 2.5) in females; 21–29 (25.1 ± 2.4) ventral scales contained in 1 head length in males, 19–30 (23.6 ± 2.8) in females; 104–110 (107.8 ± 2.8) scales around midbody in males, 92–108 (99.2 ± 3.8) in females; tubelike axillary pocket absent; precloacal scales not keeled; pair of greatly enlarged postcloacal scales in males; tail slightly to distinctly compressed, TH/TW 1.06–1.68 in 25; all subcaudal scales keeled; lateral caudal scales keeled, homogeneous, al-though indistinct division in segments discernable; dorsal medial caudal scale row enlarged, keeled, not forming crest; scales on anterior surface of antebrachium distinctly keeled, unicarinate; 21–27 (24.1 ± 1.6) subdigital lamellae on Phalanges II–IV of Toe IV of hind limbs; 5–8 (7.0 ± 0.8) subdigital scales on Phalanx I of Toe IV on 54 sides of hind limbs; SVL 42.0–47.5 (44.6 ± 1.8) mm in males, 38.0–56.0 (49.1 ± 4.9) mm in females; TAL/SVL 2.21–2.69 in five males, 2.16–2.49 in nine females; HL/SVL 0.26–0.32 in males, 0.25–0.30 in females; SHL/SVL 0.21–0.27 in males, 0.21–0.25 in females; SHL/HL 0.72–0.96 in males, 0.75–0.92 in females; longest toe of adpressed hind limb usually reaching between ear opening and posterior margin of eye.

Color in life of an adult male paratype (USNM 521919) was described by Köhler et al. (1999: 291, 294): "dorsal and lateral surfaces of head and body pale yellowish gray; dorsal surfaces of limbs pale yellowish gray with slightly darker crossbands; venter dirty cream; dewlap Flame Scarlet (color 15), with tan to dark brown gorgetal scales; iris pale metallic green." Those authors (p. 294) also described an adult female paratype (SMF 78377) as having "Raw Umber (color 123) lateral surfaces, a dirty white middorsal stripe, and Orange-Yellow (color 18) dewlap."

Color in alcohol: dorsal surfaces of head and body some shade of brown; middorsal pattern variable, some lack pattern, some have dark brown chevrons; most females have thin vertebral pale line, occasional females have broad pale middorsal stripe bordered by dark brown or pattern of dark brown diamonds; lateral surface of head pale grayish brown; dorsal surfaces of limbs brown, usually with indistinct dark brown crossbands; dorsal surface of tail brown, some with darker brown chevrons or crossbands, those markings more evident proximally.

Hemipenis: the partially everted hemipenis of SMF 87153 is a medium-sized, bilobed organ; asulcate processus divided, with medial flap; sulcus spermaticus bordered by

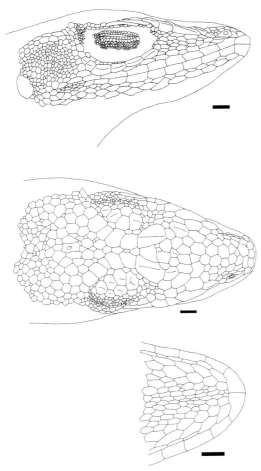

Figure 37. *Norops muralla* head in lateral, dorsal, and ventral views. SMF 78093, adult male from trail to Cerro de Enmedio, Olancho. Scale bar = 1.0 mm. Drawing by Gunther Köhler.

well-developed sulcal lips, opening into broad concave area at base of apex; lobes calyculate; truncus with transverse folds.

*Diagnosis/Similar Species. Norops muralla* is distinguished from all other Honduran *Norops*, except *N. heteropholidotus* and *N. sminthus*, by the combination of having about 6–8 distinctly enlarged middorsal scale rows, heterogeneous lateral scales, smooth and imbricate ventral scales, a red male dewlap in life, and a pair of greatly enlarged postcloacal scales in males. *Norops muralla* differs from *N. heteropholidotus* by having small scales irregularly interspersed among the enlarged medial dorsal scales (medial dorsal scales uniform in size, without interspersed small scales in *N. heteropholidotus*). *Norops muralla* differs from *N. sminthus* by having completely smooth ventral scales (midventrals weakly keeled in *N. sminthus*). *Norops amplisquamosus* shares most of the characters listed above with *N. muralla*, but differs most notably in having an orange-yellow male dewlap in life (male dewlap red in life in *N. muralla*).

*Illustrations* (Figs. 37, 38, 102). Köhler et al., 1999 (adult, head scales, body scales); Wilson and McCranie, 2004a (adult); Köhler, 2014 (dorsal scales; as *Anolis*).

*Remarks. Norops muralla* is in the *N. crassulus* species subgroup of the *N. auratus* species group of Nicholson et al. (2012; also see Remarks for *N. amplisquamosus*). *Norops muralla* tissues were not utilized in the Nicholson et al. (2012) phylogenetic analyses.

*Natural History Comments. Norops muralla* is known from 1,440 to 1,740 m elevation in the Lower Montane Wet Forest formation and peripherally in the Premontane Wet Forest formation. Females sleep at night on low vegetation, whereas males were rarely encountered at night. Both sexes are active on the ground, low on tree trunks, and in low vegetation. Specimens were collected during February, April, and from July to September and is likely active throughout the year. We have heard several reports from biologists that *N. muralla* could not be found during the last five years in the areas where it was collected during the 1990s. Nothing has been reported on its diet or reproduction.

*Etymology.* The specific name *muralla* refers to Refugio de Vida Silvestre Muralla. This species is currently known only from the environs of that wildlife reserve.

*Specimens Examined* (29 [0]; Map 19). OLANCHO: along trail to Cerro de Enmedio, SMF 78093, 78373–75, USNM 521910, 521913–17; Monte Escondido, SMF 87001, USNM 521912; Monte Escondido Campground and along trail to Cerro de Enmedio, SMF 78378–83, 87000,

Figure 38.    *Norops muralla.* (A) Adult male (USNM 521919) from Monte Escondido Campground, Olancho; (B) adult male dewlap (USNM 521919). Photographs by James R. McCranie.

USNM 521919–23; Parque Nacional La Muralla Centro de Visitantes, SMF 78372, USNM 521918; Quebrada del Monte Escondido, SMF 78376, USNM 521911; Quebrada La Habana, SMF 78377.

### *Norops nelsoni* (Barbour)

*Anolis nelsoni* Barbour, 1914: 287 (holotype, MCZ R-7892; type locality: "Swan Islands, Caribbean Sea"), 1930: 134; Regan, 1916: 16; Barbour and Loveridge, 1929b: 222; Peters, 1952: 28; Cochran, 1961: 90; Duellman and Berg, 1962: 197; Powell and Henderson, 2012: 91, 92.

*Anolis sagrei nelsoni*: Etheridge, 1959: 209; Ruibal, 1964: 490; Schwartz and Thomas, 1975: 101; MacLean et al., 1977: 4; Kluge, 1984: 12; Morgan, 1985: 43; Schwartz and Henderson, 1991: 335; Losos and de Queiroz, 1997: 463.

*Anolis sagrei*: Williams, 1969: 363; Lister, 1976a: 662, 1976b: 678; Lee, 1992: 953;

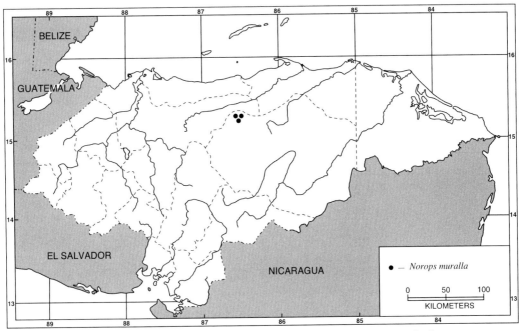

Map 19.   Localities for *Norops muralla*. Solid symbols denote specimens examined.

Rodríguez Schettino, 1999: 319; Henderson and Powell, 2009: 267.

*Norops sagrei nelsoni*: Schwartz and Henderson, 1988: 160.

*Norops nelsoni*: Savage and Guyer, 1989: 111; McCranie, 2011: 277.

*Norops* spp.: Ferrari, 2007: A14, *in* Anonymous, 2007.

*Geographic Distribution*. *Norops nelsoni* is endemic to the Islas del Cisne (Swan Islands), Honduras.

*Description*. The following is based on 10 males (USNM 494751, 494781, 494785, 494787, 494838, 494843, 494845, 494849, 494856–57) and six females (SMF 90449; USNM 494754, 494839, 494841, 494847, 494854). *Norops nelsoni* is a medium-sized anole (SVL 76 mm in largest male examined [MCZ R-192072, not used in description below], 50 mm in largest female examined [USNM 494847]); dorsal head scales keeled in internasal, prefrontal, and frontal areas, smooth to faintly keeled in parietal area; shallow frontal depression present; parietal depression absent; 3–6 (3.8 ± 0.8) postrostrals; anterior nasal single, contacting rostral, usually not in contact with first supralabial; 4–7 (4.9 ± 0.8) internasals; canthal ridge sharply defined; scales comprising supraorbital semicircles usually keeled or strongly ridged, largest scale in semicircles larger than largest supraocular scale; supraorbital semicircles well defined; 0–2 (1.0 ± 0.4) scales separating supraorbital semicircles at narrowest point; 2–3 (2.4 ± 0.5) scales separating supraorbital semicircles and interparietal at narrowest point; interparietal well defined, greatly enlarged relative to adjacent scales, surrounded by scales of moderate size, longer than wide, smaller than ear opening; 2 rows of about 5–7 (total number) enlarged, smooth to faintly keeled supraocular scales; enlarged supraoculars completely separated from supraorbital semicircles by 1 row of small scales; 2–3 elongate superciliaries, posteriormost shortest; usually 3 enlarged canthals; 4–7 (5.4 ± 0.8) scales between second canthals; 5–8 (6.9 ± 0.9) scales between posterior canthals; loreal region slightly concave, 17–35 (24.5 ± 4.9) keeled or rugose loreal scales in maximum of 4–6 (5.0 ± 0.6) horizontal rows; 5–7 (5.8 ± 0.5) supralabials and 5–6 (5.1 ± 0.3) infralabials to level below center of eye; suboculars distinctly keeled, in broad contact with supralabials; ear opening vertically oval; scales anterior to ear opening distinctly larger than those posterior to ear opening; 4–8 (5.8 ± 0.9) postmentals, outer pair much enlarged, their lengths greater than greatest length of mental scale, much larger than first infralabial; gular scales not keeled; male dewlap extending to level equal to middle of forelimb insertion; male dewlap with about 6 horizontal gorgetal-sternal scale rows, about 6–9 scales per row ($n = 2$); 3 (modal number) anterior marginal pairs in male dewlap; female dewlap rudimentary; nuchal crest and dorsal ridge present in breeding males; about 8–9 middorsal scale rows slightly enlarged, smooth to faintly keeled, dorsal scales lateral to middorsal series grading into granular lateral scales; no enlarged scales among laterals; 63–91 (81.7 ± 8.3) dorsal scales along vertebral midline between levels of axilla and groin in males, 67–101 (80.5 ± 11.7) in females; 42–59 (50.6 ± 6.4) dorsal scales along vertebral midline contained in 1 head length in males, 41–59 (48.7 ± 6.8) in females; ventral scales on midsection much larger than largest dorsal scales; ventral body scales distinctly keeled, imbricate; 52–63 (55.9 ± 3.2) ventral scales along midventral line between levels of axilla and groin in males, 48–63 (53.3 ± 5.6) in females; 28–33 (30.5 ± 1.6) ventral scales contained in 1 head length in males, 28–41 (31.8 ± 5.2) in females; 160–190 (175.7 ± 10.4) scales around midbody in males, 151–170 (160.8 ± 6.8) in females; tubelike axillary pocket absent; precloacal scales not keeled; enlarged to slightly enlarged postcloacal scales usually present in males; tail distinctly compressed, TH/TW 1.68–2.92 in males, 1.13–1.44 in females; basal subcaudal scales keeled; lateral caudal scales keeled, homogeneous; dorsal medial caudal scale row enlarged, keeled, forming

crest, especially in males; most scales on anterior surface of antebrachium unicarinate; 29–36 (33.1 ± 2.1) subdigital lamellae on Phalanges II–IV of Toe IV of hind limbs; 7–10 (8.8 ± 0.9) subdigital scales on Phalanx I of Toe IV of hind limbs; SVL 51.0–61.0 (55.9 ± 3.9) mm in males, 43.0–50.0 (46.7 ± 2.8) mm in females; TAL/SVL 1.65–2.07 in males, 1.59–2.12 in females; HL/SVL 0.26–0.27 in both males and females; SHL/SVL 0.25–0.30 in males, 0.26–0.28 in females; SHL/HL 0.98–1.14 in males, 0.97–1.06 in females; longest toe of adpressed hind limb usually reaching between posterior and anterior borders of eye.

Color in life of an adult male (MCZ R-192113; Little Swan): top and side of head Dark Yellow Buff (54 of Köhler, 2012) with Dark Brownish Olive (127) stripes and flecking; dorsal and lateral surfaces of body Olive Clay Color (85) with Dark Brownish Olive longitudinal stripes and flecking; tail Olive Yellow (117) with Dark Brownish Olive flecking suggesting banding on posterior third, some crest scales also Dark Brownish Olive; ventral surfaces of head, body, and limbs pale brown with olive-green scales laterally on body; subdigital scales Dark Brownish Olive; dewlap Buff (18) along outer edge, Clay Color (20) centrally and basally, gorgetal scales and skin between marginal scales Light Buff (2); marginal scales on anterior and posterior portions of dewlap pale yellow, those on central portion of margin yellow with sparse brown flecking. Color in life of another adult male (MCZ R-192114; Big Swan): dorsal and lateral surfaces of head and body Smoke Gray (267 of Köhler, 2012) with Dark Drab (45) flecking and lines; ventral surfaces of head, body, and limbs pale brown, except subdigital lamellae Antique Brown (24); dewlap Burnt Umber (48) with Orange Yellow (8) skin between anterior and posterior marginal scales; central marginal scales white, remaining marginal scales white with dark brown mottling; gorgetal scales white with dark brown mottling. Dewlap color in life of another adult male

(MCZ R-192108; Big Swan): skin Burnt Umber (48 of Köhler, 2012); gorgetal scales white with Burnt Umber flecking and mottling; anterior and posterior marginal scales white, separated by Light Yellow Ocher (13) skin, skin separating central marginal scales Light Yellow Ocher with dark brown infusion.

Color in alcohol: dorsal and lateral surfaces of head brown, usually without distinct markings in those from Big Swan Island and dorsal and lateral surfaces of head yellowish brown in those from Little Swan Island, especially those of juveniles and subadults; dorsal surface of body brown, usually with darker brown blotches or chevrons, some females with paler brown broad middorsal stripe; dorsal surfaces of limbs brown, usually without darker brown markings; dorsal surface of tail brown, usually without distinct markings; ventral surface of head brown with darker brown flecking; ventral surface of body white to cream, with varying amounts of brown flecking; subcaudal surface brown with traces of darker brown crossbands.

Hemipenis: unknown.

*Diagnosis/Similar Species. Norops nelsoni* is distinguished from all other Honduran *Norops*, except *N. sagrei*, by having a distinctly compressed tail with a dorsal crest (especially in large males) and the outer postmental scale on each side greatly enlarged with its length greater than that of the mental scale. *Norops nelsoni* from Big Swan Island differs from *N. sagrei* by having a dark brown male dewlap in life, and usually five or more postmentals (dewlap orange or orange-red in life, and usually four postmentals in *N. sagrei*). *Norops nelsoni* from Little Swan Island differs from *N. sagrei*, in addition to the five or more postmentals, in that adults have considerable amounts of yellow pigment on the dorsal and lateral surfaces of the head, with the juveniles and subadults having even more dense yellow pigment on those surfaces and males have a brownish-yellow dewlap in life (head color same shade of brown as body and male dewlap dark orange in *N. sagrei*).

*Illustrations* (Figs. 39, 40). None previously published.

*Remarks.* Barbour (1914: 287), in describing *Norops nelsoni*, designated MCZ 7892 as the holotype and also stated that there was "a large series of paratypes" in the MCZ. Barbour (1914: 287) did not specify on which of the two Swan Islands the holotype was collected saying only that *N. nelsoni* was "excessively abundant in all situations on both islands." Barbour and Loveridge (1929b) also did not say from which island the holotype was collected, nor does the MCZ catalogue have such information. Ruibal (1964; as *Anolis*) opined that *N. nelsoni* might be a full species, as opposed to a subspecies of *N. sagrei* where it is usually placed. A UPGMA cluster analysis based on meristic and morphometric characters showed the Swan Island population to be sister to the clade containing the Mexican and Central American populations of *N. sagrei* (Lee, 1992). The CAS specimens used by Lee (1992) are apparently a mixture of specimens from both Big and Little Swan, whereas those MCZ specimens listed by Lee are apparently from Big Swan. Based on several unique characters, we prefer to recognize *N. nelsoni* as a species distinct from *N. sagrei* (Powell and Henderson, 2012, listed *N. nelsoni* as a full species based on the suggestion of McCranie). However, recognizing *N. nelsoni* as a species might render the remaining *N. sagrei* paraphyletic. A trip to the Swan Islands by McCranie and colleagues during December 2012, revealed significant color differences of the head and male dewlap between the *N. "nelsoni"* populations on Big and Little Swan Islands. Those color differences are so significant as to suggest that two species might be involved. A study involving both molecular and morphological data of these two *Norops* populations plus *N. sagrei* from the Central American mainland is underway.

*Norops nelsoni* is a member of the *N. sagrei* species group (Savage and Guyer, 1989, Nicholson et al., 2012). The Honduran

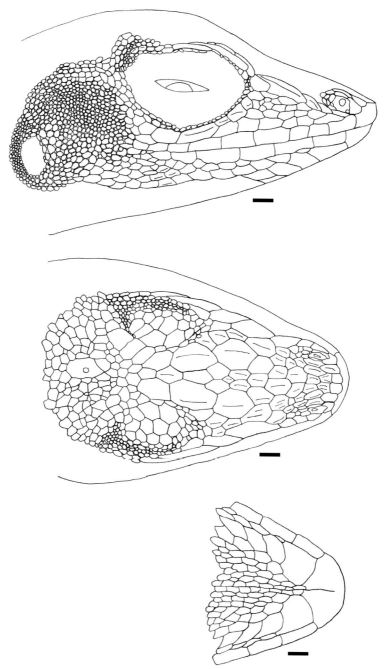

Figure 39.    *Norops nelsoni* head in lateral, dorsal, and ventral views. SMF 90450, adult male from Isla Grande, Islas del Cisne, Gracias a Dios. Scale bar = 1.0 mm. Drawing by Gunther Köhler.

Figure 40.   *Norops nelsoni.* (A) Adult male (MCZ R-192108) from Isla Grande, Islas del Cisne, Gracias a Dios; (B) adult male dewlap (MCZ R-192108). Photographs by James R. McCranie.

members of the *N. sagrei* species group (*N. nelsoni* and *N. sagrei*) are defined by having a distinctly compressed tail with a dorsal crest (especially in adult males) and greatly enlarged outer postmental scales. *Norops nelsoni* tissues were not available to Nicholson et al. (2005, 2012).

*Natural History Comments. Norops nelsoni* is known from near sea level to 10 m elevation in the Lowland Dry Forest (West Indian Subregion) formation. This species was said to be "abundant in all situations"

on both Isla Grande and Isla Pequeña on the Islas del Cisne (Barbour, 1914: 287). A visit to those islands during December 2012 confirmed the species to be abundant on both islands. *Norops nelsoni* is seen on tree trunks and branches of shrubby vegetation on all parts of those two islands visited. It is also abundant on the big island on fence posts, wooden beams of dilapidated buildings, on brush piles, on concrete walls of abandoned buildings, on karsted limestone rocks and associated grassy vegetation, and

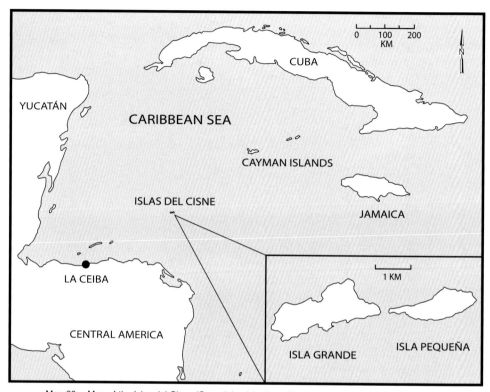

Map 20.   Map of the Islas del Cisne (Swan Islands) where *Norops nelsoni* occurs on both islands.

among leaf litter. It sleeps at night on vegetation, wooden fence posts, and beams and walls of dilapidated buildings. This species is easily the most commonly observed species of anole in all of Honduras. Lister (1976a) reported that *N. nelsoni* on Isla Grande feeds primarily on ants, but also takes a wide variety of other food. Henderson and Powell (2009: 271) summarized Lister's diet data as "ants (51.2%), other hymenopterans, flies, beetles (adults and larvae), termites (12.3%), hemipterans, lepidopterans (adults and larvae), earwigs, orthopterans, spiders, mites, isopods, gastropods, an anole [*N. nelsoni*], and plant matter." Nothing has been reported on its reproduction.

*Etymology.* The name *nelsoni* is a patronym honoring George Nelson, formerly the Chief Taxidermist at the Museum of Comparative Zoology, Harvard University, who also collected the type series.

*Specimens Examined* (260 [0] + 8 skeletons, 1 C&S; Map 20). GRACIAS A DIOS: Islas del Cisne, Isla Grande, MCZ R-160889–922, 160924–36, 192108–12, 192114, 192031–42, 192059–78, SMF 90448–55, USNM 75957, 142271–72, 494749–98, 494863–65; Islas del Cisne, Isla Grande or Isla Pequeña, AMNH 80066 (formerly part of UMMZ 60225), ANSP 31910–11, KU 47168 (formerly in MCZ collection), MCZ R-7892, 9958, 10013 (skeleton), 10014–18, 10020 (skeleton), 10022 (skeleton), 10023–24, 10026–33, 10034 (skeleton), 10036, 10037 (skeleton), 10038–41, 10043–45, 10046 (skeleton), 10047–55, 10057, 10058 (skeleton), 10059–62, 171429 (skeleton), 171430–43, SMF 22982–85 (formerly part of untagged paratype series); Islas del Cisne, Isla Pequeña, MCZ R-192043–58, 192091–92, 192113, USNM 76939–44, 142270, 494862 (C&S), 494834–57.

*Other Records.* GRACIAS A DIOS: Islas del Cisne, Isla Grande, UMMZ S-560 (skeleton); Islas del Cisne, Isla Grande or Isla Pequeña, CAS 39410–15 (Lee, 1992), UMMZ 60225 (9; Peters, 1952).

*Norops ocelloscapularis* Köhler, McCranie, and Wilson

*Norops* new species: Nicholson, 2001: 64.

*Norops ocelloscapularis* Köhler, McCranie, and Wilson, 2001: 248 (holotype, SMF 78841; type locality: "Honduras, Departamento de Copán, near Quebrada Grande off a trail to Laguna del Cerro, 15°04.82′N, 88°55.45′W, 1200 m elevation"); Nicholson, 2002: 120; Köhler, 2003: 102, 2008: 108; Wilson and McCranie, 2003: 59, 2004b: 43; McCranie, 2005: 20; McCranie et al., 2006: 217; Wilson and Townsend, 2006: 105; Nicholson et al., 2012: 12.

*Anolis ocelloscapularis*: Nicholson et al., 2005: 933; Townsend, 2006: 35; Townsend et al., 2006: 32; Köhler et al., 2007: 391; Townsend and Wilson, 2008: 160; Pyron et al., 2013: fig. 19.

*Anolis* (*Norops*) *ocelloscapularis*: Townsend et al., 2005: 466.

*Anolis cusuco*: Townsend and Wilson, 2008: 313 (in part).

*Geographic Distribution. Norops ocelloscapularis* occurs at moderate and at the lower edge of intermediate elevations in the sierras Espíritu Santo and Omoa in northwestern Honduras and adjacent northeastern Guatemala.

*Description.* The following is based on 10 males (FMNH 282581–82, 282585; MVZ 267192; SMF 78841, 79078, 88710, 91285, 91324; UF 149679) and 19 females (SMF 79077, 79090–92, 86993–94, 88711, 91322–23, 91325, 91327–28, 91330–31; UF 142458–59, 149669–70, 149673). *Norops ocelloscapularis* is a small anole (SVL 43 mm in largest Honduran male examined [FMNH 282581], 49 mm in largest Honduran female examined [UF 149673]); dorsal head scales weakly keeled in inter-nasal region, weakly keeled, rugose, or tuberculate in prefrontal, frontal, and parietal areas; deep frontal depression present; parietal depression shallow or absent; 4–9 (6.3 ± 1.1) postrostrals; anterior nasal usually entire (occasionally divided), usually contacting rostral and first supralabial, occasionally only rostral; 6–9 (7.3 ± 0.9) internasals; canthal ridge sharply defined; scales comprising supraorbital semicircles weakly keeled, largest scale in semicircles about same size as largest supraocular scale; supraorbital semicircles well defined; 1–3 (2.0 ± 0.7) scales separating supraorbital semicircles at narrowest point; 1–4 (2.3 ± 0.8) scales separating supraorbital semicircles and interparietal at narrowest point; interparietal well defined, greatly enlarged relative to adjacent scales, surrounded by scales of moderate size, longer than wide, usually larger than ear opening; 2–3 rows of about 3–10 (total number) enlarged, keeled supraocular scales; enlarged supraoculars usually completely separated by 1 row of small scales; 2 elongate superciliaries, posterior much shorter than anterior; usually 3 (occasionally 4) enlarged canthals; 3–13 (10.2 ± 1.8) scales between second canthals; 10–15 (12.2 ± 1.4) scales between posterior canthals; loreal region slightly concave, 22–54 (40.0 ± 7.6) mostly keeled (some smooth or rugose) loreal scales in maximum of 5–9 (7.1 ± 1.3) horizontal rows; 5–8 (6.6 ± 0.8) supralabials and 5–9 (6.8 ± 1.0) infralabials to level below center of eye; suboculars weakly keeled, 2 suboculars usually in broad contact with supralabials; ear opening vertically oval; scales anterior to ear opening granular, similar in size to those posterior to ear opening; 6–9 (6.3 ± 0.6) postmentals, outer pair usually largest; keeled granular scales present on chin and throat; male dewlap small, extending past level of axilla onto chest; male dewlap with 9–12 horizontal gorgetal-sternal scale rows, about 10–15 scales per row (n = 3); 2 (modal number) anterior marginal pairs in male dewlap; female dewlap absent; no nuchal crest or dorsal ridge; 2 middorsal scale rows slightly enlarged, weakly keeled, dorsal scales lateral to middorsal series grading into granular lateral scales; no enlarged scales scattered among laterals; 70–99 (84.5 ± 9.2) dorsal scales along vertebral midline between levels of axilla and groin in males, 70–94 (81.9 ± 8.8) in females; 42–64 (54.7 ± 7.0) dorsal scales along vertebral midline contained in 1 head length in males, 42–71 (49.4 ± 7.0) in females; ventral scales on midsection larger than largest dorsal scales; ventral body scales weakly keeled, subimbricate; 50–72 (61.0 ± 6.4) ventral scales along midventral line between levels of axilla and groin in males, 42–63 (52.9 ± 5.3) in females; 32–44 (39.7 ± 3.9) ventral scales contained in 1 head length in males, 26–44 (33.4 ± 4.6) in females; 109–149 (135.5 ± 11.9) scales around midbody in males, 110–139 (128.2 ± 8.4) in females; tubelike axillary pocket absent; precloacal scales weakly keeled; enlarged postcloacal scales absent in males; tail rounded to distinctly compressed, TH/TW 1.00–1.75; basal subcaudal scales keeled; lateral caudal scales keeled, homogeneous, although indistinct division in segments discernable; dorsal medial caudal scale row not enlarged, keeled, not forming crest; most scales on anterior surface of antebrachium keeled, unicarinate; 20–28 (24.3 ± 2.0) subdigital lamellae on Phalanges II–IV of Toe IV of hind limbs; 6–11 (8.0 ± 1.3) subdigital scales on Phalanx I of Toe IV of hind limbs; SVL 30.0–42.7 (39.6 ± 3.6) mm in males, 40.0–49.1 (44.6 ± 2.2) mm in females; TAL/SVL 1.45–2.45 in five males, 1.77–2.23 in 16 females; HL/SVL 0.26–0.29 in males, 0.25–0.29 in females; SHL/SVL 0.25–0.30 in males, 0.25–0.31 females; SHL/HL 0.94–1.14 in males, 0.91–1.16 in females; longest toe of adpressed hind limb usually reaching between posterior and anterior borders of eye.

Color in life of an adult male (SMF 88710): dorsum of body Olive-Brown (28) with series of widely separated, small Sepia (119) round spots along middorsal line; neck Straw Yellow (56) above, Dark Brownish Olive (129) laterally, latter color passing posteriorly over shoulders and enclosing Cream Color (54) spot above forelimb insertions and anteriorly onto temporal region of head; dorsum of tail banded Olive-Brown and Buff-Yellow (53); venter Straw Yellow with scattered dark brown flecking; dewlap Spectrum Orange (17) with Straw Yellow gorgetal scales; iris rust brown. Color in life of an adult female (SMF 86993): dorsum Yellowish Olive-Green (50) with middorsal series of 5 rectangular Olive-Brown (28) blotches, ground color grading to Olive-Yellow (52) laterally with small Olive-Brown edged pale yellow spot above forelimb insertions; head olive bronze above and on sides, with vague Olive-Brown supraocular spots; limbs Olive-Yellow with Olive-Brown transverse bars; chin cream with pale brown smudging; belly Spectrum Yellow (55). Color in life of the male holotype (SMF 78841) was described by Köhler et al. (2001: 251): "Dorsal and lateral surfaces of head and body Hair Brown (119A), with three distinct Sepia (119) chevrons; apex of chevrons extending posteriorly to form Sepia (119) vertical bars; Sepia (119) suprascapular blotch with Chamois (123D) center present; tail with Raw Umber (223) vertical longitudinal bars; dorsal surfaces of hind limbs Hair Brown (119A) with Mars Brown (223A) blotches (one each on thigh and shank), each hind limb with oblique Sayal Brown (223C) cross-stripe; ventral surface of body dirty white with a few Hair Brown (119A) spots and reticulations; ventral surfaces of limbs Grayish Olive (43); subcaudal surface Tawny Olive (223D); dewlap Chrome Orange (16), becoming Orange Yellow (18) along upper edge; gorgetal scales Chamois (123D); iris Orange Rufous (132C)." Dewlap color of another male (FMNH 282581) was Spectrum Orange (17) with dark brown scales. Color in life of an adult female (SMF 79091): dorsal surface of head pale olive green with paler irregular bar in front of supraoculars and a dark-outlined bronze interocular bar; dorsal surface of body olive-gold with series of dark brown marks on about anterior half of body; lateral surface of head mottled pale olive green and brown; white-centered black ocelli on shoulders; dorsal surfaces of forelimbs pale olive green with brown mottling; dorsal surfaces of hind limbs olive gold with broad rust brown crossbars; tail olive gold with brown crossbands; chin pinkish white with heavy brown smudging; belly white with coppery sheen; subcaudal surface pale olive with rust smudging; dewlap white. Color in life of another adult female (USNM 529976): dorsal surface of head dark olive green with dark brown interocular bar and parietal markings; dorsal surface of body dark olive green; large black ocelli with small central white spot on shoulders; dorsal surfaces of forelimbs dark olive green with narrow, indistinct, white crossbars; dorsal surfaces of hind limbs olive rust with pale olive crossbars; chin brown with white punctations; belly white with pinkish sheen; subcaudal surface pale olive green with rust red smudging; dewlap pale pink-orange; iris coppery red.

Color in alcohol: dorsal and lateral surfaces of head and body grayish brown, usually with dark brown vertebral chevrons; some specimens with dark brown interorbital bar, dark brown lines radiating out from eye, and dark brown lyriform marking in neck region; a dark brown ocellated scapular blotch usually present, ocelli incomplete dorsally in some individuals; pale brown lateral stripe visible or not on body; dorsal surfaces of limbs and tail with indistinct dark brown crossbands; ventral surfaces of head and body dirty white to cream color; ventral surface of head patternless, or with oblique narrow dark brown lines, or mottled with dark brown pigment.

Hemipenis: the completely everted hemipenis of SMF 78841 is a rather elongate organ with only rudimentary lobes; sulcus spermaticus bordered by weakly developed sulcal lips, bifurcating at base of apex, branches continuing to tips of lobes; small asulcate-side crotch flap present; no particular surface structure discernible on most of truncus; apex extensively calyculate.

*Diagnosis/Similar Species. Norops ocelloscapularis* is distinguished from all other Honduran *Norops*, except *N. carpenteri, N. limifrons, N. rodriguezii, N. yoroensis,* and *N. zeus*, by the combination of having a single elongated prenasal scale, weakly keeled ventral scales, and slender habitus. *Norops ocelloscapularis* differs from all of those species in usually having an ocellated shoulder spot (no such spot present in those species). *Norops ocelloscapularis* also differs from *N. carpenteri* by usually having a pale lateral stripe and in having a brown dorsum without paler spots in life (no pale lateral stripe and dorsum greenish brown with pale spots in life in *N. carpenteri*). *Norops ocelloscapularis* also differs from *N. limifrons* and *N. zeus* by having shorter hind legs (longest toe of adpressed hind limb usually reaching between posterior and anterior borders of eye in *N. ocelloscapularis* versus between anterior border of eye and tip of snout in *N. limifrons* and *N. zeus*), weakly keeled ventral scales, and by having a predominantly orange male dewlap in life (smooth ventrals and dewlap dirty white with or without a basal orange-yellow spot in life in *N. limifrons* and *N. zeus*). *Norops ocelloscapularis* also differs from *N. rodriguezii* in having weakly keeled, flat, imbricate (ventral scales smooth, nonimbricate, slightly conical in *N. rodriguezii*).

*Illustrations* (Figs. 41, 42). Köhler et al., 2001 (adult, head scales, hemipenis); Köhler, 2003 (adult), 2008 (adult); Wilson and McCranie, 2004a (adult); Townsend and Wilson, 2008 (adult, as *Anolis*).

*Remarks.* Köhler et al. (2001) were unable to place *N. ocelloscapularis* in any existing species group. Nicholson (2002) included this species in the *N. fuscoauratus* species group, with her nuclear DNA sequence data estimating a sister species relationship with *N. carpenteri*. In turn, her results showed the *carpenteri-ocelloscapularis* clade forming a sister clade to *N. fuscoauratus* (D'Orbigny). Further molecular analyses by Nicholson et al. (2005, 2012) presented a similar *carpenteri-ocelloscapularis* relationship, but failed to recover a sister clade relationship with *N. fuscoauratus*. Despite the conflicting data and the still

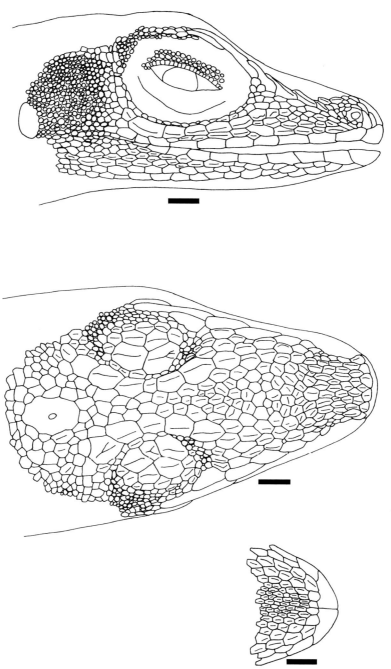

Figure 41.   *Norops ocelloscapularis* head in lateral, dorsal, and ventral views. SMF 78841, adult male from near Quebrada Grande, Copán. Scale bar = 1.0 mm. Drawing by Gunther Köhler.

Figure 42.   *Norops ocelloscapularis.* (A) Adult male (FMNH 282581) from Quebrada Las Piedras, Copán; (B) adult male dewlap (FMNH 282581). Photographs by James R. McCranie.

very incompletely understood *Norops* species relationships, we consider *N. ocelloscapularis* to belong to the *N. fuscoauratus* species subgroup of the *N. auratus* species group of Nicholson et al. (2012). Other Honduran species we include in the *N. fuscoauratus* subgroup include *N. carpenteri, N. limifrons, N. rodriguezii, N. yoroensis,* and *N. zeus.* These are all small, slender species that have smooth to weakly keeled ventral scales and a single elongated prenasal scale.

*Natural History Comments. Norops ocelloscapularis* is known from 1,040 to 1,550 m elevation in the Premontane Wet Forest formation and peripherally in the Lower Montane Wet Forest formation. It is active in the shade on tree trunks, low leafy vegetation, and on tree buttresses about 50–150 cm above the ground. It sleeps at night on stems and leaves of low vegetation. All specimens from the vicinity of the type locality found at night were adult females.

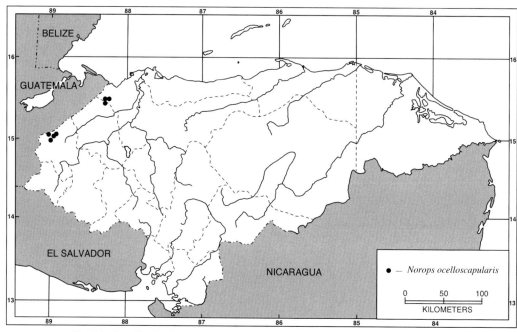

Map 21.   Localities for *Norops ocelloscapularis*. Solid symbols denote specimens examined.

Townsend and Wilson (2008: 160) reported specimens of *N. ocelloscapularis* in Parque Nacional Cusuco were "usually encountered on low (<3 m) vegetation or on the ground" and that it "can also be found at night sleeping on ferns or palm fronds less than 3 m above the ground" (those authors did not report the months of those observations). However, at least one of their observations was during July (Townsend et al., 2005). Others were collected during April, May, July, October, and November and is likely active throughout the year under favorable conditions. Nothing has been published on its diet or reproduction.

*Etymology.* The name *ocelloscapularis* is formed from the Latin *ocellatus* (marked with spots), *scapula* (shoulder-blade), and *-arius* (pertaining to). The name alludes to the distinctive ocellated shoulder spot in most specimens of this species.

*Specimens Examined* (52 [0]; Map 21). COPÁN: between Laguna del Cerro and Quebrada Grande, SMF 78841, 79077–78, UF 166246; Quebrada Cañon Oscuro, SMF

79091, USNM 529976–77; Quebrada Grande SMF 79090, 79092, 86993, USNM 529973–75; Quebrada Las Piedras, FMNH 282581–86, MVZ 267192–93; San Isidro, SMF 86994, 88710–11, 91285–87, UF 142458–59. CORTÉS: 0.5 km S of Buenos Aires, UF 149675; El Cusuco, SMF 91329; Guanales Camp, SMF 91322–23, 91327–28, 91330–31, UF 144266, 149672, 149674, 149676–77, 149680; between Guanales Camp and El Cusuco, UF 149678–79. SANTA BÁRBARA: La Fortuna, UF 149670–71; La Fortuna Camp, SMF 91324–26, UF 149669, 149673.

### *Norops oxylophus* (Cope)

*Anolis oxylophus* Cope, 1875: 123 (syntypes, USNM 30556–57 [see Cochran, 1961: 90]; type locality not given [according to Savage, 2002: 473, the syntypes were collected on the Atlantic versant in eastern Costa Rica]); Poe, 2004: 63; Köhler, 2014: 210.

*Norops oxylophus*: Savage and Villa, 1986: 118; McCranie, 1993: 255; Köhler et al.,

2000: 425; Nicholson et al., 2000: 30; Wilson et al., 2001: 136.

*Norops lionotus*: McCranie et al., 2006: 123; Wilson and Townsend, 2006: 105.

*Geographic Distribution.* *Norops oxylophus* occurs at low elevations on the Atlantic versant from northeastern Honduras to eastern Costa Rica. In Honduras, this species is restricted to the eastern portion of the country.

*Description.* The following is based on 12 males (SMF 78809–10, 80808–10, 80812, 88676–80, 88687) and 11 females (SMF 78811, 80811, 80813–14, 88681, 88683–86, 88688–89). *Norops oxylophus* is a medium-large anole (SVL 74 mm in largest Honduran male examined [SMF 80808], 61 mm in largest Honduran female examined [SMF 88688]; maximum reported SVL 85 mm [Savage, 2002]); dorsal head scales rugose or keeled in internasal region, smooth in prefrontal and frontal areas, most scales rugose in parietal area; deep frontal depression present; parietal depression shallow or absent; 5–8 (6.8 ± 1.2) postrostrals; anterior nasal divided, lower section contacting rostral and first supralabial, or only contacting first supralabial; 7–11 (9.1 ± 1.3) internasals; canthal ridge sharply defined; scales comprising supraorbital semicircles weakly keeled, largest scale in semicircles about same size as, or slightly larger than, largest supraocular scale; supraorbital semicircles usually well defined; 1–2 (1.2 ± 0.4) scales separating supraorbital semicircles at narrowest point; 1–3 (1.7 ± 0.6) scales separating supraorbital semicircles and interparietal at narrowest point; interparietal well defined, moderately enlarged relative to adjacent scales, surrounded by scales of moderate size, longer than wide, smaller than ear opening; 2–3 rows of about 5–7 (total number) enlarged, rugose or weakly keeled supraocular scales; 2–3 enlarged supraoculars in broad contact with supraorbital semicircles; single large elongate superciliary, followed posteriorly by short, less elongate scale; 2–3 enlarged canthals; 7–11 (9.6 ± 1.0) scales between second canthals; 7–12 (9.4 ± 1.4) scales between posterior canthals; loreal region slightly concave, 47–98 (72.3 ± 13.6) mostly strongly keeled loreal scales in maximum of 7–11 (9.2 ± 1.2) horizontal rows; 8–13 (9.4 ± 1.2) supralabials and 8–12 (9.6 ± 1.0) infralabials to level below center of eye; suboculars weakly keeled, separated from supralabials by 1 scale row; ear opening vertically oval; scales anterior to ear opening granular, larger than those posterior to ear opening; 6–9 (7.6 ± 0.8) postmentals, two outer pairs usually larger than median scales; keeled granular scales present on chin and throat; male dewlap moderately large, extending past level of axilla onto chest; male dewlap with 10–13 horizontal gorgetal-sternal scale rows, about 18–21 scales per row (*n* = 3); 3 (modal number) anterior marginal pairs in male dewlap; female dewlap absent; low nuchal crest present, but no dorsal ridge in adult males; about 16–20 middorsal scale rows distinctly enlarged, dorsal scales lateral to middorsal series smooth (striated in some individuals), flat, juxtaposed, hexagonal-shaped, grading into granular lateral scales; no enlarged scales scattered among laterals; 48–71 (56.8 ± 7.6) dorsal scales along vertebral midline between levels of axilla and groin in males, 52–75 (61.5 ± 6.6) in females; 32–44 (37.0 ± 3.6) dorsal scales along vertebral midline contained in 1 head length in males, 34–44 (38.0 ± 3.0) in females; ventral scales on midsection about same size as largest dorsal scales; ventral body scales faintly keeled, imbricate; 50–68 (57.4 ± 5.3) ventral scales along midventral line between levels of axilla and groin in males, 49–66 (58.2 ± 5.5) in females; 34–46 (39.7 ± 3.7) ventral scales contained in 1 head length in males, 32–46 (36.6 ± 4.3) in females; 96–120 (108.7 ± 7.5) scales around midbody in males, 98–120 (108.7 ± 6.8) in females; tubelike axillary pocket absent; precloacal scales weakly keeled; no enlarged postcloacal scales in males; tail nearly rounded to distinctly compressed, TH/TW 0.80–1.52; basal subcaudal scales keeled; lateral caudal scales smooth, homogeneous although indistinct division in segments discernable; dorsal medial caudal scale row not enlarged, weakly keeled, not forming crest; most scales on anterior surface of antebrachium weakly keeled, unicarinate; 18–24 (20.2 ± 1.6) subdigital lamellae on Phalanges II–IV of Toe IV of hind limbs; 8–11 (9.4 ± 0.9) subdigital scales on Phalanx I of Toe IV of hind limbs; SVL 39.0–74.0 (60.3 ± 11.3) mm in males, 35.5–60.5 (53.9 ± 7.9) mm in females; TAL/SVL 1.56–1.85 in eight males, 1.49–1.83 in six females; HL/SVL 0.22–0.27 in males, 0.23–0.26 in females; SHL/SVL 0.25–0.31 in males, 0.26–0.27 in females; SHL/HL 1.08–1.19 in males, 1.01–1.14 in females; longest toe of adpressed hind limb usually reaching between ear opening and posterior border of eye.

Color in life of an adult male (SMF 78810): dorsal surfaces of head and body Vandyke Brown (121); lateral stripe Straw Yellow (56); ventrolateral stripe dark brown, narrow, irregular; dorsal surfaces of fore- and hind limbs marbled brown and horn color; lip stripe Straw Yellow with Vandyke Brown spots; chin creamy white with brown smudging; belly pale yellow; dewlap Orange Yellow (18) with white scales; iris copper. Dewlap color of another male (USNM 321733) was Orange Yellow (18) with cream scales.

Color in alcohol: dorsal and lateral surfaces of head and body uniformly grayish brown with accumulation of dark brown pigment in shoulder region, dark pigment becoming indistinct posteriorly; dorsal surfaces of limbs and tail with indistinct dark brown crossbands; well-demarcated white longitudinal stripe present from subocular region to groin; supra- and infralabials usually with dark brown flecks; ventral surfaces of head, body, and limbs immaculate dirty white or with brown mottling on hind limbs; ventral surface of head immaculate white or with oblique brown lines; several dark brown lines radiating out from eye; dark brown interorbital bar absent; male dewlap dirty white with pale gray anterior portion.

Hemipenis: the completely everted hemipenis of SMF 88676 is a medium-sized, bilobed organ; sulcus spermaticus bordered by well-developed sulcal lips, bifurcating at base of apex, branches opening into broad concave areas distal to point of bifurcation on each lobe; low asulcate ridge present; lobes strongly calyculate; truncus with transverse folds.

*Diagnosis/Similar Species. Norops oxylophus* is distinguished from all other Honduran *Norops* by the combination of having flat, juxtaposed, hexagonal dorsal scales, a distinct pale lateral longitudinal stripe, a orange-yellow male dewlap in life, and weakly keeled ventral scales. The only other species of Honduran *Norops* with flattened, smooth to weakly keeled, usually juxtaposed dorsal scales is *N. capito. Norops oxylophus* differs from the latter species by not having a conspicuous short head (a conspicuous short head in *N. capito*), not having a pale crossband on the chin (pale chin crossband present, especially in juveniles and subadults in *N. capito*), and by usually having a distinct pale lateral longitudinal stripe (no such stripe present in *N. capito*).

*Illustrations* (Figs. 43, 44, 103). Williams, 1984 (scales in frontal and middorsal areas; as *Anolis*); Fläschendräger and Wijffels, 1996 (adult; as *Anolis*); Köhler, 1999c (adult), 2000 (adult, head and male dewlap; as *N. lionotus*), 2001b (head and dewlap; as *N. lionotus*), 2003 (adult, head and dewlap; as *N. lionotus*), 2008 (adult, head and dewlap; as *N. lionotus*), 2014 (dorsal scales, tail tip; as *Anolis*); Savage, 2002 (adult, male dewlap); Guyer and Donnelly, 2005 (adult, dewlap); McCranie et al., 2006 (adult; as *N. lionotus*).

*Remarks.* Savage (2002: 473) stated "The presumed differences in scutellation between what may now be called *Norops lionotus* and *N. oxylophus* are slight, and C. Guyer and I suspect that the two nominal forms represent a single species." However, unpublished data of GK support the recognition of *N. lionotus* and *N. oxylophus* as separate species. The two species differ mostly in dorsal scalation (dorsal scales large, much larger than ventral scales in *N. lionotus* versus dorsal scales of moderate

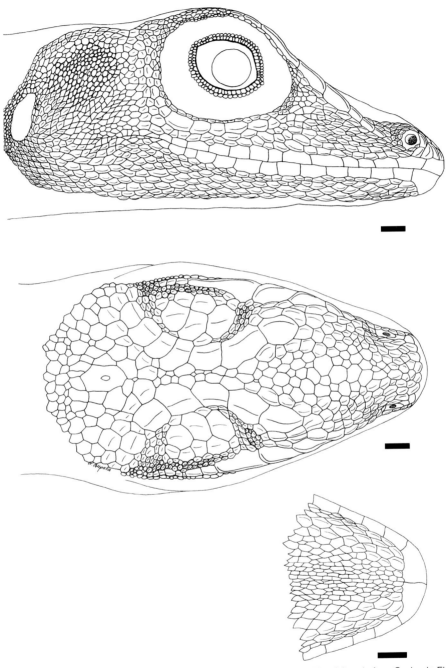

Figure 43.   *Norops oxylophus* head in lateral, dorsal, and ventral views. SMF 80809, adult male from Quebrada El Guásimo, Olancho. Scale bar = 1.0 mm. Drawing by Lara Czupalla.

Figure 44. *Norops oxylophus.* (A) Adult female (USNM 321732) from confluence of Quebrada Waskista with Río Wampú, Gracias a Dios; (B) adult male dewlap (USNM 321733) from confluence of Quebrada Siksatara with Río Wampú, Olancho. Photographs by James R. McCranie.

size and about the same size as largest ventral scales in *N. oxylophus*) and body pattern (a longitudinal lateral pale stripe usually absent in *N. lionotus* versus lateral pale stripe always present in *N. oxylophus*). Until a comprehensive study of the geographic variation in this group of anoles has been published we prefer to recognize *N. lionotus* and *N. oxylophus* as separate species. Nicholson et al. (2012) also recognized *N. oxylophus* as a valid species.

*Norops oxylophus* is in the *N. lionotus* species subgroup of the *N. auratus* species group of Nicholson et al. (2012) or the *N. lionotus* species group of Williams (1976b). However, a morphological analysis by Poe (2004) did not recover a monophyletic *N. lionotus* species group as envisioned by Savage and Guyer (1989), but *N. lionotus* and *N. oxylophus* are closely related as shown by these two species forming a clade in the molecular cladograms in Nicholson et al. (2005, 2012; also see previous paragraph). The *N. lionotus* species subgroup is characterized by having hexagonal and juxtaposed dorsal scales and a semiaquatic habitat.

*Natural History Comments. Norops oxylophus* is known from 60 to 225 m elevation in the Lowland Moist Forest formation. Fitch and Seigel (1984: 8) postulated that this species "probably does not occur farther north than Nicaragua." However, *N. oxylophus* is a common species along some streams in primary forest in eastern Honduras, but appears to be absent from many other similar streams in that region of the country. It is active on boulders and low tree trunks and is semiaquatic and will jump into the water and swim underwater in an attempt to escape. Honduran specimens were also seen to jump to the ground near the edge of a stream and then quickly run across the stream surface to the other side. Vitt *in* Pianka and Vitt (2003: 73) also noted the surface-running escape behavior in the species in southeastern Nicaragua, which he attributed to predation by fish on specimens that were dropped into the water and submerged (also see Vitt et al., 1995). Why some Honduran specimens of the same populations will swim underwater and others will cross the surface is unknown. Predation by fish was not seen in the Honduran populations. If allowed, specimens of *N. oxylophus* take refuge among tree roots at stream edges or under debris on the floor of the adjacent forest. It sleeps at night on streamside vegetation. *Norops oxylophus* has been found from July to September and during November and is probably active throughout the year. The ecology of this species is well studied in Costa Rica and southeastern Nicaragua. Savage (2002) summarized those studies and the following is taken largely from that summation. Habitat of those populations was similar to that noted in Honduras, but Savage did not mention the escape tactic of running across the surface of the stream (but see Vitt et al., 1995; Vitt *in* Pianka and Vitt, 2003). Food consists of a variety of arthropods, mostly araneid spiders, beetles (adults and larvae), homopterans, and dipterans (Guyer and Donnelly, 2005, also mentioned ants in the diet). Reproduction appears to occur year round with females depositing a single egg under moss. Several females are known to use the same nest site (also see Guyer and Donnelly, 2005). Montgomery et al. (2011) reported communal nesting in the closely related *N. lionotus* in Panama.

*Etymology.* The specific name *oxylophus* is formed from the Greek *oxys* (sharp, acute) and *lophos* (crest), and apparently alludes to the compressed tail of this species.

*Specimens Examined* (28 [0]; Map 22). GRACIAS A DIOS: Awasbila, SMF 88689; Kaska Tingni, SMF 88709; confluence of Quebrada Waskista and Río Wampú, SMF 78810–11, USNM 321732. OLANCHO: Caño El Cajón, SMF 78809; Caobita, SMF 88680; Matamoros, SMF 88676–77, USNM 565455; Quebrada El Guásimo, SMF 80808–14, 88679, 88681, 88685–88; Quebrada El Mono, SMF 88678; confluence of Quebrada Siksatara and Río Wampú, USNM 321733; Qururia, SMF 88682; Río Kosmako, SMF 88683–84.

*Norops petersii* (Bocourt)

*Anolis petersii* Bocourt, 1873: 79, *in* Dúmeril et al., 1870–1909 (two syntypes, MNHN 2479, 2479A [see Brygoo, 1989: 78; Köhler and Bauer, 2001: 122]; type locality: "la haute Vera Paz [Guatemala]"); Townsend, 2006: 35; Townsend et al., 2006: 32; Townsend and Wilson, 2008: 162.

*Norops petersi*: Savage and Talbot, 1978: 480; McCranie and Wilson, 1985: 107; Wilson et al., 2001: 136; Köhler, 2003: 109, 2008: 117; Snyder, 2011: 13.

*Norops petersii*: Wilson and McCranie, 2004b: 43; McCranie, 2005: 20.

*Anolis (Norops) petersii*: Townsend and Plenderleith, 2005: 466.

*Geographic Distribution. Norops petersii* occurs in disjunct populations at moderate

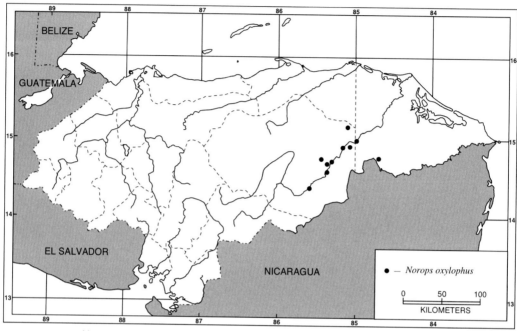

Map 22.    Localities for *Norops oxylophus*. Solid symbols denote specimens examined.

and intermediate elevations on the Atlantic versant from southeastern San Luis Potosí, Mexico, to extreme western Honduras and on the Pacific versant from Chiapas, Mexico, to Guatemala. In Honduras, this species occurs in the northwestern portion of the country.

*Description.* The following is based on one male (MVZ 263594) and four females (KU 195463; UF 142395, 144333, 144744). *Norops petersii* is a large anole (SVL 72 mm in only Honduran male examined [MVZ 263594, a subadult], 125 mm in largest Honduran female examined [UF 144333]; maximum reported SVL 135 mm [Smith and Kerster, 1955]); dorsal head scales rugose or weakly keeled in internasal region, smooth, rugose, or weakly keeled in prefrontal, frontal, and parietal areas; deep frontal and parietal depressions present; 5–8 (6.6 ± 1.1) postrostrals; anterior nasal divided, lower section contacting rostral and first supralabial, or only first supralabial; 6–10 (8.2 ± 1.6) internasals; canthal ridge sharply defined; scales comprising supraorbital semicircles rugose, strongly ridged in some, largest scale in semicircles larger than largest supraocular scale; supraorbital semicircles more or less well defined; 1–2 (1.6 ± 0.5) scales separating supraorbital semicircles at narrowest point; 2–4 (2.8 ± 0.6) scales separating supraorbital semicircles and interparietal at narrowest point; interparietal not well defined, only slightly enlarged relative to adjacent scales, surrounded by scales of moderate size, longer than wide, about equal to, or smaller than, ear opening; 2–3 rows of about 4–7 (total number) slightly enlarged, keeled supraocular scales; slightly enlarged supraoculars completely separated from supraorbital semicircles by small scales; 4–5 relatively short superciliaries, posteriormost shortest; usually 3–4 enlarged canthals; 8–11 (9.8 ± 1.3) scales between second canthals; 9–14 (11.2 ± 2.2) scales between posterior canthals; loreal region slightly concave, 40–67 (48.8 ± 10.7) rugose to faintly keeled loreal scales in maximum of 6–9 (7.0 ± 1.2) horizontal rows; 10–13 (10.7 ± 1.1) supralabials and 9–12 (10.5 ± 1.0) infra-labials to level below center of eye; subo-culars weakly to distinctly keeled, separated from supralabials by 1 row of scales; ear opening vertically oval; scales anterior to ear opening granular, slightly larger than those posterior to ear opening; 4–6 (4.6 ± 0.9) postmentals, outer pair largest; keeled granular scales present on chin and throat; male dewlap moderately large, extending about 1 orbital length posterior to level of axilla; male dewlap with 13–15 horizontal gorgetal-sternal scale rows, about 23–26 scales per row (data from 2 males from Chiapas, Mexico, and Guatemala); 3 (modal number) anterior marginal pairs in male dewlap; female dewlap well developed, but smaller than male dewlap; no nuchal crest, dorsal ridge weakly developed; about 2 middorsal scale rows slightly enlarged, faintly keeled, dorsal scales grading into granular lateral scales; no enlarged scales scattered among laterals; 101 dorsal scales along vertebral midline between levels of axilla and groin in male, 83–102 (93.5 ± 7.9) in females; 61 dorsal scales along vertebral midline contained in 1 head length in male, 52–57 (54.5 ± 2.4) in females; ventral scales on midsection much larger than largest dorsal scales; ventral body scales weakly keeled with rounded or truncate posterior margins, most subimbricate, some imbri-cate; 75 ventral scales along midventral line between levels of axilla and groin in male, 68–86 (75.0 ± 7.7) in females; 42 ventral scales contained in 1 head length in male, 39–49 (44.8 ± 4.3) in females; 116 scales around midbody in male, 121–133 (126.5 ± 5.0) in females; tubelike axillary pocket absent; precloacal scales rugose; enlarged postcloacal scales present in male; tail slightly compressed, TH/TW 1.07–1.25; all subcaudal scales keeled; lateral caudal scales keeled, homogeneous, although in-distinct division in segments discernable; 2 dorsal medial caudal scale rows slightly enlarged, keeled, not forming crest; most scales on anterior surface of antebrachium keeled, unicarinate; 39–43 (40.9 ± 1.3) subdigital lamellae on Phalanges II–IV of

Toe IV of hind limbs; 10–14 (11.9 ± 1.3) subdigital scales on Phalanx I of Toe IV of hind limbs; SVL 72.0 mm in male, 75.0–125.0 (108.8 ± 23.2) mm in females; TAL/SVL 1.36–2.38 in females (male tail incom-plete); HL/SVL 0.28 in male, 0.25–0.28 in females; SHL/SVL 0.23 in male, 0.21–0.23 in females; SHL/HL 0.80 in male, 0.81–0.88 in females; longest toe of adpressed hind limb usually reaching between shoul-der and ear opening.

Color in life of an adult female (UF 142395): dorsal surface of head Citrine (51), mottled with Olive-Brown (28); dorsal surface of body mottled Olive-Yellow (52) and Brownish Olive (29), with Olive-Brown chevrons on body; dorsal surfaces of limbs mottled Olive-Yellow and Brownish Olive; dorsal surface of tail banded with darker and paler Grayish Horn Color (91); ventral surface of body Smoke Gray (44) with Olive-Brown chevrons entering onto lateral portions; dewlap pale olive-yellow; iris copper with dark mottling. Color in life of another adult female (KU 195463) was described by McCranie and Wilson (1985: 107–108): "middorsum olive green, marbled with black and pale olive green grading to pale olive-green on the sides with black and pale green marbling; head olive green above with black marbling; side of head pale yellowish green with black marbling; tail with indistinct bands of olive gray marbled with black, separated by pale greenish gray; front limbs olive green, mottled with black; hind limbs pale olive-green mottled with black, separated by bands of pale olive-green; venter very pale green mottled with olive brown; dewlap pale greenish yellow with black spots." Color in life of dewlap of an adult male from Chiapas (kept at the zoo in Tuxtla Gutiérrez, Chiapas, Mexico; not preserved): Vinaceous Pink (221C) centrally with Olive Yellow (52) margin and with rows of Olive Yellow gorgetals.

Color in alcohol: dorsal surfaces of head, body, and limbs dark brown with paler brown mottling; dorsal surface of body dark brown with paler brown mottling, mottling concentrated in places on body to suggest about five brown crossbands, especially laterally; dorsal surface of tail dark brown; ventral surface of head pale brown with darker brown mottling, except gular region dirty white with black spots and streaks; ventral surface of body pale brown with darker brown and cream mottling; subcaudal surface dark brown with paler brown mottling, especially proximally.

Hemipenis: the completely everted hemipenis of SMF 86943 (from Departamento Baja Verapaz, Guatemala) is a medium-sized, bilobed organ; sulcus spermaticus bordered by weakly developed sulcal lips, bifurcating at base of apex, branches continuing to tips of lobes; no asulcate processus present; lobes with large calyces; truncus with transverse folds.

*Diagnosis/Similar Species.* Norops petersii is distinguished from all other Honduran *Norops*, except *N. biporcatus*, by the combination of its large size (to about 135 mm SVL), short hind legs (longest toe of adpressed hind limb usually reaching between shoulder and ear opening), and 10–13 supralabials to level below center of eye. *Norops petersii* differs from *N. biporcatus* by having a brown coloration in life, a pinkish brown male dewlap with a pale yellow margin in life, 68–86 midventrals between levels of axilla and groin, and 116–133 scales around midbody (green color in life, male dewlap with large blue basal spot and a pinkish gray to dark orange outer border in life, 34–56 midventrals, 76–120 scales around midbody in *N. biporcatus*). *Norops petersii* differs from *N. loveridgei*, another large anole, by having a pinkish brown male dewlap with a pale yellow margin in life, shorter legs with longest toe of adpressed hind limb usually reaching between shoulder and ear opening, 1–2 scales separating supraorbital semicircles at narrowest point, and 40–67 loreal scales in maximum of 6–9 horizontal rows (male dewlap orange with purple streaks in life, hind limb reaching between posterior and anterior borders of eye, 3–7 scales separating supraorbital semicircles, and 61–100 loreals in 8–12 rows in *N. loveridgei*).

*Illustrations* (Figs. 45, 46, 89). Duméril et al., 1870–1909 (adult, head scales; as *Anolis*); Álvarez del Toro, 1983 (adult; as *Anolis*); Köhler, 2000 (ventral scales), 2003 (adult, ventral scales, head and dewlap), 2008 (adult, ventral scales, head and dewlap), 2014 (ventral scales; as *Anolis*); Köhler and Vesely, 2003 (head scales); Townsend and Wilson, 2008 (adult; as *Anolis*); Snyder, 2011 (adult); Lemos-Espinal and Dixon, 2013 (adult).

*Remarks.* Norops petersii is usually placed in the *N. biporcatus* species group (Nicholson, 2002; one exception is that Williams [1976b] recognized a *N. petersi* [sic] species group that also included *N. biporcatus*). Other Honduran species sometimes placed in that group are *N. capito* and *N. loveridgei*. However, phylogenetic analyses based on molecular data, including *N. biporcatus*, *N. capito*, and *N. loveridgei* did not support that group of species as forming a clade (Nicholson, 2002; Nicholson et al. 2005, 2012). *Norops petersii* appears to be most morphologically similar to *N. biporcatus*. Poe et al. (2009) recovered a sister group relationship for *N. biporcatus* and *N. petersii* based on morphological data. Therefore, we include *N. biporcatus* and *N. petersii* in a *N. biporcatus* species subgroup within the *N. auratus* species group of Nicholson et al. (2012). These two species share a large size, relatively short hind limbs, and numerous supralabials.

Köhler and Bauer (2001) discovered that the holotype of *Dactyloa biporcata* Wiegmann represents a specimen of *N. petersii*. "Strict application of the Principle of Priority would require replacement of the name *Anolis petersii* Bocourt, 1873 by *A. biporcatus* (Wiegmann, 1834)" (Köhler and Bauer, 2001: 123). Those authors petitioned The International Commission on Zoological Nomenclature to preserve the prevailing usage of the name *A. petersii* by placing this binomen, as defined by the syntypes MNHN 2479 and 2479A, on the Official List of Specific Names in Zoology (also see Remarks for *N. biporcatus*). The International Commission on Zoological Nomenclature ruled in favor of this action (Anonymous, 2002).

*Norops petersii*

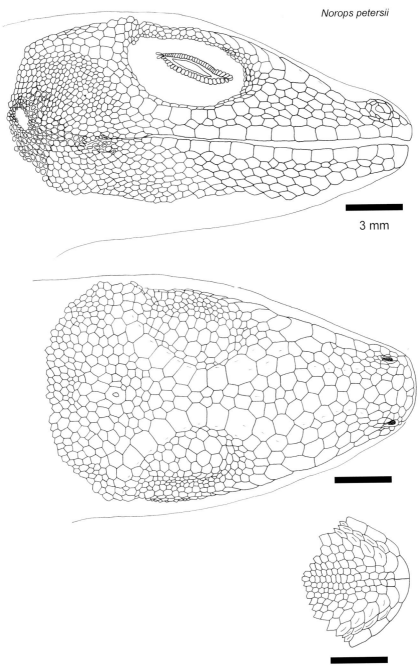

3 mm

Figure 45.    *Norops petersii* head in lateral, dorsal, and ventral views. UMMZ 120088, adult female from Barillas, Huehuetenango, Guatemala. Drawing by Gunther Köhler.

Figure 46.    *Norops petersii.* (A) Adult female (UF 142395) from Quebrada Grande, Copán; (B) adult male dewlap (not preserved) from Chiapas, Mexico. Photographs by James R. McCranie (A) and Gunther Köhler (B).

*Natural History Comments. Norops petersii* is known from 1,300 to 1,550 m elevation in the Premontane Wet Forest and Lower Montane Wet Forest formations. All Honduran specimens of this highly arboreal species were taken during July and August, but it is likely active throughout the year during suitable conditions. One was on a tree trunk about 2 m above the ground in an area of mixed pasture and forest and another "was collected on the edge of a corn field carved from cloud forest just below

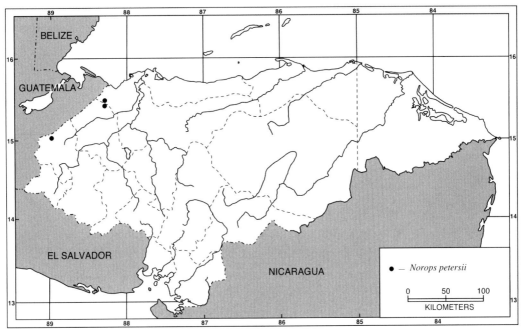

Map 23. Localities for *Norops petersii*. Solid symbols denote specimens examined.

undisturbed vegetation" (McCranie and Wilson, 1985: 107). Townsend and Plenderleith (2005: 467) reported one sleeping at night during July in a tree 3.5 m above a stream "near an edge between an agricultural clearing and broadleaf forest." Villarreal Benítez (1997) reported specimens from Veracruz, Mexico, as being crepuscular and found in slightly open areas in arboreal situations in evergreen forest. That same author reported it feeds on dipterans and coleopterans and that stomachs of some contained small fruits. Apparently nothing has been published on reproduction in this species.

*Etymology.* The name *petersii* is a patronym honoring the German naturalist Wilhelm C. H. Peters, who described several new species of anoles during his career.

*Specimens Examined* (5 [0]; Map 23). COPÁN: Quebrada Grande, KU 195463, UF 142395. CORTÉS: El Cusuco, MVZ 263594; Guanales Camp, UF 144744. SANTA BÁRBARA: La Fortuna Camp, UF 144333.

*Norops pijolense* McCranie, Wilson, and Williams

*Norops pijolense* McCranie, Wilson, and Williams, 1993: 393 (holotype, USNM 322871; type locality: "east slope of Pico Pijol [15°10′N, 87°33′W], Montaña de Pijol, northwest of Tegucigalpita, 2050 m elevation, Departamento de Yoro, Honduras"); Wilson et al., 2001: 136.

*Anolis pijolense*: Nieto-Montes de Oca, 1994b: 207, 1996: 26, 2001: 52.

*Norops pijolensis*: Köhler, 2000: 63, 2003: 102, 2008: 111; Köhler et al., 2001: 255; Wilson and McCranie, 2003: 59, 2004b: 43; McCranie et al., 2006: 217; Wilson and Townsend, 2006: 105; Köhler and Smith, 2008: 223.

*Anolis pijolensis*: Köhler et al., 2007: 391.

*Geographic Distribution. Norops pijolense* occurs at moderate and intermediate elevations in the Montaña de Pijol in southwestern Yoro, Honduras.

*Description.* The following is based on 12 males (SMF 78796–97, 86958–59, 86962;

USNM 322871, 322875–76, 322878, 322880–81, 322884) and 10 females (SMF 78798, 86960–61; UF 166254, 166261; USNM 322873–74, 322877, 322879, 322882). *Norops pijolense* is a medium-sized anole (SVL 59 mm in largest male [SMF 78797], 60 mm in largest female [USNM 322874]); dorsal head scales keeled (unicarinate) in internasal, prefrontal, and frontal regions, rugose or keeled in parietal area; deep frontal and parietal depressions present; 5–8 (6.2 ± 0.9) postrostrals; anterior nasal divided, lower section contacting rostral and first supralabial, or only first supralabial, occasionally lower anterior nasal separated from rostral and first supralabial by row of scales; 5–9 (7.0 ± 1.0) internasals; canthal ridge sharply defined; scales comprising supraorbital semicircles keeled, largest scale in semicircles about same size as, or larger than, largest supraocular scale; supraorbital semicircles usually not well defined; 2–4 (3.0 ± 0.4) scales separating supraorbital semicircles at narrowest point; 3–5 (4.0 ± 0.4) scales separating supraorbital semicircles and interparietal at narrowest point; interparietal more or less well defined, slightly to moderately enlarged relative to adjacent scales, surrounded by scales of moderate size, longer than wide, smaller than ear opening; 3–4 rows of about 9–12 (total number) enlarged, keeled supraocular scales; enlarged supraoculars in broad contact with supraorbital semicircles; 2 elongate, overlapping, subequal superciliaries, occasionally third much shorter superciliary present; usually 3 enlarged canthals; 6–12 (8.6 ± 1.6) scales between second canthals; 7–15 (9.9 ± 1.9) scales between posterior canthals; loreal region slightly concave, 33–64 (48.9 ± 8.3) keeled (stronger keeled towards anterior portion of loreal region) loreal scales in a maximum of 6–8 (6.7 ± 0.6) horizontal rows; 7–9 (8.3 ± 0.7) supralabials and 6–10 (8.0 ± 1.0) infralabials to level below center of eye; suboculars strongly keeled, usually separated from supralabials by 1 row of scales; ear opening vertically to obliquely oval; scales anterior to ear opening not granular, larger than those posterior to ear opening; 4–7 (5.5 ± 0.8) postmentals, outer pair largest; keeled granular scales present on chin and throat; male dewlap moderately large, extending posterior to level of axilla onto chest; male dewlap with 8–10 horizontal gorgetal-sternal scale rows, about 15–20 scales per row (n = 6); 2–3 (modal number) anterior marginal pairs in male dewlap; female dewlap small; low nuchal crest and dorsal ridge present in males; about 2 middorsal scale rows slightly enlarged, strongly keeled, dorsal scales grading into smaller granular lateral scales; no enlarged scales scattered among laterals; 44–62 (49.6 ± 6.5) dorsal scales along vertebral midline between levels of axilla and groin in males, 40–54 (46.1 ± 4.7) in females; 16–32 (25.7 ± 6.3) dorsal scales along vertebral midline contained in 1 head length in males, 16–43 (30.0 ± 8.5) in females; ventral scales on midsection larger than largest dorsal scales; ventral body scales smooth, flat, imbricate; 44–55 (49.4 ± 3.9) ventral scales along midventral line between levels of axilla and groin in males, 38–46 (42.9 ± 2.6) in females; 16–42 (30.1 ± 9.1) ventral scales contained in 1 head length in males, 15–34 (25.9 ± 6.1) in females; 118–138 (127.6 ± 5.7) scales around midbody in males, 120–143 (131.8 ± 8.7) in females; tubelike axillary pocket absent; precloacal scales mostly smooth, some weakly keeled; pair of greatly enlarged postcloacal scales in males; tail nearly rounded to distinctly compressed, TH/TW 0.83–2.25 in 19; basal subcaudal scales smooth, all other subcaudal scales keeled; lateral caudal scales keeled, homogeneous, although indistinct division in segments discernable; dorsal medial caudal scale row slightly enlarged, keeled, not forming crest; scales on anterior surface of antebrachium distinctly keeled, unicarinate; 21–27 (24.0 ± 1.6) subdigital lamellae on Phalanges II–IV of Toe IV of hind limbs; 8–11 (9.1 ± 0.7) subdigital scales on Phalanx I of Toe IV of hind limbs; SVL 47.9–59.0 (53.9 ± 2.7) mm in males, 33.0–59.5 (46.7 ± 10.9) mm in females; TAL/SVL 1.47–1.95 in nine males, 1.78–1.91 in four females; HL/SVL 0.26–0.30 in males, 0.24–0.32 in females; SHL/SVL 0.27–0.30 in males, 0.24–0.33 in females; SHL/HL 0.96–1.05 in males, 0.88–1.07 in females; longest toe of adpressed hind limb usually reaching between anterior border of eye and rostral.

Color in life of the adult male holotype (USNM 322871) was described by McCranie et al. (1993b: 396): "dorsal ground color Olive-Yellow (color 52), with a series of heart-shaped Grayish Horn Color (color 91) blotches on middorsal line; a Raw Sienna (color 136) anteriorly bifid blotch on parietal region and nape; dorsum of head Raw Sienna; lateral surface of head Olive-Yellow with Raw Sienna patina; dorsal surfaces of limbs Olive-Yellow with Raw Sienna crossbands; dorsal surface of tail with Sepia (color 219) bordered Raw Sienna bands separated by Olive-Yellow interspaces; iris Maroon (color 31) with copper ring around pupil; venter mottled Olive-Yellow and Light Neutral Gray (color 85); dewlap Rose (color 9) with a Spectrum Violet (color 72) central spot; scales on dewlap Spectrum Yellow (color 55)." Color in life for a subadult male paratype (USNM 322872) was also presented by McCranie et al. (1993b: 398): "Olive-Yellow (color 52) dorsum with small Sepia (color 219) spots scattered down midline of back; dorsum of head Olive-Yellow with a slight coppery patina; a pair of Chestnut (color 32) spots over sacrum; dorsal surfaces of limbs and tail Olive-Yellow with Chestnut crossbars; ventral surfaces Olive-Yellow; dewlap Rose (color 9) with a Royal Purple (color 172A) central spot; scales in dewlap Olive-Yellow." An adult female paratype (USNM 322873) was similar to USNM 322872, "except that the central dewlap spot was Cyanine Blue (color 74)" (McCranie et al., 1993b: 398).

Color in alcohol: dorsal and lateral surfaces of head and body grayish brown, uniform or with broad dark brown transverse blotches or bands, or with series of 3–4 narrow dark brown chevrons; some females with pale brown vertebral stripe; dark brown lyriform marking present in neck region in some; usually dark brown interorbital bar and dark brown lines radiating outward from eye; dorsal surfaces of limbs and tail with dark brown crossbands; ventral surfaces of head and body pale grayish brown, slightly paler brown than dorsal surfaces; male dewlap reddish brown with dark brown central portion.

Hemipenis: unknown.

*Diagnosis/Similar Species. Norops pijolense* is distinguished from all other Honduran *Norops*, except *N. johnmeyeri* and *N. purpurgularis*, by the combination of having smooth, flat and distinctly imbricate ventral scales, keeled supraocular scales, long hind limbs (longest toe of adpressed hind limb usually reaching between anterior border of eye and rostral), and a pair of greatly enlarged postcloacal scales in males. *Norops pijolense* differs from *N. johnmeyeri* by its smaller body size (SVL to 59 mm in males, 60 mm in females in *N. pijolense* versus SVL to 73 mm in males, 68 mm in females of *N. johnmeyeri*), and male dewlap rose with purple central spot in life, female dewlap small and same color as male dewlap (orange-red male dewlap with large central blue blotch in life, female dewlap well developed, yellow with large central blue blotch in life in *N. johnmeyeri*). *Norops pijolense* differs from *N. purpurgularis* by having a rose male dewlap in life with a purple central spot and unicarinate snout scales (dewlap uniformly purple in life and multicarinate snout scales in *N. purpurgularis*).

*Illustrations* (Figs. 47, 48). McCranie et al., 1993b (adult, head scales); Köhler, 2000 (ventral scales), 2003 (adult, ventral scales), 2008 (adult, ventral scales).

*Remarks. Norops pijolense*, along with three other Honduran species (*N. johnmeyeri*, *N. loveridgei*, and *N. purpurgularis*), are placed in the *N. schiedii* species subgroup of the *N. auratus* species group of Nicholson et al. (2012; see Remarks for *N. johnmeyeri*). *Norops pijolense* tissues were not utilized in the Nicholson (2002) and Nicholson et al. (2005, 2012) phylogenetic analyses.

The unjustified emendation *pijolensis* has sometimes been used for this species (see species synonymy).

*Natural History Comments. Norops pijolense* is known from 1,180 to 2,050 m elevation in the Premontane Wet Forest and Lower Montane Wet Forest formations.

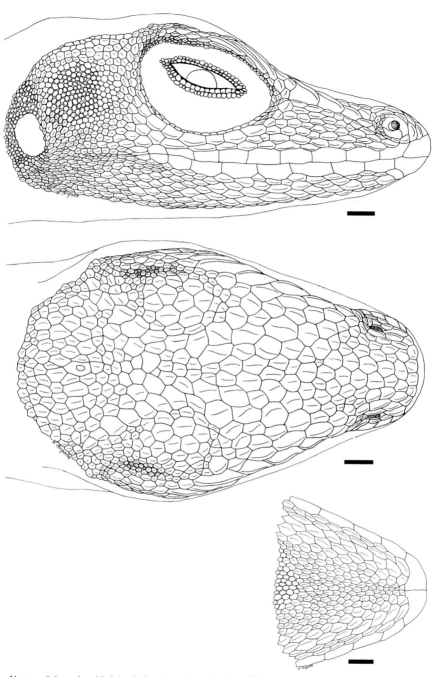

Figure 47. *Norops pijolense* head in lateral, dorsal, and ventral views. SMF 78796, adult male from Cerro de Pajarillos, Yoro. Scale bar = 1.0 mm. Drawing by Lara Czupalla.

Figure 48. *Norops pijolense.* (A) Adult female (USNM 322873); (B) adult male dewlap (USNM 322871), both from E slope of Pico Pijol, Yoro. Photographs by James R. McCranie.

This shade-tolerant species is active on the ground, on fallen logs, or on tree trunks deep within the forest. It sleeps at night on leaves and stems of low vegetation. McCranie et al. (1993b) stated that the escape behavior of this species was deliberate. We have only visited the mountains where this species is apparently isolated during August and the UF series was collected during January, February, April, and July, thus, is likely active throughout the year under favorable conditions. Nothing has been published on its diet or reproduction.

*Etymology.* The name *pijolense* refers to Montaña de Pijol, the type locality, with the Latin suffix *–ense* (denoting place, locality) added. The name is a noun in apposition.

*Specimens Examined* (38 [0]; Map 24). YORO: Cerro de Pajarillos, SMF 78796–98, 86958–62; between El Porvenir de Morazán and Quebrada La Paya, UF 166260–62; Montaña Macuzal, UF 166247–59; E slope of Pico Pijol, USNM 322871–84.

*Norops purpurgularis* McCranie, Cruz, and Holm

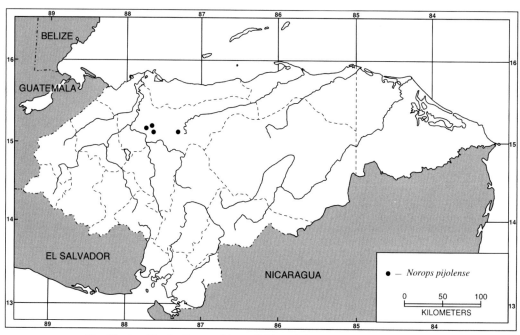

Map 24.   Localities for *Norops pijolense*. Solid symbols denote specimens examined.

*Norops purpurgularis* McCranie, Cruz, and Holm, 1993: 386 (holotype, USNM 322885; type locality: "2.5 airline km NNE La Fortuna [15°26′N, 87°18′W], 1690 m elevation, Cordillera Nombre de Dios, Departamento de Yoro, Honduras"); McCranie et al., 1993b: 394; McCranie, 1996: 32; Köhler, 2000: 63, 2003: 104, 2008: 111; Köhler et al., 2001: 255; Wilson et al., 2001: 136; Wilson and McCranie, 2003: 59, 2004b: 43; McCranie and Castañeda, 2005: 14; Köhler and Smith, 2008: 223; Nicholson et al., 2012: 12; McCranie and Solís, 2013: 242.
*Norops* sp.: Holm and Cruz D., 1994: 20.
*Anolis purpurgularis*: Nieto-Montes de Oca, 1994b: 212, 1996: 26, 2001: 52; Nicholson et al., 2005: 933; Köhler et al., 2007: 391; Townsend et al., 2010: 12, 2012: 109; Pyron et al., 2013: fig. 19.

*Geographic Distribution. Norops purpurgularis* is known from intermediate elevations in the central and western portions of the Cordillera Nombre de Dios of northern Honduras.

*Description.* The following is based on 20 males (SMF 78795, 86953, 86955; USNM 322885, 322887, 322889, 322891–92, 322895–96, 322899–901, 344806, 344808–10, 344813, 344815, 344817) and 15 females (SMF 78794, 86952, 86954, 86956–57; USNM 322890, 322893–94, 322897–98, 344807, 344811–12, 344814, 344816). *Norops purpurgularis* is a medium-sized anole (SVL 59 mm in largest male [USNM 322891], 58 mm in largest female [USNM 322898]); dorsal head scales keeled (multicarinate) in internasal, prefrontal, and frontal regions, rugose or keeled in parietal area; deep frontal and parietal depressions present; 4–7 (5.7 ± 0.9) postrostrals; anterior nasal divided, lower section contacting rostral and first supralabial, or contacting only first supralabial, occasionally lower anterior nasal separated from rostral and first supralabial by 1 row of scales; 5–7 (6.1 ± 0.6) internasals; canthal ridge sharply defined; scales comprising supraorbital semicircles keeled, largest scale in semicircles about same size as, or larger than, largest supraocular scale; supraorbital semicircles usually not well defined; 1–4 (2.7 ± 0.6) scales separating supraorbital semicircles at narrowest point; 2–4 (3.3 ± 0.5) scales separating supraorbital semicircles and interparietal at narrowest point; interparietal well defined, slightly to moderately enlarged relative to adjacent scales, surrounded by scales of moderate size, longer than wide, subequal or smaller than ear opening; 3–4 rows of about 4–8 (total number) enlarged, keeled supraocular scales; enlarged supraoculars in broad contact with supraorbital semicircles; 2 elongate, overlapping, subequal superciliaries, usually third much shorter superciliary present; usually 3 enlarged canthals; 6–9 (7.7 ± 0.8) scales between second canthals; 6–10 (8.5 ± 1.1) scales between posterior canthals; loreal region slightly concave, 40–74 (57.5 ± 8.3) keeled (stronger keeled towards anterior portion of loreal region) loreal scales in maximum of 6–9 (7.8 ± 0.6) horizontal rows; 7–9 (7.8 ± 0.7) supralabials and 6–9 (7.6 ± 0.7) infralabials to level below center of eye; suboculars strongly keeled, usually separated from supralabials by 1 row of scales; ear opening vertically to obliquely oval; scales anterior to ear opening not granular, slightly larger than those posterior to ear opening; 4–6 (5.1 ± 0.8) postmentals, outer pair usually largest; keeled granular scales present on chin and throat; male dewlap moderately large, extending posterior to level of axilla onto chest; male dewlap with 7–9 horizontal gorgetal-sternal scale rows, about 13–17 scales per row ($n = 10$); 3 (modal number) anterior marginal pairs in male dewlap; female dewlap small; low nuchal crest and dorsal ridge present in males; about 2–4 middorsal scale rows slightly enlarged, strongly keeled, dorsal scales grading into granular lateral scales; no enlarged scales scattered among laterals; 40–60 (53.8 ± 5.3) dorsal scales along vertebral midline between levels of axilla and groin in males, 37–60 (51.3 ± 6.6) in females; 29–48 (34.9 ±

4.6) dorsal scales along vertebral midline contained in 1 head length in males, 28–36 (31.6 ± 2.5) in females; ventral scales on midsection larger than largest dorsal scales; ventral body scales smooth, flat, imbricate; 40–60 (49.5 ± 5.1) ventral scales along midventral line between levels of axilla and groin in males, 34–49 (43.9 ± 3.8) in females; 25–40 (33.4 ± 3.5) ventral scales contained in 1 head length in males, 24–34 (28.4 ± 3.5) in females; 110–140 (125.8 ± 9.8) scales around midbody in males, 100–134 (119.6 ± 11.7) in females; tubelike axillary pocket absent; precloacal scales mostly smooth, some weakly keeled; pair of greatly enlarged postcloacal scales in males; tail slightly compressed, TH/TW 1.07–1.50 in 34; basal subcaudal scales smooth, all other subcaudal scales keeled; lateral caudal scales keeled, homogeneous, although indistinct division in segments discernable; dorsal medial caudal scale row slightly enlarged, keeled, not forming crest; scales on anterior surface of antebrachium distinctly keeled, unicarinate; 20–28 (24.0 ± 1.7) subdigital lamellae on Phalanges II–IV of Toe IV of hind limbs; 7–10 (8.7 ± 1.0) subdigital scales on Phalanx I of Toe IV of hind limbs; SVL 49.0–59.3 (54.4 ± 2.7) mm in males, 39.0–58.1 (52.5 ± 5.2) mm in females; TAL/SVL 1.96–2.22 in 11 males, 1.67–2.14 in five females; HL/SVL 0.27–0.31 in males, 0.25–0.31 in females; SHL/SVL 0.25–0.30 in males, 0.25–0.33 in females; SHL/HL 0.90–1.11 in males, 0.91–1.15 in females; longest toe of adpressed hind limb usually reaching between anterior border of eye and rostral.

Color in life of the adult male holotype (USNM 322885) was described by McCranie et al. (1993a: 389, 391): "dorsum Cinnamon-Brown (color 33) with faint indication of slightly paler, elongate middorsal blotches; dorsal surface of head Cinnamon-Brown with a metallic patina; dorsal surfaces of limbs and tail Cinnamon (color 39) with Buff-Yellow (color 53) crossbands; iris reddish copper; venter Salmon Color (color 6); chin dirty white; dewlap Purple (color 1) with dirty white scales." An adult male paratype (UNAH 2724), that was considerably paler than the other specimens, was described by McCranie et al. (1993a: 391): "dorsum pale brown anteriorly, golden brown, suffused with red posteriorly; a dark brown pinstripe extending posteriorly from each eye, converging on each shoulder just posterior to the axilla; middorsum with a slightly darker hourglass pattern; chin yellowish, remainder of venter with a reddish suffusion." Color in life of another adult male (USNM 344806): dorsal surface of head dark brown with indistinct darker brown interorbital bar; dorsal and lateral surfaces of body medium brown with dark brown small spots; vertebral ridge dark brown with pale brown chevrons; dorsal surfaces of limbs dark brown with pale brown crossbars; pale brown crossbands present anteriorly on otherwise dark brown dorsal surface of tail; dewlap Bluish Violet (172B); iris dark brown with pale brown outer border. Another adult male (USNM 344815) was similar to USNM 344806, except that the dewlap was Royal Purple (172A). An adult female (USNM 344816) was similar to USNM 344806, except that the dewlap was Smalt Blue (170) and the dewlap scales were dirty white.

Color in alcohol: dorsal and lateral surfaces of head and body grayish brown, uniform or with broad dark brown transverse blotches or bands; dark brown vertebral stripe present in occasional females; no interorbital bar or lines radiating outward from eye; dorsal surfaces of limbs and tail with indistinct dark brown crossbands; ventral surfaces of head and body pale grayish brown, slightly paler than dorsal surfaces; male dewlap grayish brown with dark brown anterior portion.

Hemipenis: the almost completely everted hemipenis of SMF 86955 is a stout bilobed organ; sulcus spermaticus bordered by well-developed sulcal lips, opening at base of apex into broad concave area; lobes strongly calyculate; truncus with transverse folds; fingerlike asulcate processus present.

*Diagnosis/Similar Species. Norops purpurgularis* is distinguished from all other Honduran *Norops*, except *N. johnmeyeri* and *N. pijolense*, by the combination of having smooth, flat and distinctly imbricate ventral scales, keeled supraocular scales, long hind limbs (longest toe of adpressed hind limb usually reaching between anterior border of eye and rostral), and a pair of greatly enlarged postcloacal scales in males. *Norops purpurgularis* differs from *N. johnmeyeri* by its smaller body size (SVL to 59 mm in males, 58 mm in females in *N. purpurgularis* versus SVL to 73 mm in males, 68 mm in females of *N. johnmeyeri*), in having a uniformly purple male dewlap in life and a small female dewlap similar in color to the male dewlap (orange-red male dewlap with a large central blue blotch in life, female dewlap well developed and yellow with large central blue blotch in life in *N. johnmeyeri*). *Norops purpurgularis* differs from *N. pijolense* by having a uniform purple male dewlap in life and multicarinate snout scales (dewlap rose in life with purple central spot and unicarinate snout scales in *N. pijolense*).

*Illustrations* (Figs. 49, 50). McCranie et al., 1993a (adult, head scales); Köhler, 2003 (adult), 2008 (adult).

*Remarks. Norops purpurgularis*, along with three other Honduran species (*N. johnmeyeri*, *N. loveridgei*, and *N. pijolense*), are placed in the *N. schiedii* species subgroup (McCranie et al., 1993b) of the *N. auratus* species group of Nicholson et al. (2012; see Remarks for *N. johnmeyeri* and *N. loveridgei*). *Norops purpurgularis* tissues were not utilized in the Nicholson et al. (2012) phylogenetic analysis.

*Natural History Comments. Norops purpurgularis* is known from 1,550 to 2,040 m elevation in the Lower Montane Wet Forest formation. It is active on the forest floor, on fallen logs or limbs, or low on tree trunks. Two were also on walls of an abandoned wooden house. McCranie et al. (1993a) described its escape behavior as deliberate. It sleeps at night on leaves and stems of low vegetation. *Norops purpurgularis* has been found during January and from June to August and is

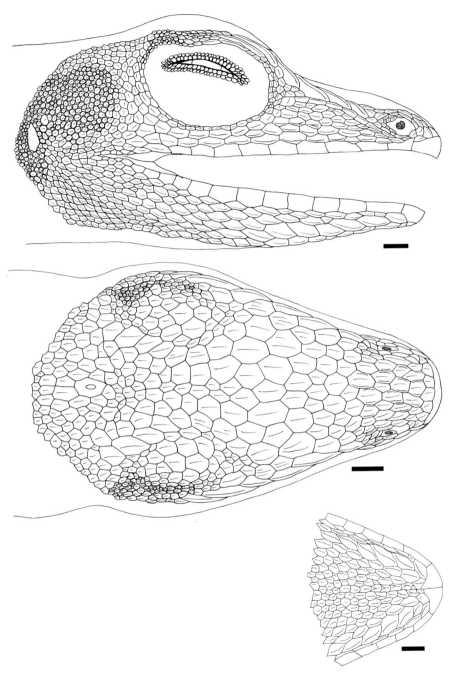

Figure 49.   *Norops purpurgularis* head in lateral, dorsal, and ventral views. SMF 78794, adult female from 2.5 airline km NNE of La Fortuna, Yoro. Scale bar = 1.0 mm. Drawing by Lara Czupalla.

Figure 50.   *Norops purpurgularis.* (A) Adult male (USNM 322889); (B) adult male dewlap (USNM 322885), both from 2.5 airline km NNE of La Fortuna, Yoro. Photographs by James R. McCranie.

probably active throughout the year under favorable conditions. Nothing has been reported on its diet or reproduction.

*Etymology.* The name *purpurgularis* is derived from the Latin *purpureus* (purple), *gula* (throat), and *-aris* (pertaining to), and refers to the purple dewlap of males of this species.

*Specimens Examined* (48 [0]; Map 25). ATLÁNTIDA: S slope of Cerro Búfalo, SMF 86952, USNM 344806–17, 508433–36. YORO: 2.5 airline km NNE of La Fortuna, SMF 78794–95, 86953–57, UF 166263, USNM 322885–901. ATLÁNTIDA or YORO: 0.4–1.0 km N and 0.1 km W–2.1 km E of Cerro San Francisco, UNAH 2684, 2687, 2724–26, 2907.

*Norops quaggulus* (Cope)
*Anolis quaggulus* Cope, 1885: 391 (holotype, USNM 24979 [see Cochran, 1961: 90]; type locality: "San Juan river, Nicaragua").

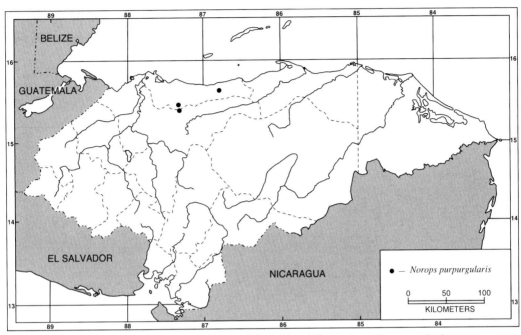

Map 25.　Localities for *Norops purpurgularis*. Solid symbols denote specimens examined.

*Anolis humilis*: Meyer, 1969: 219; Meyer and Wilson, 1972: 107, 1973: 17; Cruz Díaz, 1978: 25; Fitch and Seigel, 1984: 7.

*Norops humilis*: Wilson and McCranie, 1994: 418; Köhler et al., 2000: 425; Nicholson et al., 2000: 30; Wilson et al., 2001: 136; Castañeda, 2002: 15; McCranie et al., 2002a: 27.

*Norops quaggulus*: Köhler et al., 2003: 215, 2006: 243; McCranie et al., 2006: 124; Wilson and Townsend, 2006: 105.

*Geographic Distribution. Norops quaggulus* occurs at low and moderate elevations on the Atlantic versant from eastern Honduras to northern Costa Rica. In Honduras, this species occurs only in the eastern portion of the country.

*Description.* The following is based on 12 males (SMF 78803, 80817, 80819; UF 144608, 144610; USNM 549359–60, 563056–57, 563061, 563066–67) and 12 females (SMF 79927, 80818; UF 144609; USNM 549361–62, 563055, 563059–60, 563065, 563069–71). *Norops quaggulus* is a small anole (SVL 37 mm in largest Honduran male examined [USNM 563067], 41 mm in largest Honduran female examined [USNM 563060]; maximum reported SVL 40 mm in males, 44 mm in females [Köhler et al., 2006]); dorsal head scales strongly keeled in internasal, prefrontal, and frontal areas, most scales keeled in parietal area; deep frontal depression present; parietal depression absent; 4–9 (6.4 ± 1.2) postrostrals; anterior nasal divided, lower section contacting rostral and first supralabial; 5–9 (6.6 ± 1.0) internasals; canthal ridge sharply defined; scales comprising supraorbital semicircles keeled, largest scale in semicircles about same size as largest supraocular scale; supraorbital semicircles usually not well defined; 1–3 (1.8 ± 0.5) scales separating supraorbital semicircles at narrowest point; 2–4 (2.6 ± 0.5) scales separating supraorbital semicircles and interparietal at narrowest point; interparietal not well defined, only slightly enlarged relative to adjacent scales, surrounded by scales of moderate size, longer than wide, smaller than ear opening;

2–3 rows of about 3–8 (total number) enlarged, keeled supraocular scales; enlarged supraoculars varying from completely separated from supraorbital semicircles by 1 row of small scales to 1–2 enlarged supraoculars in broad contact with supraorbital semicircles; 2–3 elongate superciliaries, posteriormost shortest; 3 enlarged canthals; 7–11 (8.5 ± 1.3) scales between second canthals; 7–12 (9.8 ± 1.4) scales between posterior canthals; loreal region slightly concave, 27–47 (35.6 ± 4.6) mostly strongly keeled (some smooth or rugose) loreal scales in maximum of 5–7 (6.1 ± 0.6) horizontal rows; 6–8 (7.3 ± 0.5) supralabials and 7–9 (7.9 ± 0.5) infralabials to level below center of eye; suboculars weakly to strongly keeled, separated from supralabials by 1 row of scales; ear opening vertically oval; scales anterior to ear opening granular, similar in size to those posterior to ear opening; 4–7 (5.3 ± 0.7) postmentals, outer pair usually largest; keeled granular scales present on chin and throat; male dewlap moderately large, extending to level of axilla; male dewlap with 6–10 horizontal gorgetal-sternal scale rows, about 10–17 scales per row ($n = 6$); 3 (modal number) anterior marginal pairs in male dewlap; female dewlap absent; no nuchal crest or dorsal ridge; about 7–11 middorsal scale rows distinctly enlarged, strongly keeled, dorsal scales lateral to middorsal series abruptly larger than granular lateral scales; no enlarged scales scattered among laterals; 30–40 (36.0 ± 3.5) dorsal scales along vertebral midline between levels of axilla and groin in males, 32–44 (38.1 ± 3.8) in females; 20–32 (25.6 ± 2.8) dorsal scales along vertebral midline contained in 1 head length in males, 19–32 (23.8 ± 3.7) in females; ventral scales on midsection about 3–4 times smaller than largest dorsal scales; ventral body scales keeled, imbricate; 35–61 (51.3 ± 7.1) ventral scales along midventral line between levels of axilla and groin in males, 48–59 (50.9 ± 3.5) in females; 30–44 (37.9 ± 4.4) ventral scales contained in 1 head length in males, 31–40 (35.0 ± 3.6) in females; 102–115 (106.7 ± 5.0) scales around midbody in males, 97–121 (104.7 ± 6.9) in females; tubelike, scaleless

axillary pocket present; precloacal scales not keeled; no enlarged postcloacal scales in males; tail slightly compressed, TH/TW 1.08–1.44; basal subcaudal scales keeled; lateral caudal scales keeled, homogeneous, although indistinct division in segments discernable; dorsal medial caudal scale row not enlarged, keeled, not forming crest; most scales on anterior surface of antebrachium distinctly keeled, unicarinate; 17–22 (19.5 ± 1.2) subdigital lamellae on Phalanges II–IV of Toe IV of hind limbs; 5–10 (7.8 ± 1.2) subdigital scales on Phalanx I of Toe IV of hind limbs; SVL 33.0–37.2 (35.0 ± 1.3) mm in males, 25.0–40.9 (37.1 ± 4.2) mm in females; TAL/SVL 1.08–1.55 in six males, 1.26–1.44 in seven females; HL/SVL 0.26–0.29 in males, 0.25–0.29 in females; SHL/SVL 0.26–0.29 in males, 0.24–0.29 in females; SHL/HL 0.96–1.03 in males, 0.93–1.00 in females; longest toe of adpressed hind limb usually reaching between posterior and anterior borders of eye.

Color in life of an adult male (USNM 563057): region of enlarged middorsal scales coppery brown with series of dark brown chevrons, chevrons increasing in size posteriorly; sides of body Buff (24); dorsal surfaces of front limbs Buff; dorsal surfaces of hind limbs Buff on thigh with Antique Brown (37) crossbars flanked by pale grayish-brown crossbars; head brown with dark brown shield-shaped mark between eyes, dark brown spot posterior to eye, pale bars extending below eye; tail with alternating Buff and Cinnamon-Rufous (40) crossbands; venter pale gray; dewlap Flame Scarlet (15) with Orange Yellow (18) marginal scales, gorgetal and sternal dewlap scales pale gray.

Color in alcohol: dorsal surface of head dark brown; dorsal surface of body brown to dark brown; lateral surface of body brown; some specimens have small darker brown dorsal spots or chevrons on body; dorsal surfaces of limbs dark brown without distinct markings; dorsal surface of tail dark brown with indistinct darker brown crossbands; ventral surface of head cream centrally, cream with brown flecking or

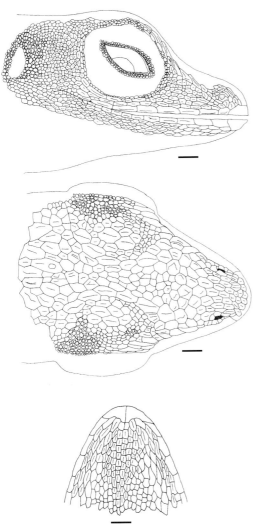

Figure 51. *Norops quaggulus* head in lateral, dorsal, and ventral views. SMF 77480, adult male from Selva Negra, Matagalpa, Nicaragua. Scale bar = 1.0 mm. Drawing by Gunther Köhler.

(length of lobes less than half length of truncus); sulcus spermaticus bifurcating at base of apex, branches continuing to tips of lobes; truncus and lobes not calyculate, but tiny papillae present in many specimens, these papillae frequently black; no asulcate processus; truncus without transverse fold.

*Diagnosis/Similar Species.* *Norops quaggulus* is distinguished from all other Honduran species of *Norops*, except *N. tropidonotus*, *N. uniformis*, and *N. wampuensis*, by having a deep tubelike and scaleless axillary pocket. *Norops quaggulus* differs from *N. tropidonotus* and *N. wampuensis* in having the scales anterior to the ear opening more-or-less subequal in size to those posterior to the ear opening (scales anterior to ear opening distinctly larger than those posterior to ear opening in *N. tropidonotus* and *N. wampuensis*). *Norops quaggulus* differs from *N. uniformis* in lacking pale vertical lines in the flank region (1–3 pale vertical lines usually present in *N. uniformis*) and having a reddish-orange male dewlap with a yellow margin (rose-colored male dewlap with large central purple spot in *N. uniformis*).

*Illustrations* (Figs. 51, 52). Köhler, 1999c (adult, scales anterior and posterior to ear opening; as *N. humilis*), 2000 (adult, head and dewlap, scales anterior and posterior to ear opening; as *N. humilis*), 2001b (adult, head and dewlap, scales anterior and posterior to ear opening; as *N. humilis*), 2003 (adult, head and dewlap), 2008 (adult, head and dewlap, superciliary scales); Savage, 2002 (adult, dewlap; as *N. humilis*); Köhler et al., 2003 (axillary pocket region, hemipenis), 2006 (adult, head scales); Guyer and Donnelly, 2005 (adult; as *N. humilis*); McCranie et al., 2006 (adult); Beckers, 2009 (adult; as *Anolis*).

*Remarks.* Köhler et al. (2003) resurrected *N. quaggulus* from the synonymy of *N. humilis* (Peters), based largely on hemipenial differences. The phylogenetic analysis of molecular data by Nicholson (2002) and Nicholson et al. (2005, 2012) rejected the monophyly of the *N. humilis* species group as envisioned by Savage and Guyer (1989). The Honduran species of this subgroup

mottling along anterior and lateral edges; ventral surface of body cream to pale brown, with darker brown flecking or small spots; subcaudal surface pale brown with darker brown flecking proximally, becoming mostly dark brown distally.

Hemipenis: the completely everted hemipenis of SMF 78803 is a relatively small, bilobed organ with short and stout lobes

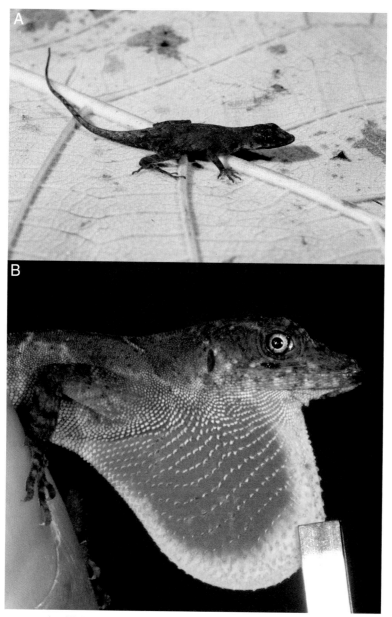

Figure 52.   *Norops quaggulus.* (A) Adult male (USNM 563067) from Quebrada El Guásimo, Olancho; (B) adult male dewlap (in UCR collection) from Heliconias Lodge, Volcán Tenorio, Alajuela, Costa Rica. Photographs by James R. McCranie (A) and Gunther Köhler (B).

(Savage and Guyer, 1989) are *N. quaggulus, N. tropidonotus, N. uniformis,* and *N. wampuensis* (Williams, 1976b also placed one South American species in the *N.* *humilis* group), all of which have a deep axillary pocket (an unique character among the anoles), distinctly enlarged middorsal scales, relatively poorly defined dorsal head scales, and strongly keeled ventral scales. However, because those species are not recovered as a monophyletic group in those phylogenetic analyses, we include all four species as incertae sedis within the *N. auratus* species group of Nicholson et al. (2012). Tissues of *N. quaggulus* were not included in the phylogenetic analyses of Nicholson et al. (2005, 2012).

*Natural History Comments. Norops quaggulus* is known from 60 to 840 m elevation in the Lowland Moist Forest and Premontane Wet Forest formations. This shade-tolerant species is usually active in leaf litter. The clearing of low vegetation and ground debris to set up campsites in several heavily forested areas revealed the species to be abundant when otherwise not very noticeable. Unlike most anoles, it apparently does not sleep on low vegetation at night, rather it sleeps in leaf litter on the ground (JRM, pers. observ). *Norops quaggulus* is a deep forest species that has been found during February, March, and from May to November and is likely active throughout the year. Based on our experience, it does not survive in deforested areas. Extensive ecological studies of this species have been carried out in northern Costa Rica, with those studies summarized by Savage (2002, for more information see the literature cited by him). The following is taken from that summation. Habitats in Costa Rica are similar to those for Honduran populations. This species is a trunk-litter forager that perches head downward low on tree trunks, tree stumps, and logs. It preys on a wide variety of arthropods, principally araneid spiders and isopods (see Savage, 2002, for a list of other prey items). Females lay a single egg in leaf litter as often as every seven days with the greatest activity during the rainy season (Guyer and Donnelly, 2005, apparently erroneously stated that reproductive activity peaks during the dry season). Paemelaere et al. (2011, 2013) studied dorsal pattern morphology and microhabitat selection in female *N. quaggulus* at La Selva, Costa Rica, in relation to their survival rates.

*Etymology.* The name *quaggulus* is likely derived from *quagga*, a native name for an extinct member of the zebra family found in South Africa until the late 19th century, which had stripes only on the area around the head and shoulders, with the rest of its body being a yellowish-brown color and the Latin suffix *–ule* (little). The name probably alludes to the general yellowish-brown body color of this small species.

*Specimens Examined* (62 [1]; Map 26). COLÓN: El Ocotillal, SMF 85951, 85953–57. GRACIAS A DIOS: Bodega de Río Tapalwás, UF 144608–10, USNM 549361–62, 563055–63; vicinity of Crique Unawás, SMF 85958–60; Kakamuklaya, USNM 573140; Pomokir, SMF 85949–50, 85952; Raudal Kiplatara, SMF 85942–45; vicinity of Río Cuyamel, SMF 85946–48; Rus Rus, USNM 563054; San San Hil, USNM 570087; San San Hil Kiamp, USNM 570086; Urus Tingni Kiamp, USNM 565460, 570088–91, 573141; Warunta Tingni Kiamp, USNM 563064–66. OLANCHO: about 40 km E of Catacamas, TCWC 23628; Montaña de Malacate, KU 194272–73; Quebrada El Guásimo, SMF 80819, USNM 563067–68; Matamoros, SMF 80817–18, USNM 549359–60; Quebrada de Las Marías, SMF 78803, USNM 563070; near Quebrada El Mono, USNM 563071–72; Río Kosmako, USNM 563069; Yapuwás, SMF 79927.

*Other Records* (Map 26). COLÓN: Cañon del Chilmeca, UNAH 5465, 5468, Empalme Río Chilmeca, UNAH 5475 (Cruz Díaz, 1978, specimens now lost).

*Norops roatanensis* Köhler and McCranie
*Anolis lemurinus*: Meyer, 1969: 222 (in part); Meyer and Wilson, 1973: 17 (in part); Wilson and Hahn, 1973: 110 (in part); Hudson, 1981: 377; O'Shea, 1986: 68.
*Anolis allisoni*: Meyer and Wilson, 1973: 15 (in part).
*Anolis sagrei*: Meyer and Wilson, 1973: 19 (in part); Wilson and Hahn, 1973: 111 (in part).

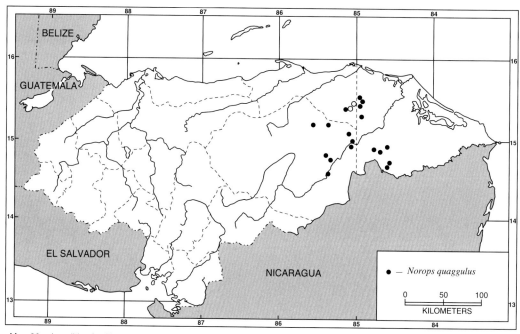

Map 26.   Localities for *Norops quaggulus*. Solid symbols denote specimens examined and open symbols accepted records.

*Norops lemurinus*: Wilson and Cruz Díaz, 1993: 17 (in part); Wilson et al., 2001: 136 (in part).

*Norops bicaorum* Köhler, 1996b: 21 (in part), 1999a: 49 (in part); Monzel, 1998: 159 (in part); Lundberg, 2001: 25; Klütsch et al., 2007: 1125 (in part).

*Norops* sp.: Köhler, 1998b: 377.

*Norops roatanensis* Köhler and McCranie, 2001: 240 (holotype, SMF 79953; type locality: "between West End Point and Flowers Bay [16°17.98′N, 86°34.82′W], 30 m elevation, Isla de Roatán, Departamento de Islas de la Bahía, Honduras"); Köhler, 2003: 104, 2008: 113; Wilson and McCranie, 2003: 60; McCranie et al., 2005: 106; Wilson and Townsend, 2006: 105; McCranie and Valdés Orellana, 2014: 45.

*Norops* cf. *bicaorum*: Monzel, 2001: 32.

*Geographic Distribution.* *Norops roatanensis* occurs at low elevations on Isla de Roatán, Islas de la Bahía, Honduras.

*Description.* The following is based on 10 males (FMNH 282571; SMF 79222, 79224, 79979–80, 79984–85, 79988, 79990; USNM 541021) and 11 females (SMF 79223, 79974–78, 79981, 79983, 79986–87, 79989). *Norops roatanensis* is a moderately large anole (SVL 65 mm in largest male [FMNH 282571], 62 mm in largest female [SMF 79223]); dorsal head scales keeled in internasal and prefrontal regions, smooth or tuberculate in frontal and parietal areas; weak to moderate frontal and parietal depressions present; 5–8 (5.4 ± 0.8) postrostrals; anterior nasal usually divided, lower section contacting rostral and first supralabial; 6–11 (8.0 ± 1.0) internasals; canthal ridge sharply defined; scales comprising supraorbital semicircles keeled, largest scale in semicircles about same size as, or larger than, largest supraocular scale; supraorbital semicircles well defined; 1–5 (1.9 ± 1.0) scales separating supraorbital semicircles at narrowest point; 1–5 (3.3 ± 1.0) scales separating supraorbital semicircles and interparietal at narrowest point; interparietal well defined, greatly enlarged relative to adjacent scales, surrounded by scales of moderate size, longer than wide, usually slightly smaller than ear opening; 2–3 rows of about 5–9 (total number) enlarged, keeled supraocular scales; enlarged supraoculars completely separated from supraorbital semicircles by 1–2 rows of small scales; 2 elongate, overlapping superciliaries, posterior much shorter than anterior; usually 3 enlarged canthals; 8–11 (9.8 ± 0.9) scales between second canthals; 8–13 (10.4 ± 1.1) scales between posterior canthals; loreal region slightly concave, 50–75 (62.9 ± 7.7) mostly keeled (some smooth or rugose) loreal scales in maximum of 8–10 (8.7 ± 0.8) horizontal rows; 6–10 (8.1 ± 1.1) supralabials and 7–9 (7.9 ± 0.7) infralabials to level below center of eye; suboculars strongly keeled, separated from supralabials by 1 row of scales; ear opening vertically oval; scales anterior to ear opening granular, similar in size to those posterior to ear opening; 6–8 (7.7 ± 0.7) postmentals, outer pair largest; keeled granular scales present on chin and throat; male dewlap moderately large, extending onto chest; male dewlap with 8–9 horizontal gorgetal-sternal scale rows, about 7–12 scales per row (*n* = 5); 3 (modal number) anterior marginal pairs in male dewlap; female dewlap small or absent; low nuchal crest and dorsal ridge present in largest males; about 2 middorsal scale rows slightly enlarged, strongly keeled, dorsal scales grading into smaller granular lateral scales; no enlarged scales scattered among laterals; 52–101 (72.6 ± 17.4) dorsal scales along vertebral midline between levels of axilla and groin in males, 58–85 (70.3 ± 7.0) in females; 42–66 (49.8 ± 7.4) dorsal scales along vertebral midline contained in 1 head length in males, 36–48 (43.8 ± 3.8) in females; ventral scales on midsection much larger than largest dorsal scales; ventral body scales keeled, mucronate, imbricate; 38–53 (47.5 ± 4.7) ventral scales along midventral line between levels of axilla and groin in males, 39–54 (43.5 ± 4.4) in females; 30–36 (32.8 ± 2.7) ventral scales contained in 1

head length in males, 21–38 (29.5 ± 4.7) in females; 122–190 (148.5 ± 18.4) scales around midbody in males, 124–160 (147.6 ± 9.2) in females; tubelike axillary pocket absent, but shallow and scaled axillary pocket present; precloacal scales weakly to strongly keeled; no enlarged postcloacal scales in males; tail slightly to distinctly compressed, TH/TW 1.10–1.71; all subcaudal scales keeled; lateral caudal scales keeled, homogeneous, although indistinct division in segments discernable; dorsal medial caudal scale row enlarged, keeled, not forming crest; scales on anterior surface of antebrachium distinctly keeled, unicarinate; 24–30 (27.2 ± 1.8) subdigital lamellae on Phalanges II–IV of Toe IV of hind limbs; 6–9 (8.2 ± 1.0) subdigital scales on Phalanx I of Toe IV of hind limbs; SVL 44.0–65.2 (56.2 ± 6.8) mm in males, 56.0–62.0 (58.0 ± 1.9) mm in females; TAL/SVL 1.93–2.23 in four males, 1.91–2.24 in six females; HL/SVL 0.25–0.27 in males, 0.26–0.32 in females; SHL/SVL 0.25–0.29 in males, 0.26–0.28 in females; SHL/HL 0.96–1.12 in males, 0.86–1.06 in females; longest toe of adpressed hind limb usually reaching between posterior and anterior borders of eye.

Color in life of an adult male (USNM 565461): lateral surface of body with Dark Brownish Olive (129) swath; middorsum pale brown with dark brown blotches; area between middorsum and lateral swath with cream stripe; cream stripe also present ventrolaterally below lateral swath; dorsal surface of head pale brown with dark brown lyriform mark, interorbital bar, and snout area; dorsal surfaces of limbs pale brown with dark brown crossbands; belly pale brown; iris brown with copper rim; dewlap Scarlet (14) with white scales. Color in life of the dewlap of the adult male holotype (SMF 79953) was "Geranium Pink (color 13) with suffusion of black pigment centrally and white gorgetal scales" (Köhler and McCranie, 2001: 243).

Color in alcohol: dorsal surfaces of head, body, and tail some shade of brown with a variety of possible dorsal markings (e.g., broad dark brown transverse bands, dia-

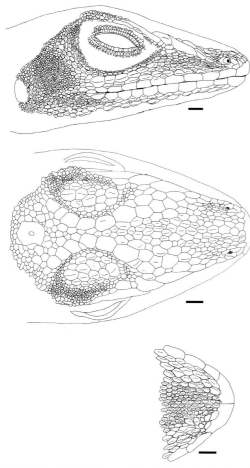

Figure 53. *Norops roatanensis* head in lateral, dorsal, and ventral views. SMF 79953, adult male from Flowers Bay, Isla de Roatán, Islas de la Bahía. Scale bar = 1.0 mm. Drawing by Gunther Köhler.

monds, a pale vertebral stripe) or patternless; dorsal surfaces hind limbs with oblique dark bands; tail with faint to distinct dark brown bands; distinct dark brown lines radiate outward from eye; distinct dark brown interorbital bar present; dark brown lyriform nuchal mark usually present.

Hemipenis: the completely everted hemipenis of SMF 81124 is a medium-sized bilobed organ; sulcus spermaticus bifurcating at base of apex, branches continuing to tips of lobes; asulcate processus absent; lobes strongly calyculate; truncus with transverse folds.

*Diagnosis/Similar Species. Norops roatanensis* is distinguished from all other Honduran *Norops*, except *N. lemurinus* and *N. bicaorum*, by the combination of having a dark brown lyriform mark in the nuchal region (usually present, obscure or absent in some specimens), long hind legs (longest toe of adpressed hind limb usually reaching between posterior and anterior borders of eye), keeled, mucronate, imbricate ventral scales, about two dorsal scale rows slightly enlarged, a shallow and scaled axillary pocket, and no enlarged postcloacals in males. *Norops roatanensis* differs from *N. lemurinus* by having a pink-red male dewlap in life, with suffusion of black pigment centrally and white gorgetal scales (dewlap red to red-orange in life, without suffusion of black pigment, often with black or dark brown-edged gorgetal scales in *N. lemurinus*). *Norops roatanensis* is most similar to *N. bicaorum* from which it differs by male dewlap coloration (dewlap pink-red in *N. roatanensis* versus orange-red in *N. bicaorum*), and body size (males average about 56 mm SVL, females about 58 mm in *N. roatanensis* versus males average about 64 mm SVL, females about 66 mm in *N. bicaorum*), in having the sulcal branches of the hemipenis continuing to tips of lobes and an asulcate processus absent (sulcal branches opening into broad concave area distal to point of bifurcation on each lobe, and asulcate processus present in *N. bicaorum*).

*Illustrations* (Figs. 53, 54). Köhler and McCranie, 2001 (adult, head scales, head and dewlap); Lundberg, 2001 (adult; as *N. bicaorum*); Monzel, 2001 (head and dewlap; as *N.* cf. *bicaorum*); Köhler, 2003 (adult), 2008 (adult); McCranie et al., 2005 (adult).

*Remarks. Norops roatanensis* is a member of the *N. lemurinus* species subgroup of the *N. auratus* species group of Nicholson et al. (2012). The three Honduran species of this subgroup (*N. bicaorum*, *N. lemurinus*, and *N. roatanensis*) form a closely related group based on their similar external morphology. Members of this subgroup are characterized by having distinct dark lines radiating

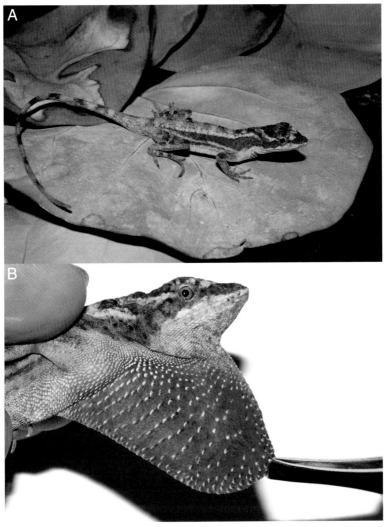

Figure 54.   *Norops roatanensis.* (A) Adult male (FMNH 282571) from Palmetto Bay, Isla de Roatán, Islas de la Bahía; (B) adult male dewlap (FMNH 282571). Photographs by James R. McCranie.

outward from the eye, a red or pink male dewlap in life, strongly keeled midventral scales, a shallow and scaled axillary pocket, and by usually having a distinct dark-colored lyriform nuchal mark. The recovered phylogenetic analyses in Nicholson et al. (2005, 2012) showed *N. bicaorum* and *N. lemurinus* to be sister to a clade containing *N. limifrons* and *N. zeus. Norops roatanensis* tissues were not included in those two phylogenetic analyses. Klütsch et al. (2007) found significant genetic differentiation

among *N. bicaorum, N. lemurinus,* and *N. roatanensis*; however, those authors were not aware that the population on Roatán Island had been described as *N. roatanensis* (Köhler and McCranie, 2001).

*Natural History Comments. Norops roatanensis* is known from near sea level to 30 m elevation in the Lowland Moist Forest formation. It is active on tree trunks near the ground and on the ground near small trees. Those on the ground attempt to escape by running to and climbing the

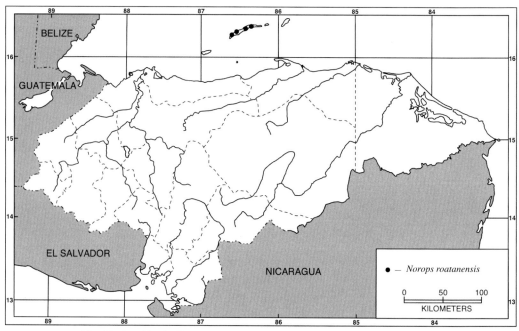

Map 27.   Localities for *Norops roatanensis*. Solid symbols denote specimens examined.

nearest tree. It sleeps at night on leaves of low vegetation, on rock walls of caves, and on various elevated human trashed items. The species is probably active throughout the year, as specimens have been collected during February, from May to September, and during November and December, during both the dry and wet seasons.

*Etymology.* The name *roatanensis* refers to Isla de Roatán (where the species is endemic), with the Latin suffix *-ensis* (denoting place, locality) added.

*Specimens Examined* (96 [27]; Map 27). ISLAS DE LA BAHÍA: Isla de Roatán, N of Coxen Hole, SMF 79223; Isla de Roatán, N of Flowers Bay, SMF 79222, USNM 565461–64; Isla de Roatán, Key Hole, SMF 79983–90; Isla de Roatán, near Oak Ridge, TCWC 10709–11; Isla de Roatán, Oak Ridge, CM 27599, 27603, FMNH 53828–29, MCZ R-150950; Isla de Roatán, Palmetto Bay, FMNH 282571–72, 283594; Isla de Roatán, 1 km E of Pollytilly Bight, FMNH 282573; Isla de Roatán, Port Royal, USNM 578760; Isla de Roatán, Port Royal

Harbor, UTA R-10714–15; Isla de Roatán, about 1.5 km N of Roatán, KU 203136; Isla de Roatán, about 0.5–2.0 km N of Roatán, LSUMZ 21353–59; Isla de Roatán, about 0.5–1.0 km N of Roatán, LSUMZ 21360–66; Isla de Roatán, near Roatán, BMNH 1985.1111, CM 64593, LSUMZ 22314–15, UF 28463–70, 28500–01; Isla de Roatán, 1.2 km E and 0.4 km S of Sandy Bay, KU 203137, 203146–48; Isla de Roatán, Sandy Bay, KU 203138–43, LSUMZ 33810–11; Isla de Roatán, Tabyana Beach, SMF 79202–03; Isla de Roatán, between West End Point and Flowers Bay, SMF 79953, 79974–82, 80764, 81124–26, USNM 541021; Isla de Roatán, near West End Town, USNM 520276; Isla de Roatán, West End Town, SMF 79224, TCWC 21949; "Isla de Roatán," IRSNB 12.484 (2), MCZ R-191104, UNAH 5322–23, 5325.

*Norops rodriguezii* (Bocourt)
*Anolis rodriguezii* Bocourt, 1873: 62, *in* Duméril et al., 1870–1909 (holotype, MNHN 2411 [see Brygoo, 1989: 87];

type locality: "à Pansos sur le Polochic [Amérique Centrale]"); Townsend, 2006: 35; Wilson and Townsend, 2007: 145; Köhler et al., 2007: 391; Townsend and Wilson, 2010b: 697 (in part); Köhler 2014: 210.

*Anolis limifrons*: Meyer, 1969: 225 (in part); Meyer and Wilson, 1973: 18 (in part).

*Norops rodriguezi*: Villa et al., 1988: 51; Wilson and McCranie, 1994: 417, 1998: 16; Wilson et al., 2001: 136.

*Norops rodriguezii*: Köhler et al., 2001: 254; Köhler, 2003: 105; McCranie, 2005: 20, 2011: 372; McCranie and Castañeda, 2005: 14; Castañeda and Marineros, 2006: 3–8; McCranie et al., 2006: 217; Wilson and Townsend, 2006: 105.

*Geographic Distribution.* *Norops rodriguezii* occurs at low and moderate elevations on the Atlantic versant from Tabasco, Mexico, to northwestern Honduras. In Honduras, this species occurs in the northwestern portion of the country.

*Description.* The following is based on seven males (SMF 79079, 79082, 79084, 79086, 88691, 88693; UF 142546) and 15 females (SMF 77746, 79080–81, 79085, 79087, 88690, 88692, 88694–95, 88702–07). *Norops rodriguezii* is a small anole (SVL 44 mm in largest Honduran male [UF 142546] and female [SMF 88690] examined; maximum reported SVL 49 mm [Köhler, 2008]); dorsal head scales weakly keeled in internasal region, rugose, weakly keeled, or tuberculate in prefrontal, frontal, and parietal areas; deep frontal depression present; parietal depression shallow; 4–7 (5.3 ± 1.2) postrostrals; anterior nasal usually entire, usually contacting rostral and first supralabial, occasionally only rostral; 4–9 (6.3 ± 1.2) internasals; canthal ridge sharply defined; scales comprising supraorbital semicircles weakly keeled, largest scale in semicircles about same size as largest supraocular scale; supraorbital semicircles well defined; 1–3 (1.8 ± 0.6) scales separating supraorbital semicircles at narrowest point; 1–4 (2.7 ± 0.8) scales separating supraorbital semicircles and interparietal at narrowest point; interparietal well defined, greatly enlarged relative to adjacent scales, surrounded by scales of moderate size, longer than wide, usually larger than ear opening; 2–3 rows of about 5–10 (total number) enlarged, keeled supraocular scales; enlarged supraoculars usually completely separated by 1 row of small scales; 2 elongate superciliaries, posterior much shorter than anterior; 2–3 enlarged canthals; 8–13 (10.8 ± 1.5) scales between second canthals; 9–15 (12.0 ± 1.6) scales between posterior canthals; loreal region slightly concave, 30–51 (40.1 ± 6.0) mostly strongly keeled (some smooth or rugose) loreal scales in maximum of 5–8 (6.1 ± 0.9) horizontal rows; 6–8 (7.1 ± 0.7) supralabials and 6–9 (7.5 ± 0.9) infralabials to level below center of eye; suboculars weakly keeled, usually 1–4 suboculars in broad contact with supralabials; ear opening vertically oval; scales anterior to ear opening granular, similar in size to those posterior to ear opening; 4–8 (5.8 ± 0.8) postmentals, outer pair usually largest; keeled granular scales present on chin and throat; male dewlap moderately small, extending posterior to level of axilla onto chest; male dewlap with 10–12 horizontal gorgetal-sternal scale rows, about 15–20 scales per row (n = 4); 2 (modal number) anterior marginal pairs in male dewlap; female dewlap absent; no nuchal crest or dorsal ridge; 2–3 middorsal scale rows slightly enlarged, weakly keeled, dorsal scales lateral to middorsal series grading into granular lateral scales; no enlarged scales scattered among laterals; 52–96 (73.6 ± 17.6) dorsal scales along vertebral midline between levels of axilla and groin in males, 49–103 (76.3 ± 17.9) in females; 24–66 (48.3 ± 16.5) dorsal scales along vertebral midline contained in 1 head length in males, 23–56 (34.8 ± 12.5) in females; ventral scales on midsection larger than largest dorsal scales; ventral body scales smooth, nonimbricate, slightly conical; 48–76 (57.7 ± 9.9) ventral scales along midventral line between levels of axilla and groin in males, 54–76 (63.7 ± 7.5) in females; 19–58 (39.6 ± 16.4) ventral scales contained in 1 head length in males, 16–46 (26.6 ± 10.0) in females; 111–158 (129.3 ± 14.7) scales around midbody in males, 123–160 (138.2 ± 12.3) in females; tubelike axillary pocket absent; precloacal scales not keeled; pair of only slightly enlarged post-cloacal scales in most males; tail rounded to slightly compressed, TH/TW 0.79–1.40 in 16; basal subcaudal scales keeled; lateral caudal scales keeled, homogeneous, although indistinct division in segments discernable; dorsal medial caudal scale row not, or only slightly enlarged, keeled, not forming crest; most scales on anterior surface of antebrachium keeled, unicarinate; 23–27 (25.4 ± 1.4) subdigital lamellae on Phalanges II–IV of Toe IV of hind limbs; 6–10 (7.7 ± 1.0) subdigital scales on Phalanx I of Toe IV of hind limbs; SVL 30.0–43.8 (35.5 ± 4.2) mm in males, 30.0–44.0 (39.5 ± 4.1) mm in females; TAL/SVL 1.78–2.18 in five males, 1.69–2.03 in five females; HL/SVL 0.25–0.28 in males, 0.23–0.29 in females; SHL/SVL 0.27–0.28 in males, 0.25–0.31 in females; SHL/HL 0.99–1.07 in males, 0.92–1.29 in females; longest toe of adpressed hind limb usually reaching between ear opening and center of eye.

Color in life of an adult male (USNM 330172): dorsal surface of head pale olive brown with some darker brown smudging along supraorbital semicircles and parietal region; dorsal surface of body pale olive brown; dorsal surfaces of forelimbs pale olive green with obscure darker crossbars; dorsal surfaces of hind limbs pale olive green with rust patina and rust brown spot at knee; dorsal surface of tail pale olive green with scattered brown spots; supralabials cream; pale color of supralabials continues onto side of neck to end at anterior point of insertion of forelimb; chin and belly cream; dewlap pale orange with dull white scales and darker orange central blotch. Color in life of another adult male (SMF 91289): dorsal ground color Brownish Olive (29) with dark brown middorsal spots along back; dorsal surface of head in front of eyes rust brown, that posterior to eyes Brownish Olive; dark brown interorbital stripe and V-shaped mark posterior to eyes present; dorsal surface of tail Brownish Olive with Burnt Umber (22) crossbands; dewlap Yellow Ocher (123C) with pinkish brown tinge, also with slightly darker orange central blotch.

Color in alcohol: dorsal and lateral surfaces of head and body grayish brown and patternless, or with dark brown vertebral chevrons; some females with pale brown, dark brown–bordered vertebral stripe; dark brown interorbital bar occasionally present; dorsal surfaces of limbs and tail with indistinct dark brown crossbands; ventral surfaces of head and body usually pale grayish brown, dark grayish brown at midventer in some; chin region usually mottled with dark brown.

Hemipenis: the completely everted hemipenis of SMF 83230 (from Cayo District, Belize) is a medium-sized, bilobed organ; sulcus spermaticus bordered by well-developed sulcal lips, bifurcating at base of apex; branches opening into broad concave areas distal to point of bifurcation on each lobe; low asulcate ridge present; lobes strongly calyculate; truncus with transverse folds.

*Diagnosis/Similar Species.* *Norops rodriguezii* is distinguished from all other Honduran *Norops*, except *N. carpenteri*, *N. limifrons*, *N. ocelloscapularis*, *N. yoroensis*, and *N. zeus*, by the combination of having a single elongated prenasal scale, smooth to slightly keeled ventral scales, and slender habitus. *Norops rodriguezii* differs from *N. carpenteri* by having brown dorsal surfaces without paler spots in life (dorsum greenish brown with paler spots in *N. carpenteri*). *Norops rodriguezii* differs from *N. limifrons* and *N. zeus* by having shorter hind legs (longest toe of adpressed hind limb usually reaching between ear opening and center of eye in *N. rodriguezii* versus between anterior border of eye and tip of snout in *N. limifrons* and *N. zeus*), and by having a predominantly orange male dewlap in life (dewlap dirty white and with or without a basal orange-yellow spot in life in *N.*

*limifrons* and *N. zeus*). *Norops rodriguezii* differs from *N. ocelloscapularis* and *N. yoroensis* by lacking a pale lateral stripe (pale lateral stripe usually present in *N. ocelloscapularis* and *N. yoroensis*). Also, *N. rodriguezii* differs from *N. ocelloscapularis* in lacking an ocellated shoulder spot (such a spot usually present in *N. ocelloscapularis*).

*Illustrations* (Figs. 55, 56, 83). Lee, 1996 (adult; as *Anolis*), 2000 (adult, head and dewlap); Campbell, 1998 (adult); Stafford and Meyer, 1999 (adult); Köhler, 2000 (head and dewlap), 2003 (adult, head and dewlap), 2008 (adult, head and dewlap), 2014 (ventral scales; as *Anolis*); D'Cruze, 2005 (adult, head and dewlap); Calderón-Mandujano et al., 2008 (adult; as *Anolis*).

*Remarks*. The relationships of *N. rodriguezii* remain uncertain (Nicholson, 2002). Savage and Guyer (1989) regarded the species as of uncertain relationship status in the *N. laeviventris* subseries. Köhler (2008) regarded it as a member of the *N. fuscoauratus* species group. The monophyly of the *N. fuscoauratus* species group as envisioned by Savage and Guyer (1989) was rejected by both molecular (Nicholson, 2002) and morphological (Poe, 2004) analyses. Nicholson (2002) included this species in her molecular analysis, but Nicholson et al. (2005, 2012) did not include *N. rodriguezii* in their subsequent analyses. Herein, we retain *N. rodriguezii* in a more restricted *N. fuscoauratus* species subgroup than that envisioned by Savage and Guyer (1989; also see Remarks for *N. ocelloscapularis*).

*Norops rodriguezii* was not reported from Honduras until 1994, although it was collected in the country as early as 1923. Meyer and Wilson (1973) did not distinguish *N. rodriguezii* from *N. limifrons*.

*Natural History Comments. Norops rodriguezii* is known from near sea level to 1,200 m elevation in the Lowland Moist Forest, Lowland Dry Forest, and Premontane Wet Forest formations and peripherally in the Premontane Moist Forest formation. It is usually found in open situations both on the ground and in vegetation to at least 2 m above the ground, along streams and other forest edge situations, and in forested areas well away from streams. It basks in the sun and is also active in shaded areas. It sleeps at night on leaves and stems of low vegetation. *Norops rodriguezii* was found from April to August and during October and November and is likely active throughout the year. D'Cruze (2005) reported the species to be highly arboreal in Belize, with an average perch height of 885.5 mm, and that it was most abundant in dense vegetation. Villarreal Benítez (1997) reported that in Veracruz, Mexico, the species occurs in coniferous forest, oak and pine-oak forests, dry forest, and evergreen forest. Lee (1996) said populations on the Yucatán Peninsula occur in a wide variety of forest situations, but are most abundant in seasonally dry forests and forest edges. Lee (1996) noted that this species feeds on small invertebrates, especially insects and spiders. Apparently nothing has been published on its reproduction.

*Etymology*. The name *rodriguezii* is a patronym honoring Juan J. Rodriguez, the "conservateur du Musée zoologique de la Société économique de Guatemala" (Bocourt, 1873: 63, *in* Duméril et al., 1870–1909).

*Specimens Examined* (74 [15]; Map 28). COPÁN: 1 km W of Copán, SMF 88702, USNM 573142; Copán, FMNH 28522, SMF 93362, UMMZ 83048 (1 of 3 under that number is *N. rodriguezii*, the remaining two are *N. tropidonotus*), 83049 (2 of 3 under that number are *N. rodriguezii*, the remaining one is *N. tropidonotus*), 83050 (11), USNM 573143; 12.9 km ENE of Copán, LACM 47269–70; El Gobiado, USNM 573145; La Playona, USNM 330169–74; Laguna del Cerro, SMF 79086; near Quebrada Grande, SMF 79085; Río Amarillo, SMF 91767, USNM 573144; San Isidro, SMF 88703–04, 93363–67, UF 142546. CORTÉS: near Cofradía, MCZ R-32522; El Paraíso, UF 149663, 149666–68; between La Fortuna and Buenos Aires, SMF 77746; Naranjito, SMF 88694; W of San Pedro Sula, FMNH 5253; Santa Teresa, SMF 91289–93; 4.7 km

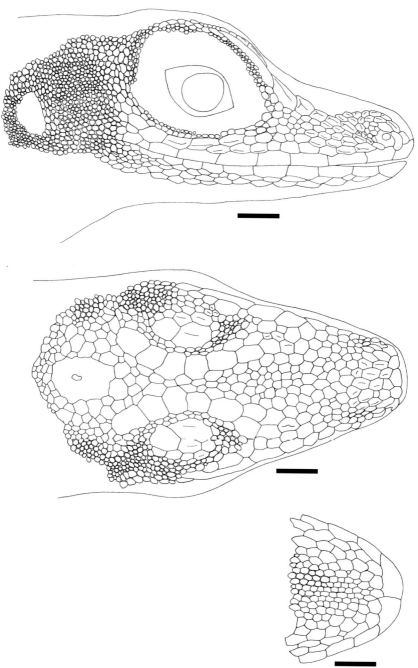

Figure 55.   *Norops rodriguezii* head in lateral, dorsal, and ventral views. SMF 79086, adult male from Laguna del Cerro, Copán. Scale bar = 1.0 mm. Drawing by Gunther Köhler.

Figure 56. *Norops rodriguezii.* (A) Adult male (in UNAH collection); (B) adult male dewlap (SMF 93363), both from San Isidro, Copán. Photographs by James R. McCranie.

N of Tegucigalpita, SMF 88690–93; about 1 km SSE of Tegucigalpita, SMF 79079–84, 79087, 88695. SANTA BÁRBARA: Cerro Negro, SMF 88705–07, USNM 565465, 573146–47; La Cafetalera, SMF 91294–96.

*Norops rubribarbaris* Köhler, McCranie, and Wilson
*Norops rubribarbaris* Köhler, McCranie, and Wilson, 1999: 280 (holotype, UF 90206; type locality: "Honduras, Departamento

de Santa Bárbara, N slope of Montaña de Santa Bárbara, 4 km S of San Luís de los Planes, elevation 1700 m"); Köhler, 2000: 63, 2003: 104, 2008: 112; Köhler et al., 2001: 254; Wilson and McCranie, 2003: 60, 2004b: 43.
*Norops* sp. 1: Wilson et al., 2001: 147.
*Anolis rubribarbaris*: Köhler et al., 2007: 391; Townsend et al., 2008: 40; Townsend and Wilson, 2009: 69; Wilson et al., 2013: 57.

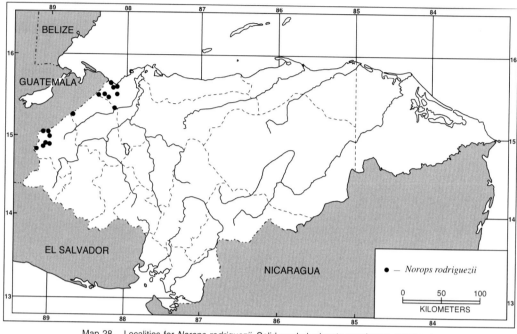

Map 28.   Localities for *Norops rodriguezii*. Solid symbols denote specimens examined.

*Geographic Distribution. Norops rubri-barbaris* occurs at intermediate elevations on Montaña de Santa Bárbara, Departamento de Santa Bárbara, Honduras.

*Description.* The following is based on four males (UF 90206, 152660; USNM 573148, 578762) and three females (SMF 92839; UF 152661–62). *Norops rubribarbaris* is a medium-sized anole (SVL 52 mm in largest male [USNM 578762] and female [UF 152661]); dorsal head scales keeled in internasal, prefrontal, frontal, and parietal areas; deep frontal depression present, parietal depression absent; 4–7 (5.0 ± 1.3) postrostrals; anterior nasal divided, lower scale contacting rostral and first supralabial; 4–7 (5.9 ± 0.9) internasals; canthal ridge sharply defined; scales comprising supraorbital semicircles smooth, rugose, or weakly keeled, largest scale in semicircles larger than largest supraocular scale; supraorbital semicircles well defined; 0–2 (0.9 ± 0.7) scales separating supraorbital semicircles at narrowest point; 1–3 (2.1 ± 0.7) scales separating supraorbital semicircles and in-

terparietal at narrowest point; interparietal well defined, slightly to distinctly enlarged relative to adjacent scales, surrounded by scales of moderate size, longer than wide, smaller than ear opening; 2–3 rows of about 3–5 (total number) enlarged, faintly to strongly keeled supraocular scales; 1–2 enlarged supraoculars in broad contact with supraorbital semicircles; 2–3 elongate superciliaries, posteriormost only slightly shorter than anteriormost; usually 3 enlarged canthals; 3–7 (5.7 ± 1.5) scales between second canthals; 6–9 (7.7 ± 1.1) scales between posterior canthals; loreal region slightly concave, 16–37 (23.9 ± 6.3) mostly strongly keeled (some smooth or rugose) loreal scales in maximum of 4–5 (4.7 ± 0.5) horizontal rows; 6–7 (6.7 ± 0.5) supralabials and 5–7 (6.1 ± 0.7) infralabials to level below center of eye; suboculars weakly to strongly keeled, usually 2–3 suboculars in broad contact with supralabials; ear opening vertically oval; scales anterior to ear opening not granular, slightly larger than those posterior to ear opening;

4–7 (5.1 ± 1.2) postmentals, outer pair usually largest; weakly keeled granular scales present on chin and throat; male dewlap moderately small, extending to level of axilla; male dewlap with 5–10 horizontal gorgetal-sternal scale rows, about 5–8 scales per row (*n* = 1); 2 (modal number) anterior marginal pairs in male dewlap; female dewlap small to rudimentary; low nuchal crest and low dorsal ridge present in males; about 8–11 middorsal scale rows distinctly enlarged, keeled; dorsal scales lateral to middorsal series grading into keeled, non-granular lateral scales; flank scales heterogeneous, solitary enlarged keeled or elevated scales scattered among laterals; 34–48 (40.3 ± 6.1) dorsal scales along vertebral midline between levels of axilla and groin in males, 30–65 (44.0 ± 18.5) in females; 21–34 (27.8 ± 5.4) dorsal scales along vertebral midline contained in 1 head length in males, 26–45 (33.7 ± 10.0) in females; ventral scales on midsection slightly larger than largest dorsal scales; ventral body scales strongly keeled, mucronate, imbricate, 36–43 (39.8 ± 2.9) ventral scales along midventral line between levels of axilla and groin in males, 30–52 (38.7 ± 11.7) in females; 21–29 (25.5 ± 3.7) ventral scales contained in 1 head length in males, 21–24 (22.7 ± 1.5) in females; 99–120 (110.3 ± 9.0) scales around midbody in males, 111–115 (113.0 ± 2.0) in females; tubelike axillary pocket absent; precloacal scales strongly keeled; pair of greatly enlarged postcloacal scales in males; tail slightly to distinctly compressed, TH/TW 1.08–1.67; all subcaudal scales keeled; lateral caudal scales keeled, homogeneous, although indistinct division in segments discernable; dorsal medial caudal scale row enlarged, keeled, not forming crest; scales on anterior surface of antebrachium distinctly keeled, unicarinate; 23–28 (25.6 ± 1.7) subdigital lamellae on Phalanges II–IV of Toe IV of hind limbs; 5–9 (7.4 ± 1.3) subdigital scales on Phalanx I of Toe IV of hind limbs; SVL 36.7–51.9 (46.3 ± 6.6) mm in males, 40.1–51.5 (45.4 ± 5.7) mm in females; TAL/SVL 2.54–2.68 in two males, 2.24–2.25 in females; HL/SVL 0.25–0.28 in males, 0.26–0.41 in females; SHL/SVL 0.22–0.25 in males, 0.19–0.26 in females; SHL/HL 0.89–0.90 in males, 0.64–0.81 in females; longest toe of adpressed hind limb usually reaching between ear opening and posterior margin of eye.

Color in life of an adult male (USNM 573148): dorsal surface of body Smoke Gray (44) with Grayish Olive (43) lateral chevron-shaped (with apex pointed dorsally) markings; body also with Grayish Olive lines connecting lateral chevrons across back; slightly paler gray than Smoke Gray indistinct lateral stripe present below bottom edges of chevrons; dorsal surface of head Smoke Gray with Sepia (119) line along inner edge of upper eyelids and paired Sepia lines on snout above nostrils that extend posteriorly to about level of first canthal; dorsal surface of tail Smoke Gray with Grayish Olive crossbands anteriorly, tail becoming darker brown distally with indistinct darker brown crossbands; belly pale gray; dewlap Chrome Orange (16) with white gorgetal scales, although some central gorgetal scales with slight brown edging. Köhler et al. (1999: 285) described the dewlap color of the male holotype (UF 90206) as "red with rows of tan gorgetal scales." Color in life of a young adult male (UF 152660) was described by Townsend et al. (2008: 41): "dorsum rust brown on enlarged middorsal scale rows, smudged middorsally with dark gray; lateral region of body yellow-brown; anterior limbs yellow-brown; posterior limbs yellow-brown with narrow brown crossbars on lower limb; dorsum of head rust brown mottled with dark gray; tail yellow-brown with dark gray crossbars; venter peach-cream; dewlap red with slight orange tinge; iris rust brown." Color in life of an adult female (UF 152661) also described by Townsend et al. (2008: 42): "dorsum uniform rust brown, lateral regions same; anterior limbs yellow-tan; posterior limbs rust brown; dorsum of head rust brown with dark gray smudging; tail rust brown; x-shaped dark brown mark at base of tail; venter pale peach with scattered black punctations; small dewlap orangish

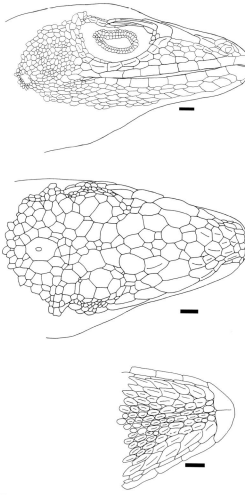

Figure 57.   *Norops rubribarbaris* head in lateral, dorsal, and ventral views. UF 90206, adult male from 4 km S of San Luís de Los Planes, Santa Bárbara. Scale bar = 1.0 mm. Drawing by Gunther Köhler.

red; iris rust brown." Color in life of another adult female (UF 152662) also described by Townsend et al. (2008: 42): "enlarged middorsal scale rows gray brown with four dark brown chevrons; lateral regions yellow-gray with scattered black punctations; anterior limbs yellow-tan; posterior limbs yellow-tan with brown crossbands on lower limb; dorsum of head gray-brown; tail brown with slightly dark crossbands; venter yellowish cream; small dewlap orangish red; iris rust brown."

Color in alcohol: dorsal and lateral surfaces of head and body grayish brown, with dark brown crossbands or chevrons usually present on body; dark brown inter-orbital bar usually present; pale longitudinal lateral stripes absent; dorsal surfaces of limbs and tail with narrow dark brown crossbands; ventral surfaces of head and body pale grayish brown.

Hemipenis: according to Townsend et al. (2008: 42), "the everted hemipenis of UF 152660, a subadult male, is a somewhat stout organ; asulcate processus undivided; sulcus spermaticus bounded by moderately well-developed sulcal lips, terminates at the base of the apex; truncus bearing some shallows [sic] folds, otherwise lacking surface structures; apical region appearing relatively smooth, slightly calyculate."

*Diagnosis/Similar Species. Norops rubribarbaris* is distinguished from all other Honduran *Norops*, except *N. crassulus* and *N. morazani*, by the combination of having distinctly enlarged middorsal scale rows, heterogeneous lateral scales, strongly keeled, mucronate, imbricate ventral scales, and a pair of greatly enlarged postcloacal scales in males. *Norops rubribarbaris* differs from *N. crassulus* by having about 8–11 rows of enlarged dorsal scales (about 14–18 in *N. crassulus*), and by the absence of, or having a relatively small female dewlap (female dewlap well-developed in *N. crassulus*). *Norops rubribarbaris* differs from *N. morazani* by having a hemipenis with an undivided asulcate processus (processus divided in *N. morazani*).

*Illustrations* (Figs. 57, 58). Köhler et al., 1999 (adult, head scales, body scales); Townsend et al., 2008 (adult, subadult, dewlap; as *Anolis*).

*Remarks. Norops rubribarbaris* is in the *N. crassulus* species subgroup (see Remarks for *N. amplisquamosus*). *Norops rubribarbaris* tissues were not used in the phylogenetic analyses of Nicholson (2002) or Nicholson et al. (2005, 2012).

*Natural History Comments. Norops rubribarbaris* is known from 1,600 to 1,800 m

Figure 58.   *Norops rubribarbaris*. (A) Adult female (SMF 92839); (B) adult male dewlap (UF 152660), both from near El Cedral, Santa Bárbara. Photographs by James R. McCranie (A) and Josiah Townsend (B).

elevation in the Lower Montane Wet Forest formation. The holotype was active during March on a limestone boulder along a stream, and another was active in leaf litter in forest being converted to a coffee field during August. Another was sleeping at night on a vine about 2 m above the ground on a coffee farm during October, and two others were sleeping at night during November. The three specimens reported by Townsend et al. (2008) were sleeping on vegetation at night during January. The species is likely active throughout the year under favorable conditions. Nothing has been reported on its diet or reproduction. However, a female collected during October contained an unshelled egg.

*Etymology.* The name *rubribarbaris* is formed from the Latin *ruber* (red), *barba* (beard), and *-arius* (pertaining to). The name refers to the red dewlap of the adult male holotype.

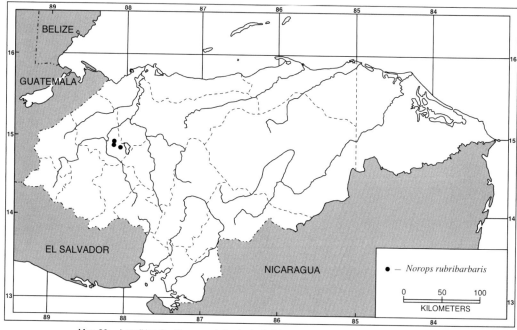

Map 29.   Localities for *Norops rubribarbaris*. Solid symbols denote specimens examined.

*Specimens Examined* (8 [0]; Map 29). SANTA BÁRBARA: Cañon del Cielo, USNM 578761–62; 1.2 W of El Cedral, UF 152660–62, USNM 573148; El Cedral, SMF 92839; 4 km S of San Luís de Los Planes, UF 90206.

*Norops sagrei* (Cocteau, *in* Duméril and Bibron)

*Anolis sagrei* Cocteau, *in* Duméril and Bibron, 1837: 149 (eight syntypes, MCZ R-2172, MNHN 2430, 2430A, 2430B, 6795, 6795A, 6797, 6797A [see Barbour and Loveridge, 1929b: 223, Brygoo, 1989: 90]; type locality: "l'île de Cuba" [see Remarks]); Fugler, 1968: 97; Meyer, 1969: 231; Meyer and Wilson, 1973: 19 (in part); Wilson and Hahn, 1973: 111 (in part); Hudson, 1981: 377; Schoener, 1988: 22; Monzel, 1998: 157, 2001: 30; Lundberg, 2001: 27; McCranie and Nuñez, 2014: 91.

*Anolis sagrei sagrei*: MacLean et al., 1977: 4; Garrido and Jaume, 1984: 73.

*Norops sagrei*: Schwartz and Henderson, 1988: 159; Köhler, 1998b: 377; Köhler et al., 2001: 255; McCranie et al., 2005: 108, 2006: 217; Wilson and Townsend, 2006: 105; McCranie and Valdés Orellana, 2014: 45.

*Norops sagrai*: Wilson et al., 2001: 136.

*Geographic Distribution.* The native range of *Norops sagrei* includes low elevations on Cuba, the islands in the Bahamas, Little Cayman Island, and Cayman Brac. It has been introduced and established on Bermuda, Jamaica (but see below), Grand Cayman, the Grenada Bank, the Barbados Bank, the St. Vincent Bank, Hawaii, California, and the southeastern USA, Taiwan, Singapore, and many other places, with the list of localities constantly growing. It is not certain if the populations occurring along the coast of the mainland of Middle America from south-central Veracruz, Mexico, to north-central Honduras and those on Roatán Island and many other islands off the coast of Belize and Quintana Roo,

Mexico, and those on Jamaica are introduced or native. The population on Isla de Utila, Islas de la Bahía, Honduras, has apparently been recently introduced, likely from La Ceiba or Roatán. An introduced and established population occurs in east-central Costa Rica. Another population has been established in the central portion of the Yucatán Peninsula, Mexico. Many of these introduced populations are expanding their ranges. In Honduras, this species is known to occur at Puerto Cortés and San Pedro Sula, Cortés, and at La Ceiba and Tela, Atlántida, on the northern mainland and on islas de Roatán and Utila in the Islas de la Bahía.

*Description.* The following is based on 10 males (KU 203144; SMF 77743; UF 28547–49, 28551; UIMNH 66634, 66641; USNM 573151–52) and eight females (CM 64598; LSUMZ 21369, 22394–96; SMF 77744; UF 28552; USNM 573150). *Norops sagrei* is a medium-sized anole (SVL 70 mm in largest Honduran male examined [MCZ R-192030; not used in description], 53 mm in largest Honduran female examined [LSUMZ 21369]); dorsal head scales keeled in internasal, prefrontal, and frontal areas, smooth to faintly keeled in parietal area; shallow frontal depression present; parietal depression absent; 4–6 (4.8 ± 0.8) postrostrals; anterior nasal single, contacting rostral, usually not in contact with first supralabial; 4–6 (5.0 ± 0.8) internasals; canthal ridge sharply defined; scales comprising supraorbital semicircles usually keeled or strongly ridged, largest scale in semicircles larger than largest supraocular scale; supraorbital semicircles well defined; 0–2 (0.9 ± 0.5) scales separating supraorbital semicircles at narrowest point; 1–4 (2.5 ± 0.6) scales separating supraorbital semicircles and interparietal at narrowest point; interparietal well defined, greatly enlarged relative to adjacent scales, surrounded by scales of moderate size, longer than wide, smaller than ear opening; 2 rows of about 4–8 (total number) enlarged, smooth to faintly keeled supraocular scales; enlarged supraoculars completely separated from supraorbital semicircles by 1 row of small scales; 2–3 elongate superciliaries, posteriormost shortest; usually 3 enlarged canthals; 4–8 (5.9 ± 1.1) scales between second canthals; 5–12 (6.8 ± 1.6) scales between posterior canthals; loreal region slightly concave, 16–30 (22.6 ± 3.6) keeled or rugose loreal scales in maximum of 4–6 (4.9 ± 0.6) horizontal rows; 4–8 (5.8 ± 0.9) supralabials and 4–6 (5.0 ± 0.6) infralabials to level below center of eye; suboculars distinctly keeled, in broad contact with supralabials; ear opening vertically oval; scales anterior to ear opening distinctly larger than those posterior to ear opening; 4–6 (4.3 ± 0.6) postmentals, outer pair much enlarged, their lengths greater than greatest length of mental scale, much larger than first infralabial; gular scales not keeled; male dewlap small, extending to level of forelimb insertion; male dewlap with about 6 horizontal gorgetal-sternal scale rows, about 6–9 scales per row (n = 3); 3 (modal number) anterior marginal pairs in male dewlap; female dewlap rudimentary; nuchal crest and dorsal ridge present in breeding males; about 2–3 middorsal scale rows slightly enlarged, smooth to faintly keeled, dorsal scales lateral to middorsal series grading into granular lateral scales; no enlarged scales among laterals; 61–85 (69.1 ± 7.7) dorsal scales along vertebral midline between levels of axilla and groin in males, 67–84 (76.1 ± 8.5) in females; 24–51 (42.9 ± 7.4) dorsal scales along vertebral midline contained in 1 head length in males, 28–50 (42.6 ± 11.2) in females; ventral scales on midsection much larger than largest dorsal scales; ventral body scales distinctly keeled, imbricate; 39–51 (46.1 ± 4.0) ventral scales along midventral line between levels of axilla and groin in males, 41–54 (48.5 ± 6.5) in females; 29–34 (31.4 ± 1.6) ventral scales contained in 1 head length in males, 22–39 (30.5 ± 8.5) in females; 141–164 (151.2 ± 8.3) scales around midbody in males, 134–149 (143.0 ± 7.5) in females; tubelike axillary pocket absent; precloacal scales not keeled; enlarged postcloacal scales usually absent in males, sometimes

1–2 enlarged to slightly enlarged postcloacals present; tail strongly compressed, TH/TW 1.38–2.40 in males, 1.18–2.31 in seven females; basal subcaudal scales keeled; lateral caudal scales keeled, homogeneous; dorsal medial caudal scale row enlarged, keeled, forming crest, especially in large males; most scales on anterior surface of antebrachium unicarinate; 20–33 (30.2 ± 2.9) subdigital lamellae on Phalanges II–IV of Toe IV of hind limbs; 6–9 (7.6 ± 0.7) subdigital scales on Phalanx I of Toe IV of hind limbs; SVL 47.0–62.3 (55.6 ± 6.1) mm in males, 42.5–53.3 (48.6 ± 5.4) mm in females; TAL/SVL 1.47–1.96 in six males, 1.53–1.74 in five females; HL/SVL 0.25–0.28 in males, 0.24–0.31 in females; SHL/SVL 0.24–0.30 in males, 0.23–0.28 in females; SHL/HL 0.96–1.06 in males, 0.90–1.00 in females; longest toe of adpressed hind limb usually reaching between posterior and anterior borders of eye.

Color in life of an adult male (MCZ R-192005): dorsal ground color Raw Umber (22 of Köhler, 2012) with lateral Pale Pinkish Buff (3) flecking, longitudinal lines, and vertical lines; top of head Fawn Color (258); tail Raw Umber with Pale Pinkish Buff (3) on original portion; belly yellowish brown; ventral surfaces of limbs brown; subcaudal surface on original portion Tawny Olive (17) with dark brown mottling; iris dark brown; dewlap Dark Salmon Color (59), except basal portion and skin along outer portion (except margin) Dark Yellow Buff (54); gorgetal scales white with varying amounts of dark brown flecking; anterior marginal scales pale yellow, posterior marginal scales white with dark brown flecking, flecking becoming more dense on central marginal scales; skin between marginal scales on central portion Dark Yellowish Buff. Color in life of another adult male (USNM 573152): middorsum Clay Color (26); slightly paler brown longitudinal stripe present on dorsolateral surface of body; dorsolateral and lateral areas mottled with Sepia (119), Cream Color (54), and Army Brown (219B); dorsal surfaces of limbs Dark Drab (119B) with slightly paler crossbands;

dorsal surfaces of head and tail Clay Color; venter of body Light Drab (119C); chin pale gray with gray-brown lineate markings; dewlap Ferruginous (41) with brown stippled white gorgetal scales; iris dark brown. Dewlap color in another adult male (USNM 573151): dark orange with gray scales; marginals yellowish white anteriorly, becoming orange-white posteriorly, central marginals separated by dark orange skin.

Color in alcohol: dorsal surface of head brown, usually without distinct markings; dorsal surface of body brown, usually with darker brown blotches or chevrons, some females with paler brown broad middorsal stripe; dorsal surfaces of limbs brown, usually without darker brown markings; dorsal surface of tail brown, usually without distinct markings; ventral surface of head brown with darker brown flecking; ventral surface of body white to cream, with varying amounts of brown flecking; subcaudal surface brown with traces of darker brown crossbands.

Hemipenis: the completely everted hemipenis of SMF 83344 (from Stann Creek District, Belize) is a small, slightly bilobed organ; sulcus spermaticus bordered by well-developed sulcal lips, opening into single broad concave area at base of apex; lobes calyculate; truncus with transverse folds; asulcate processus absent.

*Diagnosis/Similar Species. Norops sagrei* is distinguished from all other Honduran *Norops*, except *N. nelsoni*, by having a very strongly compressed tail with a dorsal crest (especially in large males) and the outer postmental scale on each side greatly enlarged with its length greater than that of the mental scale. *Norops sagrei* differs from *N. nelsoni* (but see Remarks for *N. nelsoni*) by having an orange male dewlap in life (dewlap dark brown in life in *N. nelsoni* from Big Swan Island, dewlap brownish-yellow in *N. nelsoni* from Little Swan Island). *Norops sagrei* differs further from *N. nelsoni* from Little Swan Island in having the dorsal and lateral surfaces of the head similar in color to that of the body (those surfaces of head yellowish-brown in life in

*N. nelsoni* from Little Swan Island). *Norops sagrei* differs further from *N. nelsoni* from both Little and Big Swan islands in usually having four postmentals (usually five or more in *N. nelsoni*).

*Illustrations* (Figs. 59, 60, 81). Campbell, 1998 (adult); Stafford and Meyer, 1999 (adult, head and dewlap); Lee, 2000 (adult [male only]); Monzel, 2001 (adult; as *Anolis*); McCranie et al., 2005 (adult,); Calderón-Mandujano et al., 2008 (adult and male dewlap; as *Anolis*); Köhler, 2008 (adult, head and dewlap).

*Remarks. Norops sagrei* is a member of the *N. sagrei* species group of Williams (1976a) and Nicholson et al. (2012; also see Remarks for *N. nelsoni*). Ruibal (1964: 490) stated "It appears reasonable to restrict the type locality [of *N. sagrei*] to the city of La Habana, Habana." The species was first collected on the Honduran mainland during 1963 at Puerto Cortés (Fugler, 1968) and was first collected on Roatán Island in 1940 (see Wilson and Hahn, 1973) and again in 1947. However, Meyer (1969) was the first to report *N. sagrei* from Roatán. Thus, Williams (1969) had called the *N. sagrei* colonization of the Honduran Bay Islands a "failure."

Fugler (1968) showed that Boulenger's (1885) records of *Norops sagrei* from "Honduras" are based on specimens from British Honduras (= Belize). We herein consider *Norops nelsoni* of the Islas del Cisne to be a full species, instead of a subspecies of *N. sagrei* as proposed by some workers (see synonymy in *N. nelsoni* account).

*Norops sagrei* is speculated to have reached the Honduran Bay Island of Roatán by recent overwater dispersal from Cuba (see Schoener, 1988). On the other hand, Kraus (2009: 364) considered it "Unclear if introduced by humans" for the Central American populations and also (p. 78) stated that the population of *N. sagrei* on "the coast of northern Central America" was native. Molecular analyses of Honduran populations of *N. sagrei*, as well as those from other parts of Central America are needed to test those scenarios. Such a study using recently collected tissues from Honduras is underway. The phenetic analysis by Lee (1992) resulted in *N. sagrei* from Mexico and Central America forming a separate clade within the *N. sagrei* populations studied by him.

Fieldwork during December 2012 resulted in the collection of long series' of *N. sagrei* from La Ceiba, Atlántida, and from Isla de Utila, Islas de la Bahía. Color pattern of the body and male dewlap of these two populations could be interpretable as that they might not be conspecific with *N. sagrei* populations from Cuba and south Florida, USA. Smith and Burger (1949) described *Anolis sagrei mayensis* (type locality in Campeche, Mexico) for "the form occurring in the eastern coastal areas of southern Mexico and Central America." One of the distinctive characters given by those authors for *N. sagrei mayensis* "is the peculiarity that the median line of the dewlap of males is broken by grey mottling" (p. 408). Apparently, Smith and Burger (1949) were referring to the outer marginal scales not being a continuous white or yellow as is the case with some Cuban individuals (see illustrations in Rodríguez Schettino, 1999, 2003) and Florida (see photograph in Bartlett and Bartlett, 1999) specimens. However, much variation does occur in male dewlap color on Cuba (Rodríguez Schettino, 1999) as well as on the Bahama Islands (Vanhooydonck et al., 2008). Color notes in life of an adult male from La Ceiba indicated the anterior margin of the dewlap had yellow scales, that of the posterior margin were white with dark brown flecking, and those of the central margin scales were white with dense dark brown flecking, thus agreeing with *N. s. mayensis* in that character. Unfortunately, so much variation occurs in dewlap color on Cuba and in the various introduced populations in Florida, that using dewlap color as a diagnostic character for any *N. sagrei* complex population (including isolated island populations) might not be informative. However, it must be remembered that the various Honduran

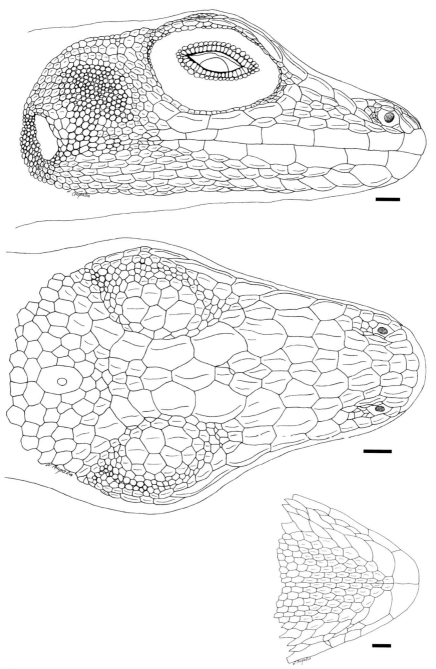

Figure 59.    *Norops sagrei* head in lateral, dorsal, and ventral views. SMF 77743, adult male from San Pedro Sula, Cortés. Scale bar = 1.0 mm. Drawing by Lara Czupalla.

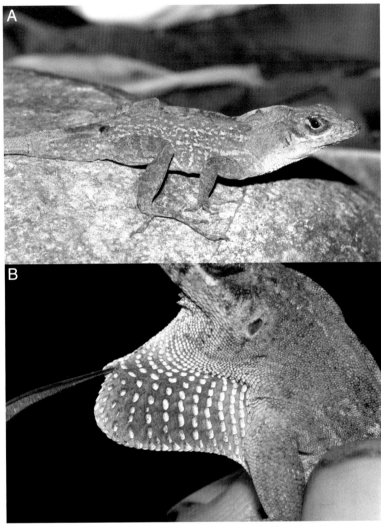

Figure 60.    *Norops sagrei*. (A) Adult male (MCZ R-192005) from La Ceiba, Atlántida; (B) adult male dewlap (MCZ R-192005). Photographs by James R. McCranie.

populations we have seen in life all have population specific (i.e., apparently nonvariable in each population) dewlap color and pattern.

*Natural History Comments. Norops sagrei* is known from near sea level to 100 m elevation in the Lowland Moist Forest and Lowland Dry Forest formations. This species is edificarian on the mainland and on islas de Roatán and Utila, being found on low vegetation, coconut palms, mangrove trees, concrete walls, wooden fences, rock piles, rock walls, and on the ground near those objects. We have not searched for it at night, but it probably sleeps at night on low vegetation as close relatives in south Florida are known to do (JRM pers. observ.) It has been collected during February, July, November, and December and is likely active throughout the year. West Indian populations were classified as "cosmopolitan" (see Henderson and Powell, 2009). *Norops*

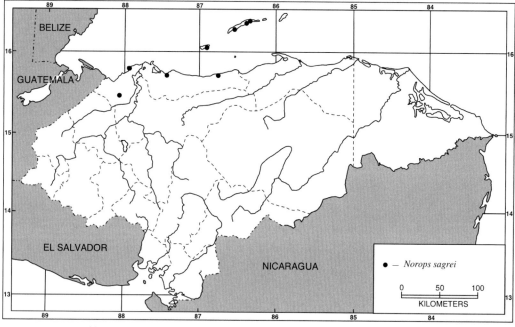

Map 30.   Localities for *Norops sagrei*. Solid symbols denote specimens examined.

*sagrei* from outside Honduras are reported to primarily feed on insects, but it is a diet generalist that will eat nearly anything that it can overpower (including other anoles), and also will occasionally take plant matter and fruit (Henderson and Powell, 2009; Holbrook, 2012; and references cited in those studies). A seasonal reproductive pattern (peaking during the rainy season) has been reported for populations in Belize (Lee, 1996; Stafford and Meyer, 1999).

*Etymology.* The name *sagrei* is a patronym honoring Ramon de la Sagra, who collected part of the type series, and also authored the extensive work "Historia Fisca, Politica y Natural de la Isla de Cuba" during the 1800s.

*Specimens Examined* (99 [12]; Map 30). ATLÁNTIDA: La Ceiba, CM 29010, MCZ R-192002–30, USNM 573152; Tela, AMNH 124039. CORTÉS: Puerto Cortés, UIMNH 66632–41; San Pedro Sula, SMF 77743–44. ISLAS DE LA BAHÍA: Isla de Roatán, Coxen Hole, LSUMZ 33812, USNM 573149–51; Isla de Roatán, near French Harbor, CM 64598, UF 28547–52; Isla de Roatán, French Harbor, KU 203144, LSUMZ 21369–70, 22394–97; Isla de Roatán, Oak Ridge, MCZ R-150951, UTA R-10712–13, 55236–38; "Isla de Roatán," UF 149697; Isla de Utila, Utila Town, MCZ R-191093, 192079–80, 192093–107, UNAH 5567–68.

*Norops sminthus* (Dunn and Emlen)
*Anolis sminthus* Dunn and Emlen, 1932: 26 (holotype, ANSP 22946 [formerly ANSP 19878, see Malnate, 1971: 360, Köhler, 1996c: 33 erroneously listed MCZ R-22946 as the holotype]); type locality: "San Juancito, Honduras, 6900 feet"); Smith, 1933: 35; Barbour, 1934: 150; Barbour and Loveridge, 1946: 73; Etheridge, 1959: 216; Meyer, 1969: 232 (in part); Peters and Donoso-Barros, 1970: 66; Echelle et al., 1971: 359; Malnate, 1971: 360; Meyer and Wilson, 1972: 108 (in part), 1973: 20 (in part); Fitch and Hillis, 1984: 321; Fitch and Seigel, 1984: 9; Poe, 2004: 64; Wilson and Townsend,

2007: 145; Köhler et al., 2007: 391; Townsend and Wilson, 2009: 63; Pyron et al., 2013: fig. 19; Köhler, 2014: 210.

*Norops sminthus*: Villa et al., 1988: 51; McCranie et al., 1992: 208 (in part); Caceres, 1993: 116; Köhler, 1996c: 33, 2003: 104, 2008: 113; Köhler and Obermeier, 1998: 136; Köhler et al., 1999: 298, 2001: 254; Wilson et al., 2001: 136 (in part); Lundberg, 2003: 27; Wilson and McCranie, 2003: 60 (in part), 2004b: 43; Diener, 2008: 15.

*Anolis* cf. *sminthus*: Townsend and Wilson, 2009: 63 (in part).

*Geographic Distribution*. *Norops sminthus* occurs at intermediate and the upper extreme of moderate elevations on both versants in Francisco Morazán and in southern Comayagua, in southern Honduras.

*Description*. The following is based on 11 males (KU 193100–02, 193121, 219961, 219963, 219965–66; SMF 77177, 78801; USNM 573160) and 11 females (KU 193111–12, 193114, 219962, 219967–68; SMF 77178, 77181, 78802; UF 89456–57). *Norops sminthus* is a moderate-sized anole (SVL 49 mm in largest male [KU 193100], 56 mm in largest female [UF 89457]); dorsal head scales smooth, rugose, or keeled in internasal, prefrontal, frontal, and parietal regions; frontal depression present; parietal depression absent; 4–7 (5.4 ± 0.7) postrostrals; anterior nasal usually divided, occasionally entire, lower section contacting rostral and first supralabial; 4–7 (5.1 ± 1.0) internasals; canthal ridge sharply defined; scales comprising supraorbital semicircles faintly keeled, largest scale in semicircles larger than largest supraocular scale; supraorbital semicircles well defined; 0–2 (1.0 ± 0.5) scales separating supraorbital semicircles at narrowest point; 2–4 (3.0 ± 0.5) scales separating supraorbital semicircles and interparietal at narrowest point; interparietal well defined, slightly to distinctly enlarged relative to adjacent scales, surrounded by scales of moderate size, longer than wide, smaller than ear opening; 2–3 rows of about 5–8 (total number) enlarged, smooth to faintly keeled supraocular scales; 1–2 enlarged supraoculars in broad contact with supraorbital semicircles; 2 elongate superciliaries, posterior only slightly shorter than anterior; usually 3 enlarged canthals; 4–7 (5.5 ± 0.7) scales between second canthals; 5–10 (6.6 ± 0.8) scales between posterior canthals; loreal region slightly concave, 12–23 (17.7 ± 2.7) mostly strongly keeled loreal scales in maximum of 4–5 (4.3 ± 0.4) horizontal rows; 6–8 (6.6 ± 0.6) supralabials and 5–7 (6.3 ± 0.5) infralabials to level below center of eye; suboculars weakly to strongly keeled, usually 2 suboculars in broad contact with supralabials; ear opening vertically oval; scales anterior to ear opening not granular, slightly larger than those posterior to ear opening; 4–6 (5.7 ± 0.7) postmentals, outer pair usually largest; keeled granular scales present on chin and throat; male dewlap small, extending to level of axilla; male dewlap with 6–8 horizontal gorgetal-sternal scale rows, about 5–8 scales per row (n = 8); 1–2 (modal number) anterior marginal pairs in male dewlap; female dewlap relatively well developed, smaller than adult male dewlap; low nuchal crest present in males, no dorsal ridge; about 9–12 middorsal scale rows distinctly enlarged, keeled, without small scales irregularly interspersed among enlarged dorsal scales; dorsal scales lateral to middorsal series abruptly larger than granular lateral scales; flank scales heterogeneous, solitary enlarged keeled or elevated scales scattered among laterals; 35–49 (42.8 ± 4.2) dorsal scales along vertebral midline between levels of axilla and groin in males, 38–53 (45.9 ± 4.8) in females; 21–30 (26.1 ± 2.8) dorsal scales along vertebral midline contained in 1 head length in males, 23–41 (29.3 ± 6.4) in females; ventral scales on midsection about same size as largest dorsal scales; midventral body scales weakly keeled, flat, imbricate; 31–48 (41.2 ± 5.3) ventral scales along midventral line between levels of axilla and groin in males, 37–54 (40.9 ± 5.2) in females; 21–28 (24.1 ± 2.6) ventral scales contained in 1 head length in males, 14–29 (22.1 ± 4.2) in females; 98–

118 (105.4 ± 5.9) scales around midbody in males, 92–118 (104.7 ± 8.7) in females; tubelike axillary pocket absent; precloacal scales faintly keeled; pair of greatly enlarged postcloacal scales in males; tail rounded to distinctly compressed, TH/TW 1.00–1.58 in 21; all subcaudal scales keeled; lateral caudal scales keeled, homogeneous, although indistinct division in segments discernable; dorsal medial caudal scale row enlarged, keeled, not forming crest; scales on anterior surface of antebrachium distinctly keeled, unicarinate; 23–29 (25.7 ± 1.7) subdigital lamellae on Phalanges II–IV of Toe IV of hind limbs; 5–10 (7.2 ± 1.4) subdigital scales on Phalanx I of Toe IV of hind limbs; SVL 40.5–49.0 (44.4 ± 3.3) mm in males, 41.0–55.5 (48.5 ± 4.2) mm in females; TAL/SVL 1.43–2.27 in six males, 1.81–2.52 in 10 females; HL/SVL 0.25–0.29 in both males and females; SHL/SVL 0.23–0.27 in males, 0.22–0.27 in females; SHL/HL 0.84–0.98 in males, 0.82–1.02 in females; longest toe of adpressed hind limb usually reaching between ear opening and mideye.

Color in life of an adult male (USNM 573160): dorsum of body Olive-Brown (28) with Drab (27) outlined rhomboid shaped blotches, blotches with Olive-Brown centers; top of head Olive-Brown; venter of head and body dirty white; dewlap Scarlet (14) with white scales. Color in life of another adult male (KU 219961): dorsal surface of head brown with rust wash; dorsal surface of body brown; dorsal surfaces of limbs rust brown with vague crossbars; tail alternately banded with pale rust brown and pale gray-brown; belly cream; dewlap red-orange. Color in life of a series: (KU 194288; adult male)-dorsum brown with tan middorsal stripe; belly pale yellow; dewlap orange with yellow scales; (KU 194289; subadult male)-dorsum brown with paler chevron markings; dewlap bright red-orange with pale yellow scales; (KU 194290; adult female)-dorsum brown; dewlap orange with yellow scales.

Color in alcohol: dorsal surfaces of head and body some shade of brown; middorsal pattern variable, some lack pattern, some have dark brown chevrons; most females with thin vertebral pale line, occasional females with broad pale middorsal stripe bordered by dark brown, or pattern of dark brown diamonds; lateral surface of head pale grayish brown; dorsal surfaces of limbs brown, usually with indistinct dark brown crossbands; dorsal surface of tail brown, some have darker brown chevrons or crossbands, those markings more evident proximally.

Hemipenis: the completely everted hemipenis of SMF 77177 is a medium-sized, slightly bilobed organ; sulcus spermaticus bordered by well-developed sulcal lips, opening into single broad concave area at base of apex; lobes calyculate; truncus with transverse folds; single asulcate processus present.

*Diagnosis/Similar Species. Norops sminthus* is distinguished from all other Honduran *Norops*, by the combination of having about 9–12 distinctly enlarged middorsal scale rows, heterogeneous lateral scales, weakly keeled and imbricate ventral scales, a red male dewlap in life, and a pair of greatly enlarged postcloacal scales in males. *Norops sminthus* differs from its apparent closest relatives, *N. heteropholidotus* and *N. muralla*, by having weakly keeled ventral scales (ventral scales perfectly smooth in *N. heteropholidotus* and *N. muralla*).

*Illustrations* (Figs. 61, 62, 99). Köhler, 1996c (dorsal, lateral, and ventral scales), 2003 (adult), 2008 (adult), 2014 (ventral scales; as *Anolis*); Köhler and Obermeier, 1998 (hemipenis); Lundberg, 2003 (adult); Diener, 2008 (adult).

*Remarks. Norops sminthus* is a member of the *N. crassulus* species subgroup (see Remarks for *N. amplisquamosus*).

The specimen identified as *N. sminthus* that was sequenced for the phylogenetic analyses in Nicholson (2002) and Nicholson et al. (2005, 2012) actually represents an undescribed species that appears most closely related to *N. crassulus* (see Remarks for *N. crassulus*). We plan to describe that population as a new species. Tissues of *N.*

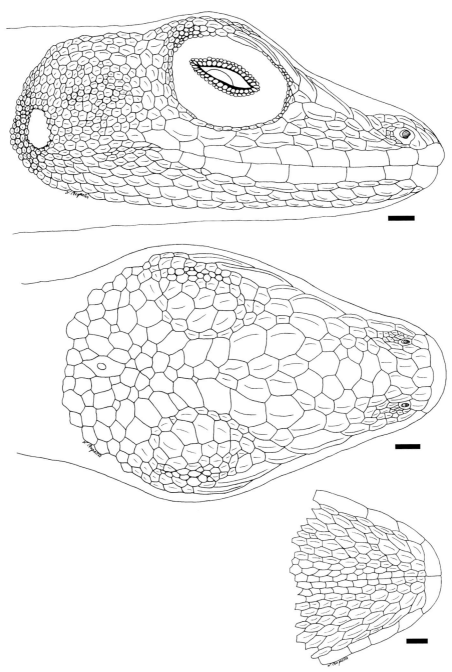

Figure 61.  *Norops sminthus* head in lateral, dorsal, and ventral views. SMF 77181, adult female from Cerro La Tigra NNE of El Hatillo, Francisco Morazán. Scale bar = 1.0 mm. Drawing by Lara Czupalla.

Figure 62.  *Norops sminthus.* (A) Adult male (SMF 91299) from Parque Nacional La Tigra Centro de Visitantes, Jutiapa, Francisco Morazán; (B) adult male dewlap (SMF 91299). Photographs by James R. McCranie.

*sminthus* were not utilized in the Nicholson (2002) and Nicholson et al. (2005, 2012) studies.

*Natural History Comments. Norops sminthus* is known from about 1,450 to 1,900 m elevation in the Lower Montane Moist Forest formation and peripherally in the Premontane Moist Forest formation. It is usually found in low vegetation or on the ground in edge situations, such as along dirt roads and streams. It is common in the clearing at the visitors' center of Parque Nacional La Tigra at Jutiapa. The species basks in the sun and is active for short periods in shaded areas, but quickly disappears during periods of cloudy conditions, even relatively short ones. It has not been found to sleep at night in exposed conditions, probably because of the cold climate where it occurs. *Norops sminthus* has been found from January to March and during June and September, but is probably active throughout the year during sunny periods. Nothing has been reported on its diet or reproduction.

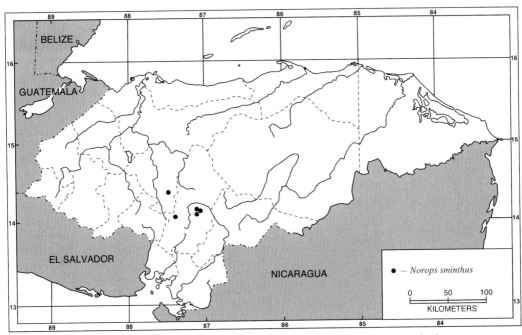

Map 31.   Localities for *Norops sminthus*. Solid symbols denote specimens examined.

*Etymology.* The name *sminthus* is formed from the Greek word *sminthos* (mouse), and refers to the hopping habits of this species according to Dunn and Emlen (1932).

*Specimens Examined* (103 [20] + 1 C&S; Map 31). COMAYAGUA: Cerro La Granadilla, UF 166266–77. FRANCISCO MORAZÁN: Cerro Cantagallo, KU 219961–64, SMF 77801–02; Cerro La Tigra NNE of El Hatillo, CM 64600, KU 194288–98, 219965–68, LSUMZ 24200–01, 24212, SMF 77176 (C&S), 77177–81 (77179–80 now in UNAH collection); El Rosario, USNM 578768; La Caballeriza, SMF 91297; La Cascada, USNM 578767; Monte Crudo, AMNH 70356, 70392–93; Parque Nacional La Tigra Centro de Visitantes, Jutiapa, SMF 91298–300, USNM 573160; Parque Nacional La Tigra near San Juancito, KU 192311–14, MCZ R-32348; Rancho Quemado, AMNH 70357–58, 70390–91, 70394; 7.2 km SW of San Juancito, KU 193100–21; 11.3 km SW of San Juancito, LACM 47676–78, LSUMZ 21415–16; San Juancito Mountains, AMNH 70396–97,

ANSP 22946–49, CM 34054, UF 89456–57, UMMZ 112330–31; near Talanga, UTA R-53991–92 (imprecise and questionable locality). "HONDURAS": AMNH 70457.

*Norops tropidonotus* (Peters)

*Anolis tropidonotus* Peters, 1863: 135 (lectotype, ZMB 382 [designated by Stuart, 1955: 27]; type locality: "aus Huanusco in Mexico"); Werner, 1896: 346; Dunn and Emlen, 1932: 26; Smith, 1950: 55; Campbell and Howell, 1965: 132; Meyer, 1969: 235; Wilson and Meyer, 1969: 146; Hahn, 1971: 111; Jackson, 1973: 309; Meyer and Wilson, 1973: 20 (in part); Fitch and Hillis, 1984: 318; O'Shea, 1986: 36, 1989: 16; Poe, 2004: 64; Nicholson et al., 2005: 933; Townsend, 2006: 35; Townsend et al., 2007: 10; Wilson and Townsend, 2007: 145; Köhler et al., 2007: 391; Townsend and Wilson, 2010b: 697; Köhler, 2014: 210.

*Norops tropidonotus*: O'Shaughnessy, 1875: 277; Espinal, 1993: table 3; Wilson and McCranie, 1998: 16, 2004b: 43; Köhler, 2000: 65; Espinal et al., 2001: 106; Köhler et al., 2001: 255, 2005b: 126; McCranie and Köhler, 2001: 232; Nicholson, 2001: 64, 2002: 120; Wilson et al., 2001: 136 (in part); McCranie and Castañeda, 2005: 14; Castañeda and Marineros, 2006: 3–8; McCranie et al., 2006: 126; Wilson and Townsend, 2006: 105; Diener, 2008: 16; Nicholson et al., 2012: 12; McCranie and Solís, 2013: 242.

*Anolis humilis*: Brattstrom and Howell, 1954: 117.

*Anolis tropidonotus tropidonotus*: Meyer, 1966: 175.

*Anolis crassulus*: Hahn, 1971: 111 (in part); Meyer and Wilson, 1973: 16 (in part).

*Norops crassulus*: Caceres, 1993: 117; Lundberg, 2003: 27.

*Geographic Distribution.* *Norops tropidonotus* occurs at low, moderate, and intermediate elevations on the Atlantic versant from central Veracruz, Mexico, to north-central Nicaragua. It also occurs on the Pacific versant in extreme northern El Salvador and southwestern to south-central Honduras. In Honduras, this species is found throughout much of the country.

*Description.* The following is based on 35 males (SMF 77505, 77729, 77736–38, 78039–40, 78396, 78705–06, 78710–12, 78831, 79059, 79095–96, 79098–100, 79102, 79249–50, 79929, 79931, 79933, 79936, 80890, 80892; UMMZ 77855 [1–6]) and 35 females (SMF 43898, 78117, 77730–31, 77735, 78427–28, 78704, 78707–09, 78804, 78806, 79055–58, 79097, 79101, 79103–04, 79152, 79248, 79251–55, 79928, 79932, 79934–35, 79947, 80891; UMMZ 77855 [1]). *Norops tropidonotus* is a medium-sized anole (SVL 59 mm in largest Honduran male examined [SMF 78040], 55 mm in largest Honduran female examined [SMF 79947]; maximum reported SVL 60 mm [McCranie and Köhler, 2001]); dorsal head scales strongly keeled in internasal, prefrontal, frontal, and parietal areas; frontal depression absent or weak; parietal depression absent; 5–8 (7.1 ± 0.6) postrostrals; anterior nasal divided, lower section contacting rostral and first supralabial; 6–9 (7.2 ± 0.8) internasals; canthal ridge sharply defined; scales comprising supraorbital semicircles keeled, largest scale in semicircles about same size as largest supraocular scale; supraorbital semicircles usually not well defined; 0–2 (1.4 ± 0.5) scales separating supraorbital semicircles at narrowest point; 1–3 (2.3 ± 0.5) scales separating supraorbital semicircles and interparietal at narrowest point; interparietal not well defined, only slightly enlarged relative to adjacent scales, surrounded by scales of moderate size, longer than wide, smaller than ear opening; 2–3 rows of about 5–10 (total number) enlarged, keeled supraocular scales; enlarged supraoculars usually in broad contact with supraorbital semicircles, occasionally completely separated from supraorbital semicircles by 1 row of small scales; 2–3 elongate superciliaries, posteriormost shortest; usually 3 enlarged canthals; 6–11 (8.0 ± 0.9) scales between second canthals; 8–16 (11.1 ± 2.0) scales between posterior canthals; loreal region slightly concave, 16–37 (26.5 ± 5.0) strongly keeled loreal scales in maximum of 4–6 (5.0 ± 0.7) horizontal rows; 4–7 (5.8 ± 0.5) supralabials and 4–7 (5.6 ± 0.6) infralabials to level below center of eye; suboculars weakly to strongly keeled, separated from supralabials by 1 row of scales; ear opening vertically oval; scales anterior to ear opening enlarged, keeled, much larger than those posterior to ear opening; 4–8 (5.5 ± 0.9) postmentals, outer pair usually largest; keeled granular scales present on chin and throat; male dewlap moderately large, extending to level of axilla; male dewlap with 6–8 oblique gorgetal-sternal scale rows, about 6–9 scales per row (n = 3); 3 (modal number) anterior marginal pairs in male dewlap; female dewlap absent; no nuchal crest or dorsal ridge; about 9–14 middorsal scale rows greatly enlarged, strongly keeled, paramedian scales larger than vertebral scales, dorsal scales lateral to middorsal series abruptly larger than granular lateral scales; no enlarged scales scattered among laterals; 15–24 (19.5 ± 2.1) dorsal scales along vertebral midline between levels of axilla and groin in 29 males, 17–26 (20.7 ± 2.1) in 34 females; 12–18 (14.8 ± 1.5) dorsal scales along vertebral midline contained in 1 head length in males, 11–16 (13.2 ± 1.3) in females; ventral scales on midsection smaller than largest dorsal scales; ventral body scales keeled, mucronate, imbricate; 22–30 (25.5 ± 2.3) ventral scales along midventral line between levels of axilla and groin in 29 males, 20–32 (24.9 ± 2.5) in 34 females; 17–22 (19.5 ± 1.7) ventral scales contained in 1 head length in 34 males, 14–22 (16.6 ± 1.9) in 34 females; 66–96 (79.0 ± 7.5) scales around midbody in 31 males, 66–88 (75.9 ± 5.5) in females; tubelike, scaleless axillary pocket present; precloacal scales not keeled; no enlarged postcloacal scales in males; tail slightly to distinctly compressed, TH/TW 1.14–1.68 in 61; all subcaudal scales keeled, mucronate; lateral caudal scales keeled, mucronate, homogeneous although indistinct division in segments discernable; dorsal medial caudal scale row not enlarged, keeled, not forming crest; most scales on anterior surface of antebrachium strongly keeled, unicarinate; 22–31 (26.5 ± 1.9) subdigital lamellae on Phalanges II–IV of Toe IV of hind limbs; 6–9 (7.8 ± 0.7) subdigital scales on Phalanx I of Toe IV of hind limbs; SVL 40.3–59.2 (50.8 ± 4.5) mm in males, 41.0–55.0 (50.2 ± 2.9) mm in females; TAL/SVL 1.34–1.84 in 17 males, 1.42–1.77 in 24 females; HL/SVL 0.25–0.28 in males, 0.24–0.30 in females; SHL/SVL 0.28–0.33 in males, 0.25–0.33 in females; SHL/HL 1.01–1.27 in males, 1.00–1.18 in females; longest toe of adpressed hind limb usually reaching between anterior border of eye and nostril.

Color in life of an adult male (USNM 565477): dorsal surface of body Brownish Olive (29) with darker brown posteriorly directed chevrons; dorsal surface of head Brownish Olive; dorsal surface of tail brown with pale brown crossbands; ventral and subcaudal surfaces dirty white, mottled with brown; dewlap Chrome Orange (16) with Geranium (12) streak, dewlap scales yellow; iris brown with pale brown rim. Color in life of another adult male (KU 220123): dorsum pale rust brown with tan stripe from axilla to groin; dorsal surfaces of limbs and tail rust brown with tan crossbars; venter white; dewlap reddish orange with burnt orange central streak and yellow scales, including marginals. Color in life of a third adult male (USNM 570095): dorsal surface of head brown with obscure darker broad interorbital bar; lateral surface of head brown; middorsal area of enlarged scales gray-brown with middorsal series of about six spots (tan with dark brown posterior edging) from just posterior to level of axilla to above vent, increasing in size posteriorly by expansion of posterior edging, each connected to pair of obscure oblique lines creating vague chevrons; lateral surface of body brown with small scattered cream spots, obscure narrow gray-brown band between axilla and groin; dorsal surfaces of forelimbs brown with rust patina; dorsal surfaces of hind limbs brown with rust patina and rust-tan crossbars on shanks; tail pale brown with obscure narrow darker brown crossbands; chin and belly cream with vague brown smudging; dewlap reddish orange with burnt orange central streak and pale yellow scales, including marginals; iris copper red with gold ring around pupil. Color in life of a fourth adult male (KU 194307): dorsum brown with darker brown middorsal markings; dewlap reddish orange with burnt orange central streak and pale yellow scales, including marginals. Color in life of an adult female (KU 194305): dorsum pale bronze-brown with wide, dark brown-bordered tan middorsal stripe; belly cream; some scales of throat with reddish orange bases. Color in life of another adult female (KU 194306): middorsal pale area scalloped and outlined by brown; dark brown interocular bar present.

Color in alcohol: dorsal and lateral surfaces of body pale brown with dark brown middorsal blotches or chevrons, those markings especially prominent in males, females usually with pale middorsal longitudinal stripe dividing dorsal markings; dorsal and lateral surfaces of head pale brown with darker brown crossbars anterior and posterior to eyes, those markings more distinct in males; dorsal surfaces of limbs pale brown with darker brown crossbands; dorsal and lateral surfaces of tail pale brown with darker brown crossbands; ventral scales pale brown to pale gray, many with dark brown flecks or small spots; scales of chin and throat pale brown, a few to many scales with darker brown flecking, flecking more distinct in males; proximal subcaudal surface pale brown, a few scales flecked with darker brown; dark brown central spot present in male dewlap.

Hemipenis: the completely everted hemipenis of SMF 78039 is a large, bilobed organ; sulcus spermaticus bordered by well-developed sulcal lips, bifurcating at base of apex, branches continuing to tips of lobes; lobes and truncus strongly calyculate; large flaplike asulcate processus present; truncus with transverse folds.

*Diagnosis/Similar Species. Norops tropidonotus* is distinguished from all other Honduran species of *Norops*, except *N. quaggulus, N. uniformis,* and *N. wampuensis,* by having a deep tubelike, scaleless axillary pocket. *Norops tropidonotus* differs from *N. quaggulus* and *N. uniformis* in having the scales anterior to ear opening much larger than those posterior to ear opening (scales anterior to ear opening similar in size to those posterior to ear opening in *N. quaggulus* and *N. uniformis*). *Norops tropidonotus* differs from *N. wampuensis* in having a distinct dark streak present in the male dewlap and a slightly larger size reaching 59 mm SVL in males and 55 mm SVL in females (distinct dark central streak absent in male dewlap and males and females reaching 51 mm SVL in *N. wampuensis*).

*Illustrations* (Figs. 63, 64, 80). Duméril et al., 1870–1909 (head scales; as *Anolis*); de Vosjoli, 1992 (adult; as *Anolis*); Fläschendräger and Wijffels, 1996 (adult; as *Anolis t. tropidonotus*); Köhler, 1996d (adult), 1999c (scales anterior and posterior to ear open-

ing), 2000 (adult, axillary pocket, head and dewlap, scales anterior and posterior to ear opening), 2001b (adult, axillary pocket, head and dewlap, scales anterior and posterior to ear opening), 2003 (adult, axillary pocket, scales anterior and posterior to ear opening, head and dewlap), 2008 (adult, axillary pocket, scales anterior and posterior to ear opening, head and dewlap), 2014 (superciliary scales, axillary pocket, scales surrounding ear opening; as *Anolis*); Lee, 1996 (adult; as *Anolis*), 2000 (adult); Campbell, 1998 (adult, head scales); Stafford and Meyer, 1999 (adult); D'Cruze, 2005 (adult, head and dewlap); Köhler et al., 2005b (adult, head scales); McCranie et al., 2006 (adult, axillary pocket); Calderón-Mandujano et al., 2008 (adult; as *Anolis*); Diener, 2008 (adult).

*Remarks. Norops tropidonotus* is usually placed in the *N. humilis* species group (Savage and Guyer, 1989; Nicholson, 2002). However, phylogenetic analyses based on molecular data rejected the monophyly of that species group (Nicholson, 2002; Nicholson et al., 2005, 2012). Also, the phylogenetic analysis in Poe (2004), based largely on morphology, did not support a close relationship between *N. tropidonotus* and *N. humilis.* Thus, we treat *N. tropidonotus* as incertae sedis within the *N. auratus* species group of Nicholson et al. (2012; also see Remarks for *N. quaggulus*).

Fitch and Hillis (1984) reported 19 specimens of *N. tropidonotus* from "Santa Barbara, Honduras." We have been unable to locate such a large series of the species from that town or department in any museums.

*Natural History Comments. Norops tropidonotus* is known from near sea level to 1,900 m elevation in the Lowland Moist Forest, Lowland Dry Forest, Lowland Arid Forest, Premontane Wet Forest, Premontane Moist Forest, and Premontane Dry Forest formations and peripherally in the Lower Montane Wet Forest and Lower Montane Moist Forest formations. It is usually found on the ground, but can be seen on fence posts and low tree trunks, and

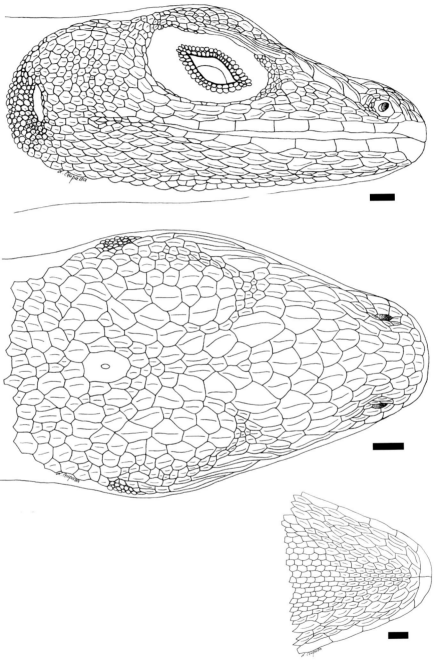

Figure 63. *Norops tropidonotus* head in lateral, dorsal, and ventral views. SMF 78705, adult male from Estación Forestal CURLA, Atlántida. Scale bar = 1.0 mm. Drawing by Lara Czupalla.

Figure 64.   *Norops tropidonotus.* (A) Adult male (KU 194310) from Finca Naranjito, Cortés; (B) adult male dewlap (KU 194310). Photographs by James R. McCranie.

is most common in open situations that have scattered low shrubby vegetation where it alternates between basking in the sun and retreating to shaded areas. Unlike many other species of *Norops*, *N. tropidonotus* does not sleep on low vegetation during the night, instead it sleeps on the ground in leaf litter. This species has been found during every month of the year. Jackson (1973) studied a midelevation pine forest population in central Honduras, and noted that densities were much higher in pine forest with extensive leaf litter from various shrubs than in pine forest that was regularly burned or grazed where abundant shrubs were absent. Males also perched higher on plant stems and spent less time on the ground than did females, with no individuals seen higher than about 2 m above the ground. Males also were seen to perch on pine tree trunks. D'Cruze (2005) studied a population in Belize and reported that individuals were

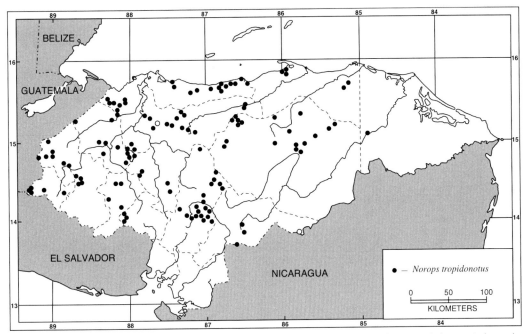

Map 32. Localities for *Norops tropidonotus*. Solid symbols denote specimens examined and open symbols accepted records.

largely terrestrial and were seen only while basking in the sun. Reproductive activity began at the end of the dry season and peaked during the rainy season. We are not aware of any published information on diet in *Norops tropidonotus*, other than the general statement that the species feeds on small insects (Villarreal Benítrez, 1997). Telford (1977) reported saurian malaria in some of the specimens collected by Jackson (1973).

*Etymology.* The name *tropidonotus* is derived from the Greek *tropos* (turn) and *notos* (back), and refers to the lateral granular scales changing abruptly to the enlarged dorsal scales in this species.

*Specimens Examined* (981 [213] + 3 skeletons; Map 32). ATLÁNTIDA: Carmelina, USNM 62970; mountains S of Corozal, LACM 47644–48, LSUMZ 21390–96, 21399; 2 km S of El Naranjal, UF 90219–21; Estación Forestal CURLA, SMF 78039–40, 78704–09, 79056–58, 79943–46, 79948–50, USNM 565476–77, 570092–94; 7.4 km SE of La Ceiba, SMF 79931–33; 8 km SE of La Ceiba, KU 101411–20; 13 km E of La Ceiba, LACM 47649, LSUMZ 21397; near La Ceiba, SMF 79152–54; La Ceiba, CM 29009, INHS 4478–84, MCZ R-32201; Lancetilla, AMNH 70432, 70435–37, ANSP 28129, FMNH 22905–09, MCZ R-29808–10 (+ 4 untagged), 31620–23, SMF 79098–102, UMMZ 70322 (3), 77841–42; about 2 km S of Nueva Armenia, KU 220123; Pico Bonito Lodge, SMF 91350–51; Piedra Pintada, LACM 47664; Quebrada de Oro, SMF 79244, 79930, USNM 539838–39, 570095; Tela, MCZ R-27577. COLÓN: Balfate, AMNH 58612–15, 58627; Barranco, ANSP 28126, 33150–54; Cerro Calentura, LSUMZ 22449, 22464, 22490; Paulaya, BMNH 1985.1125–43; Río Claura, UMMZ 58419; 2 km E of Trujillo, LACM 47650–54, LSUMZ 21398, 21400–09; 1 km SSE of Trujillo, KU 101421–22; 1 km SSW of Trujillo, KU 101423–29, MCZ R-92985; 0.5 km SW of Trujillo, LSUMZ 21410; Trujillo, CM 29367, 64615–16, 64618, 64620–21, 64623, LSUMZ 22430. COMAYAGUA: Cerro El Volcán, UF 166298;

near Río Negro, SMF 79250, 79928–29; Río Negro, UF 166306; 10 km SW of Siguatepeque, SMF 77740; 8 km N of Siguatepeque, KU 193416–40, 194893–94; Siguatepeque, UF 72116–71, 72174; Taulabé, LSUMZ 24210. COPÁN: 19.3 km ENE of Copán, LACM 47655; Copán, AMNH 63365, 70270, 124038, FMNH 28516–20, MCZ R-150314, TCWC 19191, UIMNH 52517–19, UMMZ 83042 (6), 83043 (2), 83044 (6), 83045 (4), 83048 (two of these under that number are *N. tropidonotus*, the remaining one is *N. rodriguezii*), 83049 (one of these under that number is *N. tropidonotus*, the remaining two are *N. rodriguezii*), USNM 570096; El Gobiado, USNM 573162; Quebrada Las Piedras, FMNH 282551–52; Río Amarillo, SMF 91742, USNM 570097; Río Bonito, USNM 578772; Río Higuito, ANSP 28127, 33155; San Andres, USNM 573161. CORTÉS: 6 km N of Agua Azul, AMNH 149656–57; Banaderos, UF 144105–06; Buenos Aires, SMF 77729–34, UF 144107, 149660; 3 km WSW of Cofradía, KU 67145–47; near Cofradía, MCZ R-32202, 32525; between Cofradía and Buenos Aires, SMF 77735; 1.6 km NW of El Jaral, LSUMZ 12005, 88073; 1.6 km W of El Jaral, LACM 45261–66, 47656–59; 1.6 km SE of El Jaral, LACM 47660–61, 47666; El Jaral, FMNH 5231–35; El Zapotal, SMF 78117; Finca Naranjito, KU 194305–15, LSUMZ 36592–93, 36614–15; Hacienda Santa Ana, FMNH 5229–30; Lago de Yojoa, AMNH 70289, MCZ R-29801–07, MSUM 4642–44, UF 72175, UMMZ 70323 (6); Montaña Santa Ana, MCZ R-32203–04; Naranjito, SMF 79248–49; Peña Blanca, USNM 243386; Potrerillos, MCZ R-29799–800; 3.2 km W of San Pedro Sula, LACM 47665; W of San Pedro Sula, FMNH 5227–28; San Pedro Sula, MCZ R-29798; near Santa Elena, LSUMZ 38837–38; 4.0 km ENE of Villanueva, LACM 47663. EL PARAÍSO: near Agua Fría, AMNH 70267–69; Las Manos, UTA R-41443, 52143; Mapachín, SMF 91745; Río Guayambre, AMNH 70285; Valle de Jamastrán, AMNH 70286. FRANCISCO MORAZÁN: Cantarranas, KU 192319;

Cerro Uyuca, AMNH 70287–88, 70389, 70395, CM 64602, KU 103237–38, LSUMZ 24204, 38842–44, MCZ R-49942–43, 49980–82, SMF 78427–28, 79059, 79251, UTA R-52144; El Chile, SMF 80890–93; 3.2 km NE of El Hatillo, TCWC 61761; El Hatillo, LSUMZ 24205–06, 24211, TCWC 61752–54; El Rosario, ANSP 28128; 21 km NW of El Zamorano, LSUMZ 24601; near El Zamorano, MCZ R-49978; El Zamorano, MCZ R-49758–61 (+ 1 untagged), 49979; Hacienda San Diego, AMNH 69087; La Esperanza, USNM 578771; La Montañita, AMNH 70266; La Montañuela, AMNH 70265; Los Corralitos, SMF 91301; Los Planes, USNM 581190–91; Montaña de Guaimaca, AMNH 70271; Montaña de la Sierra, UF 166307–08; Rancho Quemado, USNM 578770; Río Yeguare, AMNH 70272–79, MCZ R-48675; San Francisco, AMNH 70280–84, 142458–61; San Ignacio, SMF 82727–34; 7.2 km SW of San Juancito, KU 193441–45, 194895–99; between San Juan de Flores and Talanga, UF 90218; San Marcos de Guaimaca, SMF 79055; 15 km NW of Tegucigalpa, KU 194891–92; 9.9 km N of Tegucigalpa, LSUMZ 24181–91; 6.4 km N of Tegucigalpa, TNHC 32090; Tegucigalpa, AMNH 69088, CAS 152985, MCZ R-49922–26 (+ 2 untagged), 49933; Valle de Ángeles, UF 150155; near Zambrano, CM 59123; Zambrano, CM 59122, UTA R-53898–900. GRACIAS A DIOS: Krausirpe, LSUMZ 52494–95; Las Marías, UNAH 5436; Palacios, BMNH 1985.1123–24. INTIBUCÁ: near Jesús de Otoro, SMF 77736–39, 78396; 18.1 km NW of La Esperanza, SMF 79935–36; 15.0 km SE of La Esperanza, SMF 79938–39, 79941–42; La Rodadora, USNM 570099, 580701; Montaña de Mixcure, USNM 570098. LA PAZ: Bonilla, FMNH 283592; El Chilador, USNM 570100; 8 km S of Marcala, UF 72172, 89463–65; 13.7 km N of Marcala, SMF 79937, 79940; Marcala, CM 64603, IRSNB 12485. LEMPIRA: 4 km N and 3 km W of Gracias, UF 89466; Gracias, FMNH 40868–69, UF 135786–87; near Parque Nacional Celaque Centro de Visitantes, SMF 78710–12, 79245, 79934, 91302. OCOTEPEQUE:

El Güisayote, FMNH 283591; El Mojanal, SMF 91303; 10 km N of La Labor, UF 89467–69; Quebrada El Comatal, SMF 88674; 15 km SSW of San Marcos de Ocotepeque, UF 72173, 89470. OLANCHO: 1 km WNW of Catacamas, LACM 47622–25, LSUMZ 21381–83, 21411; 1.5–3.0 km NW of Catacamas, LACM 47626–32, 47635; 3 km N of Catacamas, LACM 47633–34; 4.5 km SE of Catacamas, LACM 45220–26, LSUMZ 21384–85; near Catacamas, KU 192315; between Catacamas and Dulce Nombre de Culmí, UTA R-53994; Cuaca, UTA R-53241–42, 53903–05, 53907–08, 53910–16; 12.1 km E of Dulce Nombre de Culmí, LACM 45153–54; El Aguacatal, UTA R-53243–66; El Carbón, SMF 91348–49; between El Díctamo and Parque Nacional La Muralla Centro de Visitantes, SMF 79096–97, USNM 342291–306; El Díctamo, SMF 79104, 79947, USNM 570101; El Vallecito, SMF 91743, USNM 342307; Guata, KU 192318; near Los Planes, USNM 342308–12; Montaña de Liquidambar, USNM 342315–17; Montaña de Las Parras, SMF 79095, USNM 342318–27; Montaña del Ecuador, USNM 342313–14; Parque Nacional La Muralla Centro de Visitantes, USNM 342345, 565473–75, 570102–05; Piedra Blanca, SMF 91744; near Piedras Negras, USNM 342360; Quebrada de Las Marías, SMF 78804–08, 78831–33; Quebrada de Las Mesetas, USNM 342328–34; Quebrada El Pinol, SMF 79096–97, 79103, USNM 342335–37; Quebrada La Calentura, USNM 342338–44; Río Cuaca, SMF 91344–47; Río de Enmedio, USNM 342351–59; 10.6 km S of San Esteban, SMF 79252; 4 km E of San José del Río Tinto, LACM 45144–45; Terrero Blanco, USNM 342346–50. SANTA BÁRBARA: Cerro Negro, USNM 565479–80, 570106; N slope of Cerro Santa Bárbara, USNM 570107; Compañia Agrícola Paradise, USNM 578778; 8.8 km SW of El Jaral, LACM 47662; El Níspero, KU 192320; El Sauce, KU 67148–50, SMF 43898; mountains W of Lago de Yojoa, KU 67151–52; Quebrada Las Cuevas, FMNH 282553–54, USNM 580304; Quimistán, USNM 128088–

89; San Pedro Zacapa, TNHC 32616–18. YORO: 2 km S of Coyoles, KU 101430–33; 5 km E of Coyoles, LACM 47636–38, LSUMZ 21386–87; Coyoles, LACM 47640–43, LSUMZ 21389; 4.8 km E of El Negrito, MVZ 171384; El Progreso, UMMZ 58387–89; 2.5 airline km NNE of La Fortuna, SMF 79253–55, UF 166302–05, USNM 565478, 570108; 2 km N of Los Guares, USNM 580305; Montañas de Mataderos, MCZ R-38798–801, UMMZ 77849 (6); Portillo Grande, FMNH 21871 (16), MCZ R-38812–15, 38816–17 (both skeletons), 38818, 38820, 38821 (skeleton), UMMZ 77847 (7); between San Francisco and La Fortuna, UF 166299–301; San Francisco, MVZ 52403–10; 38.6 km NE of Santa Rita, MVZ 171382; Sopametepe, LACM 47639, LSUMZ 21388; Subirana Valley, FMNH 21849 (25), MCZ R-38797, 38802–11, UMMZ 77855 (60); Sulaco Mountains, UMMZ 77845 (2); 10 km W of Yoro, ANSP 30689–90, 30693; 6.6 km S of Yoro, MVZ 171377, USNM 217599; 32.0 km W of Yoro, MVZ 171385, USNM 217595–98. "HONDURAS": ANSP 28125, 33111–20, 30687–88, SMF 77505, UTA R-38522–25.

*Other Records* (Map 32). EL PARAÍSO: Arenales, LACM 16850–54 (Meyer and Wilson, 1973, now lost). YORO: Mojiman Valley (Dunn and Emlen, 1932). "HONDURAS": (Werner, 1896).

*Norops uniformis* (Cope)

*Anolis uniformis* Cope, 1885: 392 (24 syntypes, MCZ R-10933, USNM 6774 (6), 24734–48, 24750, 24859 [see Barbour and Loveridge, 1929b: 224, Cochran, 1961: 92]; type locality: "Guatemala" and "Yucatan"); Townsend, 2006: 35; Köhler et al. 2007: 391.

*Norops uniformis*: Villa et al., 1988: 52; Wilson and McCranie, 1994: 418, 2004b: 43; Köhler, 2000: 64; Köhler et al., 2001: 254, 2006: 247; Wilson et al., 2001: 136; McCranie, 2005: 20; Castañeda and Marineros, 2006: 3–8; McCranie et al., 2006: 218; Wilson and Townsend, 2006:105.

*Geographic Distribution. Norops uniformis* occurs at low and moderate elevations on the Atlantic versant from southern Tamaulipas, Mexico, to north-central Honduras. In Honduras, this species occurs in the northwestern portion of the country, with an isolated population near La Ceiba in north-central Honduras tentatively assigned to this species.

*Description.* The following is based on 10 males (SMF 79130–31, 79134, 79148; UF 142460; USNM 330186, 563073–74, 563077, 565483) and 10 females (SMF 79132–33; UF 142461–62; USNM 330185, 330187–88, 563075–76, 573163). *Norops uniformis* is a small anole (SVL 41 mm in largest Honduran male examined [UF 142460], 40 mm in largest Honduran female examined [USNM 563076]; maximum reported female size 41 mm SVL [Köhler et al., 2006]); dorsal head scales keeled in internasal, prefrontal, and frontal areas, most scales keeled in parietal area; deep frontal depression present; parietal depression absent; 4–8 (6.2 ± 1.2) postrostrals; anterior nasal usually single, usually contacting only first supralabial, occasionally both rostral and first supralabial; 5–8 (6.4 ± 0.9) internasals; canthal ridge sharply defined; scales comprising supraorbital semicircles keeled, largest scale in semicircles slightly larger than largest supraocular scale; supraorbital semicircles usually not well defined; 1–3 (2.2 ± 0.6) scales separating supraorbital semicircles at narrowest point; 2–5 (2.9 ± 0.6) scales separating supraorbital semicircles and interparietal at narrowest point; interparietal usually well defined, oval in outline, smaller than ear opening; 2 rows of about 2–5 (total number) enlarged, keeled supraocular scales; enlarged supraoculars varying from completely separated from supraorbital semicircles by 1 row of small scales to 1 enlarged supraocular in broad contact with supraorbital semicircles; 2–3 elongate superciliaries, posteriormost shortest; usually 3 enlarged canthals; 6–11 (8.3 ± 1.2) scales between second canthals; 7–12 (9.1 ± 1.2) scales between posterior canthals; loreal region slightly concave, 26–49 (38.2 ± 5.9) rugose or keeled loreal scales in maximum of 5–8 (6.0 ± 0.8) horizontal rows; 7–9 (8.0 ± 0.8) supralabials and 6–10 (7.9 ± 0.9) infralabials to level below center of eye; suboculars distinctly keeled, separated from supralabials by 1 row of scales; ear opening vertically oval; scales anterior to ear opening granular, similar in size to those posterior to ear opening; 4–8 (5.9 ± 0.9) postmentals, outer pair usually largest; keeled granular scales present on chin and throat; male dewlap moderately large, extending to level of axilla; male dewlap with 6–7 horizontal gorgetal-sternal scale rows, about 7–12 scales per row (*n* = 4); 3–4 (modal number) anterior marginal pairs in male dewlap; no dewlap in females; no nuchal crest or dorsal ridge; about 8–11 middorsal scale rows distinctly enlarged, strongly keeled, dorsal scales lateral to distinctly enlarged middorsal series usually grading into granular lateral scales; no enlarged scales scattered among laterals; 23–31 (27.1 ± 2.5) dorsal scales along vertebral midline between levels of axilla and groin in males, 26–37 (29.8 ± 3.9) in females; 20–26 (22.1 ± 1.7) dorsal scales along vertebral midline contained in 1 head length in males, 17–23 (19.5 ± 2.2) in females; ventral scales on midsection much smaller than largest dorsal scales; ventral body scales keeled, imbricate; 27–52 (42.1 ± 9.0) ventral scales along midventral line between levels of axilla and groin in males, 34–49 (41.2 ± 5.2) in females; 30–42 (37.2 ± 4.6) ventral scales contained in 1 head length in males, 23–37 (29.9 ± 3.8) in females; 80–103 (93.1 ± 6.0) scales around midbody in males, 90–95 (92.7 ± 1.6) in females; deep tubelike, scaleless axillary pocket present; precloacal scales not keeled; no enlarged postcloacal scales in males; tail slightly to distinctly compressed, TH/TW 1.10–1.82 in 12; basal subcaudal scales keeled; lateral caudal scales keeled, homogeneous; dorsal medial caudal scale row not enlarged, keeled, not forming crest, although indistinct division in segments discernable; most scales on anterior surface of antebrachium distinctly keeled, unicarinate; 18–25 (21.6 ± 1.9) subdigital lamellae on Phalanges II–IV of Toe IV of hind limbs; 6–9 (7.4 ± 0.9) subdigital scales on Phalanx I of Toe IV of hind limbs; SVL 32.9–40.5 (36.9 ± 2.5) mm in males, 32.0–39.9 (37.4 ± 2.2) mm in females; TAL/SVL 1.21–1.52 in four males, 1.33–1.42 in four females; HL/SVL 0.28–0.33 in males, 0.26–0.30 in females; SHL/SVL 0.28–0.32 in males, 0.25–0.28 in females; SHL/HL 0.93–1.02 in males, 0.91–1.08 in females; longest toe of adpressed hind limb usually reaching between posterior and anterior borders of eye.

Color in life of an adult male (USNM 565482): dorsum Hair Brown (119A) with Drab-Gray (119D) narrow vertical line just posterior to midbody extending from lower body to edge of enlarged middorsal scales, irregular pale spotting also present posterior to that point; dorsum of tail Hair Brown with widely separated pale brown crossbands; forelimbs Hair Brown above; hind limbs Hair Brown above with faint slightly paler crossbands; head Hair Brown, dorsally and laterally; chin and belly Drab-Gray; dewlap Magenta (2) with central Smalt Blue (70) spot; iris rust red with thin gold line around pupil. Color in life of another adult male (USNM 330186): head olive green dorsally and laterally, with red-orange spots along lip line above and below mouth; middorsum rust brown with dark brown chevrons; lateral surface of body pale olive with scattered olive green and rust red flecks and pale yellow vertical lines posteriorly; dorsal surfaces of limbs olive green with rust patina, hind limbs also with pale rust spots; chin pale gray with scattered small white spots; belly gray-tan; dewlap Rose (9) with large, central deep purple spot; iris copper.

Color in alcohol: dorsal surface of head brown to dark brown, without distinct markings; dorsal surface of body brown to dark brown, with or without darker brown irregular blotches; lateral surface of body brown, usually with indistinct paler brown vertical stripes; dorsal surface of tail brown with indistinct darker brown crossbands; dorsal surfaces of limbs brown with indistinct darker brown crossbars; ventral surface of head white to cream, lightly to heavily flecked or mottled with brown; ventral surface of body white to cream, with brown flecking, small brown spots also sometimes present; subcaudal surface cream with dark brown flecking proximally, becoming mostly dark brown distally.

Hemipenis: the completely everted hemipenis of SMF 79131 is a medium-sized bilobed organ; sulcus spermaticus bifurcating at base of apex, branches continuing to tips of lobes; ridgelike asulcate processus present; lobes strongly calyculate; truncus with transverse folds.

*Diagnosis/Similar Species.* Norops *uniformis* is distinguished from all other Honduran species of *Norops*, except *N. quaggulus*, *N. tropidonotus*, and *N. wampuensis*, by having a deep tubelike, scaleless axillary pocket. *Norops uniformis* differs from *N. tropidonotus* and *N. wampuensis* in having the scales anterior to the ear opening similar in size to those posterior to the ear opening (scales anterior to ear opening much larger than those posterior to ear opening in *N. tropidonotus* and *N. wampuensis*). *Norops uniformis* differs from *N. quaggulus* in having a male dewlap with a large central purple spot (male dewlap without large central purple spot in *N. quaggulus*) and usually 1–3 pale vertical lines in the flank region (pale vertical lines absent in flank region in *N. quaggulus*).

*Illustrations* (Figs. 65, 66). Stafford, 1991 (adult; as *Anolis humilis uniformis*); Lee, 1996 (adult; as *Anolis*), 2000 (adult, head and dewlap); Campbell, 1998 (adult, head and dewlap); Stafford and Meyer, 1999 (adult); Köhler, 2000 (juvenile, head and dewlap), 2003 (adult, head and dewlap), 2008 (adult, head and dewlap), 2014 (scales surrounding ear opening; as *Anolis*); D'Cruze, 2005 (head and dewlap); Köhler et al., 2006 (adult, head scales).

*Remarks.* Norops *uniformis* is usually considered to be a member of the *N. humilis* species group (Savage and Guyer, 1989; Nicholson, 2002). However, phyloge-

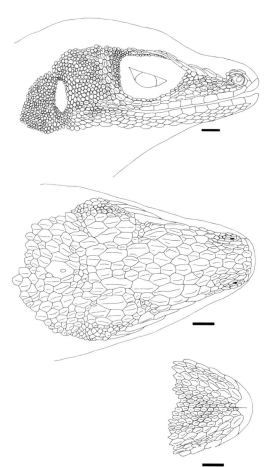

Figure 65. *Norops uniformis* head in lateral, dorsal, and ventral views. SMF 79131, adult male from 1 km SSE of Tegucigalpita, Cortés. Scale bar = 1.0 mm. Drawing by Gunther Köhler.

netic analyses based on molecular data rejected the monophyly of that group (Nicholson, 2002; Nicholson et al., 2005, 2012) as envisioned by previous workers. Thus, we place *N. uniformis* as incertae sedis in the *N. auratus* species group of Nicholson et al. (2012; also see Remarks for *N. quaggulus*).

*Natural History Comments. Norops uniformis* is known from 30 to 1,370 m elevation in the Lowland Moist Forest and Premontane Wet Forest formations and peripherally in the Lower Montane Wet Forest formation. It is usually active on the ground, but is sometimes seen perched low on tree trunks or tree buttresses. The species is usually found in deeply shaded areas, but also can be found in edge situations. *Norops uniformis* is particularly common in cacao groves SE of Tegucigalpita, Cortés. We have not seen it basking in the sun, nor have we seen it sleeping on vegetation at night. Instead, we have found it at night by disturbing leaf litter on the ground (but see Cabrera-Guzmán and Reynoso, 2010 who reported it sleeping on vegetation at night in Veracruz, Mexico). This species has been found during April, May, July, August, and October, and is probably active throughout the year. D'Cruze (2005) reported that individuals in Belize usually perched on tree buttresses or low on tree trunks. *Norops uniformis* feeds primarily on arthropods, but there is a single record of an adult feeding on a small frog (Guzmán and Reynoso, 2008, and references cited therein; also see Eifler, 1995). Female *N. uniformis* lay multiple clutches of a single egg with reproductive activity peaking during the rainy season (Lee, 1996 and references cited therein; also see Villarreal Benítez and Heras Lara, 1997).

*Etymology.* The name *uniformis* is a Latin adjective meaning "having only one shape or form, uniform." The name alludes to "the equality in size between all the scales of the frontal region, so that the supraorbitals cannot be distinguished in size" (Cope, 1885: 392).

*Specimens Examined* (51 [0]; Map 33). ATLÁNTIDA: Estación Forestal CURLA, USNM 578736–37, 578773–77. COPÁN: between Laguna del Cerro and Quebrada Grande, USNM 330185; Laguna del Cerro, SMF 79150–51; Quebrada Grande, UF 142460–61; Río Amarillo, SMF 91747; San Isidro, FMNH 282653–56, MVZ 267194, UF 142462, USNM 565481–83; Santa Rita, USNM 565484. CORTÉS: El Paraíso, UF 149656–59; Quebrada Agua Buena, SMF 79134; about 1 km SSE of Tegucigalpita, SMF 79130–34, 79148–49, USNM 330186–88, 563073–78. SANTA BÁRBARA: La Cafetalera, SMF 91304, USNM 573163–68.

Figure 66. *Norops uniformis.* (A) Adult male (FMNH 282653) from San Isidro, Copán; (B) adult male dewlap (FMNH 282653). Photographs by James R. McCranie.

*Norops unilobatus* (Köhler and Vesely)
*Anolis sallaei*: Werner, 1896: 345; Dunn and Emlen, 1932: 27 (in part); Lynn, 1944: 190.
*Anolis sericeus sericeus*: Meyer, 1966: 175 (in part).
*Anolis sericeus*: Meyer, 1969: 231 (in part); Meyer and Wilson, 1973: 19 (in part); Wilson and Hahn, 1973: 112; Lee, 1980: 311 (in part); Lee, 1983: 340.1 (in part);

O'Shea, 1986: 36; Poe, 2004: 64; Townsend, 2006: 35; Wilson and Townsend, 2007: 145 (in part); Townsend and Wilson, 2010b: 697 (in part).
*Norops sericeus*: Köhler, 1996b: 25, 1998b: 374, 1999b: 214; Wilson and McCranie, 1998: 16 (in part); Lundberg, 2000: 3; Wilson et al., 2001: 136 (in part); Castañeda, 2002: 15; McCranie et al., 2002a: 27, 2005: 110, 2006: 125; McCranie and

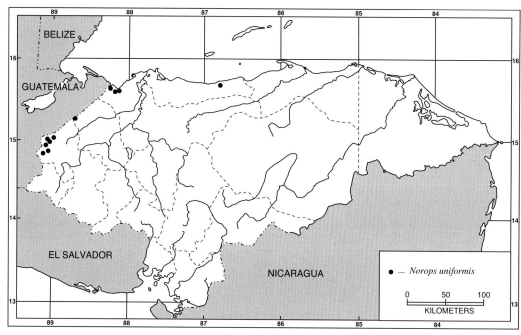

Map 33.    Localities for *Norops uniformis*. Solid symbols denote specimens examined.

Castañeda, 2005: 14; Wilson and Townsend, 2006: 105; Diener, 2008: 14;
Kramer, 2010: 5.

*Norops* (*Anolis*) *sericeus*: Monzel, 1998:
160.

*Norops* sp.: Espinal et al., 2001: 106.

*Anolis unilobatus* Köhler and Vesely, 2010:
217 (holotype SMF 87133; type locality:
"Awasbila, a village along Río Coco,
14°47′N, 84°45′W, 60 m, Departamento
Gracias a Dios, Honduras").

*Norops unilobatus*: Köhler and Vesely,
2010: 226 (in part); McCranie and Solís,
2013: 242; McCranie and Valdés Orellana, 2014: 45.

*Geographic Distribution*. *Norops unilobatus* occurs at low and moderate elevations
from Costa Rica across most of Nicaragua
(except the northwestern corner), Honduras, (except the Pacific Versant), along
the Caribbean versant of Guatemala and
Chiapas, Mexico, to the Isthmus of Tehuantepec where it crosses to the Pacific versant
and continues along the Pacific versant of

Chiapas, Mexico, to about Mazatenango,
Guatemala. In Honduras, the species occurs
in open habitats throughout the Atlantic
versant, including most interior valleys, and
on Isla de Utila.

*Description*. The following is based on 35
males (SMF 77114, 77116, 77197, 77201,
77263–64, 77763–64, 78142, 79137–38,
79366, 81495–96, 81498–500, 87125,
87127, 87129, 87131, 87133–36, 87138,
87140, 87142–43, 87150; UF 28403,
28440, 90204–05; USNM 565469) and 19
females (SMF 77112, 77115, 77198–99,
77745, 79367, 80881, 81497, 87121, 87130,
87132, 87137, 87139, 87141, 87144, 87147–
48, 87151; UF 150331). *Norops unilobatus*
is a small anole (SVL 46 mm in largest
Honduran male [SMF 87129] and female
[SMF 87147] examined; maximum reported
SVL 49 mm in males, 51 mm in females
[Köhler and Vesely, 2010]); most dorsal
head scales weakly to strongly keeled in
internasal area, smooth or rugose in prefrontal, frontal and parietal areas; deep
frontal depression present; parietal depres-

sion absent; 4–7 (5.8 ± 0.6) postrostrals; anterior nasal usually single, contacting rostral and first supralabial; 5–8 (6.6 ± 1.0) internasals; canthal ridge sharply defined; scales comprising supraorbital semicircles smooth or weakly keeled, largest scale in semicircles usually larger than largest supraocular scale; supraorbital semicircles well defined, 0–2 (0.7 ± 0.6) scales separating supraorbital semicircles at narrowest point; 1–3 (2.1 ± 0.4) scales separating supraorbital semicircles and interparietal at narrowest point; interparietal well defined, irregular in outline, longer than wide, much larger than ear opening; 2–3 rows of about 2–11 (total number) enlarged keeled supraocular scales; 1–4 enlarged supraorbitals in contact with supraorbital semicircles; a single large elongated superciliary; usually 3 enlarged canthals; 6–11 (8.3 ± 1.0) scales between second canthals; 7–12 (9.6 ± 1.2) scales between posterior canthals; loreal region slightly concave, 17–39 (26.8 ± 5.0) mostly keeled loreal scales in maximum of 4–7 (5.5 ± 0.6) horizontal rows; 6–9 (7.1 ± 0.7) supralabials and 5–8 (6.9 ± 0.6) infralabials to level below center of eye; suboculars keeled, 2–4 suboculars broadly in contact with supralabials; ear opening very small, vertically oval or round; scales anterior to ear opening not granular, slightly larger than those posterior to ear opening; 4–8 (5.9 ± 0.7) postmentals, outer pair usually largest; gular scales faintly keeled; male dewlap moderately large, extending past level of axilla onto chest; male dewlap with about 7–10 horizontal gorgetal-sternal scale rows, about 7–13 scales per row ($n$ = 4); 3 (modal number) anterior marginal pairs in male dewlap; female dewlap absent; low nuchal crest and dorsal ridge present in adult males; about 8–20 middorsal scale rows slightly enlarged, keeled, dorsal scales lateral to middorsal series grading into granular lateral scales; no enlarged scales scattered among laterals; 42–60 (52.4 ± 5.8) dorsal scales along vertebral midline between levels of axilla and groin in males, 48–60 (54.9 ± 3.9) in females; 27–34 (30.6 ±

1.9) dorsal scales along vertebral midline contained in 1 head length in males, 26–40 (30.8 ± 3.4) in females; ventral scales on midsection larger than largest dorsal scales; ventral body scales strongly keeled, mucronate, imbricate; 30–41 (35.8 ± 3.6) ventral scales along midventral line between levels of axilla and groin in males, 31–43 (36.4 ± 4.0) in females; 19–30 (25.2 ± 2.5) ventral scales contained in 1 head length in males, 16–30 (21.1 ± 3.2) in females; 92–123 (112.0 ± 9.4) scales around midbody in 23 males, 93–111 (104.1 ± 4.9) in 14 females; tubelike axillary pocket absent; precloacal scales smooth or weakly keeled; no enlarged postcloacal scales; tail rounded to distinctly compressed, TH/TW 0.69–1.53 in 50; basal subcaudal scales keeled; lateral caudal scales keeled, homogeneous; dorsal medial caudal scale row slightly enlarged, keeled, not forming crest; most scales on anterior surface of antebrachium keeled, unicarinate; 20–26 (23.4 ± 1.2) subdigital lamellae on Phalanges II–IV of Toe IV of hind limbs; 5–7 (5.9 ± 0.4) subdigital scales on Phalanx I of Toe IV on 72 sides of hind limbs; SVL 35.0–46.0 (41.6 ± 3.1) mm in males, 30.5–46.0 (41.9 ± 3.9) mm in females; TAL/SVL 2.04–2.55 in 25 males, 1.46–2.50 in 13 females; HL/SVL 0.25–0.28 in males, 0.24–0.28 in females; SHL/SVL 0.20–0.25 in males, 0.20–0.23 in females; SHL/HL 0.77–0.94 in males, 0.78–0.91 in females; longest toe of adpressed hind limb usually reaching between shoulder and ear opening, rarely beyond ear opening.

Color in life of an adult male (SMF 87127): dorsal surface of body Cinnamon (123A) with dark brown, small vertebral spots; dorsal surface of head Cinnamon; dorsal surfaces of limbs Cinnamon with dark brown mottling; belly pale yellowish brown; dewlap Trogon Yellow (153) with Ultramarine (270) central spot; iris brown with gold rim. Color in life of another adult male (KU 220122): dorsum mustard yellow; venter pale creamy yellow; lips pale creamy yellow; dewlap pale orange with royal blue spot and pale yellow scales.

Color in alcohol: dorsal surfaces of head and body brown to grayish brown, without distinct pattern in males, many females with pale brown middorsal stripe, pale middorsal stripe extending well onto tail; lateral surface of head pale brown; dorsal surfaces of limbs brown with darker brown cross-bands, at least on lower leg; dorsal surface of tail brown, usually with indistinct darker brown crossbars, at least anteriorly in those specimens in which pale middorsal stripe not extending onto tail; ventral surfaces of head and anterior third of body white or pale brown; ventral surface of midbody gray in males, gray laterally and dirty white medially in females; subcaudal surface with sparse to numerous brown flecks on scales at base, flecking becoming more prominent distally until subcaudal surface mostly brown for distal half; male dewlap with dark brown or gray central spot.

Hemipenis: the completely everted hemipenis of SMF 87133 is a small and unilobed organ; there is a distinct ridgelike structure emerging from point of bifurcation of sulcus spermaticus and reaching across tip of apex to asulcate side, this ridge with median pocket on sulcate side; sulcus spermaticus bordered by well-developed sulcal lips, bifurcating at base of apex, branches open into broad, slightly convex areas on sides of medial ridge distal to point of bifurcation; small calyculate area on asulcate surface of lower portion of apex; base of truncus without transverse folds; asulcate processus absent.

*Diagnosis/Similar Species. Norops unilobatus* differs from all Honduran species of the genus *Norops*, except *N. wellbornae*, in the following combination of characters: male dewlap yellowish orange with large blue to purple blotch; short legged (longest toe of adpressed hind limb usually reaches to between shoulder and ear opening, rarely beyond ear opening); ear opening very small (less than one-quarter size of interparietal scale); a single conspicuously large and elongate superciliary; ventral scales strongly keeled and mucronate. *Norops unilobatus* differs from *N. wellbornae* by having uni-

lobed hemipenes (hemipenes bilobed in *N. wellbornae*).

*Illustrations* (Figs. 67, 68, 112). Fitch, 1973 (adult; as *Anolis sericeus*); Campbell, 1998 (adult, head and dewlap; as *N. sericeus*); Köhler, 2000 (head and dewlap; as *N. sericeus*), 2001b (head and dewlap; as *N. sericeus*), 2003 (adult, head and dewlap; as *N. sericeus*), 2008 (adult, head and dewlap; as *N. sericeus*), 2014 (supraocular scales; as *Anolis*); Savage, 2002 (adult, dewlap; as *N. sericeus*); McCranie et al., 2005 (adult, head and dewlap; as *N. sericeus*), 2006 (adult; as *N. sericeus*); Diener, 2008 (head and dewlap; as *N. sericeus*); Köhler and Vesely, 2010 (head and dewlap, hemipenis; as *Anolis*); Kramer, 2010 (adult, male dewlap; as *Anolis sericeus*).

*Remarks.* Köhler and Vesely (2010) provided evidence that what has traditionally been considered *N. sericeus* (Hallowell) actually represents a complex of three species. *Norops sericeus* is restricted to the Atlantic versant of Mexico (Tamaulipas, Hidalgo, San Luis Potosí, Veracruz, Tabasco, Campeche, Quintana Roo, and Yucatán as well as the extreme northern portion of Oaxaca, Mexico, and Belize). *Norops wellbornae* was revalidated for the populations along the Pacific versant of Nuclear Central America (northwestern Nicaragua, extreme southern Honduras, El Salvador, and Guatemala to about Mazatenango) formerly referred to *N. sericeus*. Finally, Köhler and Vesely (2010) described the new species, *N. unilobatus*, for the remaining populations (see Geographic Distribution for *N. unilobatus*).

The single anterior nasal scale, strongly keeled ventral scales, and the conspicuously large and elongate superciliary scale along with a male dewlap that is orange-yellow with a central blue or purple blotch form a combination unique to the three named species in the *N. sericeus* subgroup of the *N. auratus* species group of Nicholson et al. (2012). Two of those species occur in Honduras. The sequence data for *N. sericeus* in Nicholson (2002) and

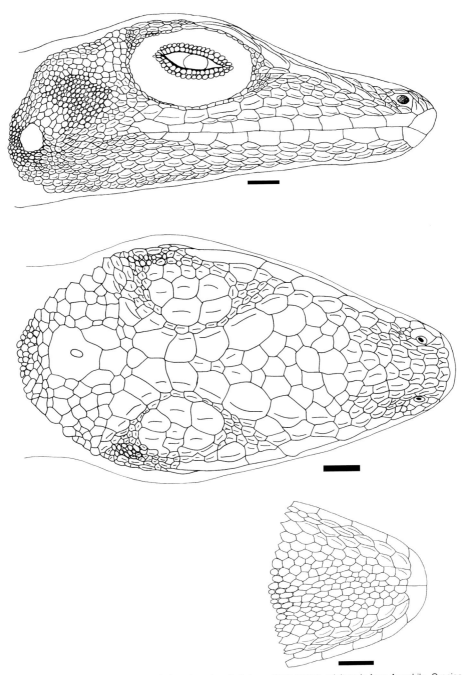

Figure 67.    *Norops unilobatus* head in lateral, dorsal, and ventral views. SMF 87133, adult male from Awasbila, Gracias a Dios. Scale bar = 1.0 mm. Drawing by Gunther Köhler.

Figure 68.  *Norops unilobatus.* (A) Adult female (SMF 87151) from Rus Rus, Gracias a Dios; (B) adult male dewlap from San Francisco, Yoro (not collected). Photographs by James R. McCranie (A) and Leonardo Valdés Orellana (B).

Nicholson et al. (2005, 2012) actually represents the species *N. unilobatus*.

*Natural History Comments. Norops unilobatus* is known from near sea level to 1,320 m elevation in the Lowland Moist Forest, Lowland Dry Forest, Lowland Arid Forest, Premontane Moist Forest, and Premontane Dry Forest formations and peripherally in the Premontane Wet Forest formation. This species occurs in open areas in the mesic forest formations. It is usually arboreal, being found in low leafy vegetation, on tree trunks up to at least 5 m above the ground, on fence posts, and rock walls used as fencerows. One was seen crossing barbed wire from one fence post to another. Occasionally, it is seen on the ground. *Norops unilobatus* sleeps at night on tree branches or limbs up to at least 5 m above the ground, and on stems, branches, and leaves of low vegetation. It has been found during every month of the year. The ecology

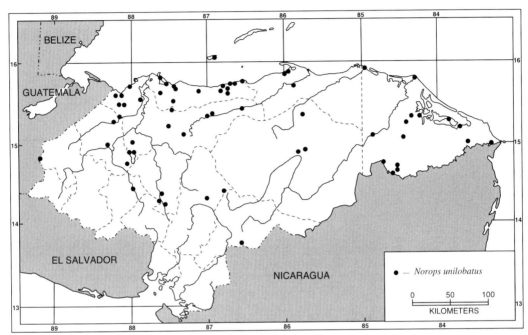

Map 34.   Localities for *Norops unilobatus*. Solid symbols denote specimens examined.

of this species has been well studied in Costa Rica, with Savage (2002) summarizing those studies. Individuals perch from about 0.3 to 3.0 m above the ground and will bask in the sun. Reproductive activity appears to be restricted to the rainy season. Apparently nothing has been published on its diet.

*Etymology.* The name *unilobatus* is formed from the Latin *unus* (one) and *lobus* (lobe) and refers to the unilobed hemipenes in this species, which distinguishes it from *Norops sericeus* and *N. wellbornae* (both with bilobed hemipenes) with which it has been confused in the past.

*Specimens Examined* (704 [101] + 1 skeleton; Map 34). ATLÁNTIDA: Corozal, LSUMZ 21432; Estación Forestal CURLA, SMF 87142; 8 km SSE of La Ceiba, KU 101406; La Ceiba, USNM 68059–60, 71721, 117607; near Lancetilla, USNM 565466; Lancetilla, AMNH 70422–29, 70431, 153436–39, MCZ R–29820–21, 29830, 31616–19, 32209, SMF 77197–98, 77200–01, UMMZ 71935 (2), 74038 (6), 172050; about 2 km S of Nueva Armenia, KU

220122; Punta Sal, SMF 81495–500; Quebrada La Muralla, SMF 77198; Ramal del Tigre, USNM 580307; Río Viejo, UTA R-41234–36; 8 km W of Sambo Creek, UF 90204; San Marcos, SMF 87128, 87134, 87140; Santiago, SMF 77199; Tela, MCZ R-27601–03, 27604 (skeleton), 27605–17, UMMZ 67696 (13). COLÓN: Jericó, LSUMZ 33684; Salamá, USNM 242057–278, 242280–443; 1–3 km W of Trujillo, KU 101407; 1 km SSW of Trujillo, KU 101408; Trujillo, CM 59121, 64613, SMF 87135. COMAYAGUA: 1.8 km SW of Comayagua, LSUMZ 24174, 24192–94; 3.7 km S of Comayagua, LSUMZ 24196–97; Las Limas, MCZ R-38824; about 7 km ESE of Villa San Antonio, LACM 72094–95. COPÁN: 1 km W of Copán, SMF 87150; Copán, FMNH 28521, 28523, UMMZ 83046 (8), 83047 (3), USNM 565467. CORTÉS: Agua Azul, AMNH 70400–01, 70503, USNM 243278–307; 3 km W of Cofradía, AMNH 149642–43, LSUMZ 24600; 1.6 km W of El Jaral, LACM 47288; 1.6 km NW of El Jaral, LSUMZ 12006–07; El Paraíso, UF 149664–

65; Hacienda Santa Ana, FMNH 5246–48; E side of Lago de Yojoa, KU 67155; Laguna Ticamaya, FMNH 5236–45; near Peña Blanca, UF 166264; Peña Blanca, USNM 243382–85; Potrerillos, MCZ R-29817–19; 27.2 km W of Puerto Cortés, SMF 79367; Río Santa Ana, FMNH 5250–52; 3.2 km W of San Pedro Sula, LACM 47289; W of San Pedro Sula, FMNH 5249; San Pedro Sula, LSUMZ 71675, USNM 24374; Santa Ana Mountains, MCZ R-32208; about 1 km SSE of Tegucigalpita, SMF 79137–38, 87139. EL PARAÍSO: between El Paraíso and Las Manos, UTA R-41264. FRANCISCO MORAZÁN: El Chile, SMF 80880–81; 1.6 km W of Talanga, TCWC 19189. GRACIAS A DIOS: Awasbila, SMF 87133, 87136; Barra Patuca, USNM 20305; Karasangkan, SMF 87147–48; Krausirpe, LSUMZ 52496–97; Mavita, SMF 91352, 91750, USNM 573155; Mocorón, UTA R-43564–65, 46169–70, 46173, 53527–30; Palacios, BMNH 1985.1122; Puerto Lempira, USNM 565471; Río Coco, USNM 24525–26; Rus Rus, SMF 87125–27, 87129, 87131, 87137, 87143–44, 87151, UF 150331, 150335, 150338, 150341–42, USNM 565468–69; Sadyk Kiamp, SMF 92846; Samil, SMF 91305, 91354; Swabila, UF 150334; Tánsin, LACM 47277–81, USNM 573156–59; Warunta, SMF 91762, USNM 565472, 573153–54. INTIBUCÁ: Valle de Otoro, SMF 77745. ISLAS DE LAS BAHÍA: Isla de Utila, on trail to Iron Bound, SMF 77114–17, 77763–64; Isla de Utila, Jake's Bight, SMF 77112; Isla de Utila, on trail to Pumpkin Hill, SMF 77113; Isla de Utila, 3.2 km N of Utila, TCWC 19190; Isla de Utila, Utila, LSUMZ 22273, SMF 78142, 79366, 79863, 87132, 87141, UF 28403, 28440. LA PAZ: La Paz, SMF 77263–64. OLANCHO: 1 km NW of Catacamas, LACM 47276; Catacamas, KU 194286; El Carbón, SMF 91353; San Esteban, UTA R-41265; Talgua, SMF 91748–49. SANTA BÁRBARA: Quimistán, USNM 128090; 1 km S of San José de Colinas, SMF 91306–07; San Pedro Zacapa, TNHC 32614–15. YORO: 0.5 km N of Coyoles, LACM 47282–83; 2 km S of Coyoles, KU 101409; Coyoles, LACM 47284–87, LSUMZ 21421–31; San José de Texíguat, USNM 580306; near San Lorenzo Abajo, USNM 565470; 4.7 km ESE of San Lorenzo Arriba, SMF 87130, 87138; San Patricio, SMF 77192, 77197; Santa Marta, MCZ R-29833; Subirana Valley, FMNH 21845 (2), MCZ R-38825–30, UMMZ 77846.

*Other Records.* "HONDURAS": (Werner, 1896).

### *Norops utilensis* Köhler

*Anolis* sp.: Köhler, 1994: 8, 1995: 97.
*Norops utilensis* Köhler, 1996a: 24 (holotype, SMF 77051; type locality: "Honduras, Islas de la Bahia, Isla de Utila, 2 km NNE of the town Utila [= Big Bight Pond]"), 1996b: 25, 1998a: 47, 1998b: 374, 2000: 64, 2003: 106, 2008: 115; Monzel, 1998: 160, 2001: 30; Lundberg, 2000: 3; Köhler et al., 2001: 254; Wilson and McCranie, 2003: 60; Gutsche et al., 2004: 297; Gutsche, 2005: 240; McCranie et al., 2005: 112; Wilson and Townsend, 2006: 105; Kramer, 2010: 9; Nicholson et al., 2012: 12; McCranie and Valdés Orellana, 2014: 45.
*Anolis utilensis*: Powell, 2003: 38; Nicholson et al., 2005: 933, 2006: 452; Köhler et al., 2007: 391; D'Angiolella et al., 2011: 38; Hallmen and Huy, 2012: 642; Pyron et al., 2013: fig. 19; Wilson et al., 2013: 66; Köhler, 2014: 210.

*Geographic Distribution.* *Norops utilensis* is known to occur only at low elevations on the eastern end of Isla de Utila, Islas de la Bahía, Honduras (see Remarks).

*Description.* The following is based on five males (SMF 77051–52, 77054, 79364–65) and four females (SMF 77053, 77055, 77983, 79866). *Norops utilensis* is a medium-sized anole (SVL 56 mm in largest male [SMF 79364], 57 mm in largest female [SMF 79866]); dorsal head scales rugose in internasal, prefrontal, and frontal areas, rugose to smooth in parietal area; shallow frontal and parietal depressions present; 5–7 (6.8 ± 0.7) postrostrals in eight; anterior nasal divided, lower scale contacting rostral and first supralabial; 5–8 (6.3 ± 1.1) internasals; canthal ridge sharply defined; scales comprising supraorbital semicircles ridged, especially anterior ones, largest scale in semicircles larger than largest supraocular scale; supraorbital semicircles well defined; 1 scale in each supraorbital semicircle in contact medially; 1–3 (1.9 ± 0.7) scales between supraorbital semicircles and interparietal; interparietal well defined, irregular in outline, longer than wide, larger than size of ear opening; 2 rows of about 3–5 (total number) enlarged, smooth to weakly keeled supraocular scales; enlarged supraoculars completely separated from supraorbital semicircles; 3 short superciliaries, posteriormost shortest; 3 enlarged canthals; 7–9 (7.7 ± 0.7) scales between second canthals; 8–11 (9.4 ± 0.9) scales between posterior canthals; loreal region concave; 22–32 (26.6 ± 3.8) smooth loreal scales in maximum of 4–5 (4.1 ± 0.3) horizontal rows; 8–10 (9.1 ± 0.9) supralabials and 7–11 (9.4 ± 1.7) infralabials to level below middle of eye; suboculars smooth to rugose, in broad contact with supralabials; ear opening vertically oval; scales anterior to ear opening slightly larger than, to about same size as, those posterior to ear opening; 4–7 (5.8 ± 0.8) postmentals, outer pair largest; gular scales smooth; male dewlap large, extending to level posterior to axilla onto body; male dewlap with 6–7 horizontal gorgetal-sternal scale rows, with about 6–12 scales per row (*n* = 3); 3–4 (modal number) anterior marginal pairs in male dewlap; female dewlap well developed, extending to level of axilla; female dewlap with 4–7 horizontal gorgetal-sternal scale rows, with about 7–15 scales per row (*n* = 4); 3–4 (modal number) anterior marginal pairs in female dewlap; no nuchal crest or dorsal ridge; about 2 middorsal scale rows slightly enlarged, dorsal scales mostly smooth, juxtaposed, dorsal scales lateral to middorsal series grading into granular lateral scales; no enlarged scales among laterals; 78–96 (90.0 ± 7.2) dorsal scales along vertebral midline between levels of axilla and groin in males,

98–110 (102.3 ± 5.7) in females; 60–68 (64.8 ± 3.0) dorsal scales along vertebral midline contained in 1 head length in males, 56–68 (61.0 ± 5.3) in females; ventral scales on midsection slightly larger than largest dorsal scales; ventral body scales smooth, obliquely conical, juxtaposed; 56–80 (70.2 ± 9.4) ventral scales along midventral line between levels of axilla and groin in males, 65–86 (75.5 ± 9.3) in females; 44–56 (50.4 ± 4.3) ventral scales contained in 1 head length in males, 48–52 (49.5 ± 1.9) in females; 140–188 (162.4 ± 18.5) scales around midbody in males, 146–160 (151.5 ± 6.0) in females; tubelike axillary pocket absent; precloacal scales not keeled; enlarged postcloacal scales absent; tail oval to distinctly compressed, TH/TW 0.90–1.83; basal subcaudal scales smooth to faintly keeled, except 2 median rows distinctly enlarged, keeled; lateral caudal scales keeled, homogeneous; dorsal medial caudal scale row enlarged, keeled, forming low crest anteriorly; 5 supracaudals present per caudal segment; most scales on anterior surface of antebrachium smooth; 18–23 (20.6 ± 1.4) subdigital lamellae on Phalanges II–IV of Toe IV of hind limbs (those proximal ones in Phalanx IV small, in more than one row); 6–8 (6.9 ± 0.5) subdigital scales on Phalanx I of Toe IV of hind limbs; SVL 51.0–56.0 (53.4 ± 1.9) mm in males, 52.0–57.0 (54.3 ± 2.2) mm in females; TAL/SVL 1.44–1.69 in four males, 1.28–1.56 in females; HL/SVL 0.23–0.28 in males, 0.24–0.28 in females; SHL/SVL 0.19–0.23 in males, 0.19–0.22 in females; SHL/HL 0.77–0.91 in males, 0.77–0.82 in females; longest toe of adpressed hind limb usually reaching shoulder region.

Color in life of an adult male (SMF 79364): dorsal surfaces of body and limbs Light Drab (119C) with dirty white blotches and longitudinal streaks, also with Hair Brown (119A) to Sepia (119) blotches; dorsal and lateral surfaces of tail Light Drab with Sepia banding; venter Pale Horn Color (92); dewlap Geranium Pink (13) with six rows of white gorgetals, color between gorgetals suffused with Maroon (31). Color in life of an adult female (USNM Herp Image 2723, specimen lost): dorsal surfaces of head and body mottled Smoke Gray (44), Drab (27), and whitish gray; dorsolateral stripes whitish gray, outlined above and below by Smoke Gray and Drab mottling; ventral surfaces whitish gray; dewlap Gem Ruby (110); iris pale brown with golden brown ring around pupil.

Color in alcohol: dorsal surface of head brown with paler brown mottling; dorsal surface of body brown with pale brown lichenose pattern, pattern especially prominent laterally, most males not as heavily patterned as females; dorsal surfaces of limbs brown with pale brown lichenose pattern; dorsal surface of tail brown with pale brown crossbands; ventral surfaces of head and body white, without distinct markings; dewlap retaining red pigment on skin; subcaudal surface white with brown mottling proximally, brown with dark brown crossbands distally; throat lining grayish black.

Hemipenis: the completely everted hemipenis of SMF 79365 is a medium-sized, bilobed organ; sulcus spermaticus bordered by well-developed sulcal lips, bifurcating at base of apex, branches continuing to tips of lobes; lobes calyculate; truncus with transverse folds; asulcate ridge present.

*Diagnosis/Similar Species. Norops utilensis* is distinguished from all other Honduran *Norops*, except *N. beckeri*, by having the proximal subdigital scales of the fourth toe of the hind limbs only slightly larger than granular, relatively short hind legs with the longest toe of adpressed hind limb usually only reaching past shoulder region, and smooth, obliquely conical, and juxtaposed ventral scales. *Norops beckeri* has four supracaudals present per caudal segment (five supracaudals present per caudal segment in *N. utilensis*).

*Illustrations* (Figs. 69, 70, 82, 104, 107, 109). Köhler, 1996a (adult, head scales, subdigital lamellae, dewlap), 2000 (adult, ventral scales, subdigital lamellae, head and dewlap), 2003 (adult, ventral scales, subdigital lamellae, head and dewlap), 2008 (adult, ventral scales, subdigital lamellae, head and dewlap), 2014 (dorsal and ventral scales; as *Anolis*); Monzel, 2001 (adult); Powell, 2003 (adult; as *Anolis*); Gutsche et al., 2004 (juvenile); Gutsche, 2005 (adult, juvenile); McCranie et al., 2005 (adult, head and dewlap, head scales, midventral scales); Kramer, 2010 (adult; as *Anolis*).

*Remarks.* The central and western portions of Isla de Utila have not been surveyed extensively for its herpetofauna. It seems likely that such a survey will reveal *N. utilensis* to occur on those portions of the island.

The phylogenetic analysis, based on molecular data, of Nicholson et al. (2005) recovered *N. utilensis* as a subclade nested between a subclade consisting of *N. loveridgei* and *N. purpurgularis* and a subclade of six mostly South American small to moderate-sized species. The phylogenetic analysis of Nicholson et al. (2012), also based on molecular data, recovered *N. utilensis* as part of a subclade of seemingly unrelated species from South America and Mexico. *Norops utilensis* is similar in external morphology to the mainland *N. beckeri* and *N. pentaprion* and we include it in the *N. pentaprion* species subgroup (see Remarks for *N. beckeri*). Poe et al. (2009), using morphological and molecular data, also did not recover a close relationship between *N. utilensis* and *N. pentaprion*. The unique characters among the Mexican and Central American anoles of having granular proximal subdigital scales and a lichenous dorsal pattern in these three species questions the results of those phylogenetic analyses. Therefore, we keep *N. beckeri* and *N. utilensis* in the *N. pentaprion* species group (Myers, 1971b) or *N. pentaprion* subgroup of the *N. auratus* species group (Nicholson et al., 2012).

*Natural History Comments. Norops utilensis* is known from near sea level to 8 m elevation in the Lowland Moist Forest formation. The type series was collected during April in mangrove trees between 1.5–6.0 m above the ground (Köhler, 1996a). It was also active in mangrove trees

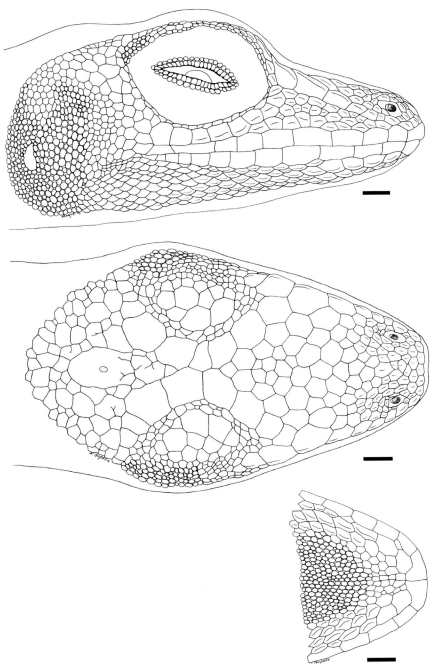

Figure 69.   *Norops utilensis* head in lateral, dorsal, and ventral views. SMF 79364, adult male from 2 km NNE of Utila, Isla de Utila, Islas de la Bahía. Scale bar = 1.0 mm. Drawing by Lara Czupalla.

Figure 70.   *Norops utilensis.* (A) Adult female from Big Bight Pond, Isla de Utila, Islas de la Bahía (not collected); (B) adult male dewlap (from series SMF 79364–65) from 2 km N of Utila town on trail to Rock Harbour, Isla de Utila, Islas de la Bahía. Photographs by James R. McCranie (A) and Gunther Köhler (B).

during May and October. One was "hugging" a mangrove limb about 4 m above the water before it slipped to the opposite side of where McCranie was standing. However, after a 5 min wait, the animal reappeared in the same general area. Hallmen and Huy (2012) reported finding a *N. utilensis* sitting on a fence post about 1.5 m above the ground at the edge of a dirt road through broadleaf forest about 250 m from the nearest mangrove swamp. Gutsche et al. (2004) presented information on an aggregate egg-laying site in a hollow limb of a red mangrove tree (*Rhizophora mangle*), first found in October, with at least one female having recently deposited an egg there. Those authors reported on a successful incubation of three of these eggs and considered *N.*

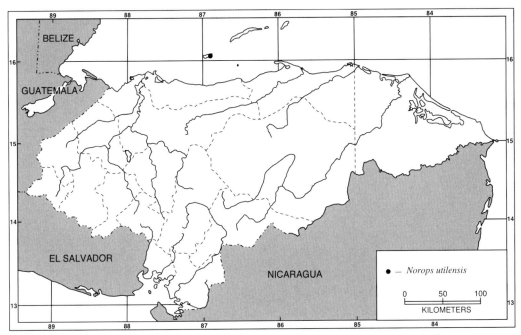

Map 35.   Localities for *Norops utilensis*. The solid symbol denotes specimens examined.

*utilensis* to be a critically endangered species. Two fertile eggs were found at that site during March two years later. Nothing has been published on its diet.

*Etymology.* The name *utilensis* is formed from Utila (the island to which the species is restricted) and the Latin suffix *-ensis* (denoting place, locality).

*Specimens Examined* (12 [0]; Map 35). ISLAS DE LA BAHÍA: Isla de Utila, 0.5 km NNE of Utila, SMF 83427; Isla de Utila 2 km NNE of Utila, SMF 77051–55, 77983, 79364–65, 79866, UNAH (2 unnumbered).

*Other Records.* ISLAS DE LA BAHÍA: Isla de Utila, Big Bight Pond, (USNM Herp Image 2723, but specimen lost by AFE COHDEFOR employee); Isla de Utila, near Pumpkin Hill (Hallmen and Huy, 2012).

**Norops wampuensis** McCranie and Köhler
*Norops tropidonotus*: Wilson et al., 2001: 136 (in part).
*Norops wampuensis* McCranie and Köhler, 2001: 228 (holotype, SMF 79902; type locality: "the confluence of Ríos Aner and Wampú, 15°03′N, 85°07′W, 110 m elevation, Departamento de Olancho, Honduras"); Köhler, 2003: 106, 2008: 114; Wilson and McCranie, 2003: 60; McCranie et al., 2006: 127; Wilson and Townsend, 2006: 105.

*Geographic Distribution.* Norops wampuensis occurs at low elevations near the Río Wampú in northeastern Olancho, Honduras.

*Description.* The following is based on seven males (SMF 79246, 79902, 79920–21; USNM 539193, 539195, 539197) and 16 females (SMF 79247, 79903, 79905–07, 79909–11, 79916–19, 79922, 79925; USMN 539194, 539196). *Norops wampuensis* is a moderate-sized anole (SVL 51 mm in largest male [SMF 79246] and largest female [USNM 539196]); dorsal head scales strongly keeled in internasal, prefrontal, frontal, and parietal areas; frontal depression absent or weak; parietal depression absent; 6–7 (6.7 ± 0.5) postrostrals; anterior nasal divided, lower section contacting rostral and first supralabial; 6–9 (6.9 ± 0.8) internasals; canthal ridge sharply defined; scales comprising supraorbital semicircles keeled, largest scale in semicircles about same size as largest supraocular scale; supraorbital semicircles usually not well defined; 1–2 (1.6 ± 0.5) scales separating supraorbital semicircles at narrowest point; 1–3 (2.0 ± 0.4) scales separating supraorbital semicircles and interparietal at narrowest point; interparietal not well defined, only slightly enlarged relative to adjacent scales, surrounded by scales of moderate size, longer than wide, smaller than ear opening; 2–3 rows of about 2–8 (total number) enlarged, keeled supraocular scales; enlarged supraoculars usually in broad contact with supraorbital semicircles, occasionally completely separated from supraorbital semicircles by 1 row of small scales; 2–3 elongate superciliaries, posteriormost shortest; 3–4 enlarged canthals; 6–15 (10.0 ± 2.6) scales between second canthals; 7–12 (9.4 ± 1.3) scales between posterior canthals; loreal region slightly concave, 16–30 (24.4 ± 4.1) strongly keeled loreal scales in maximum of 4–6 (4.8 ± 0.7) horizontal rows; 5–7 (6.3 ± 0.7) supralabials and 5–7 (6.2 ± 0.5) infralabials to level below center of eye; suboculars weakly to strongly keeled, separated from supralabials by 1 row of scales; ear opening vertically oval; scales anterior to ear opening broad, keeled, much larger than those posterior to ear opening; 4–6 (4.6 ± 0.8) postmentals, outer pair usually largest; keeled granular scales present on chin and throat; male dewlap moderately large, extending to level of axilla; male dewlap with 7–8 oblique gorgetal-sternal scale rows, about 9–11 scales per row (n = 1); 3 (modal number) anterior marginal pairs in male dewlap; female dewlap absent; no nuchal crest or dorsal ridge; about 9–12 middorsal scale rows greatly enlarged, strongly keeled, paramedian scales larger than vertebral scales, dorsal scales lateral to middorsal series abruptly larger than granular lateral scales; no enlarged scales scattered among laterals; 18–23 (21.1 ± 1.8) dorsal scales along vertebral midline between levels of

axilla and groin in males, 14–25 (19.8 ± 3.0) in females; 10–18 (14.9 ± 2.6) dorsal scales along vertebral midline contained in 1 head length in males, 12–18 (14.5 ± 1.7) in females; ventral scales on midsection smaller than largest dorsal scales; ventral body scales keeled, mucronate, imbricate; 22–29 (26.0 ± 2.3) ventral scales along midventral line between levels of axilla and groin in males, 21–30 (25.6 ± 2.9) in females; 19–26 (21.0 ± 2.4) ventral scales contained in 1 head length in males, 16–22 (19.4 ± 2.3) in females; 58–68 (64.6 ± 3.2) scales around midbody in males, 56–86 (70.8 ± 7.9) in females; deep tubelike, scaleless axillary pocket present; precloacal scales not keeled; no enlarged postcloacal scales in males; tail nearly rounded to distinctly compressed, TH/TW 0.80–1.50 in 17; all subcaudal scales keeled, mucronate; lateral caudal scales keeled, mucronate, homogeneous, although indistinct division in segments discernable; dorsal medial caudal scale row not enlarged, keeled, not forming crest; most scales on anterior surface of antebrachium strongly keeled, unicarinate; 24–30 (26.9 ± 1.7) subdigital lamellae on Phalanges II–IV of Toe IV of hind limbs; 6–9 (7.6 ± 0.8) subdigital scales on Phalanx I of Toe IV of hind limbs; SVL 25.0–51.0 (39.7 ± 9.8) mm in males, 25.0–51.3 (37.8 ± 9.0) mm in females; TAL/SVL 1.44–1.73 in six males, 1.42–1.64 in nine females; HL/SVL 0.24–0.29 in males, 0.23–0.28 in females; SHL/SVL 0.30–0.32 in males, 0.28–0.32 in females; SHL/HL 1.07–1.29 in males, 1.03–1.31 in females; longest toe of adpressed hind limb usually reaching between anterior border of eye and nostril.

Color in life of an adult male paratype (USNM 539197) was described by McCranie and Köhler (2001: 232): "dorsal and lateral surfaces of body rusty brown with dark brown chevrons across middorsal area; dorsal and lateral surfaces of head rusty brown with dark brown crossbar extending from anterior portion of supraorbital disc to about level of first canthal, dark brown pigment also present on posterior portion of supraorbital disc and adjacent supraorbital semicircles; dorsal and lateral surfaces of tail rusty brown with dark brown crossbands; dorsal surfaces of limbs brown with darker brown crossbars; belly grayish white; subcaudal surface dirty white; dewlap Pratt's Rufous (color 140) with a yellow outer edge; dewlap scales pale yellow, flecked with brown." Those same authors (p. 232) also stated "Dewlap color of another adult male (USNM 539195) was recorded as Scarlet (color 14) with a pale yellow outer edge and pale yellow dewlap scales."

Color in alcohol: dorsal and lateral surfaces of body pale brown with dark brown middorsal blotches or chevrons, those markings especially prominent in males, females with pale middorsal longitudinal stripe dividing dorsal markings; dorsal and lateral surfaces of head pale brown with darker brown crossbars anterior and posterior to eyes, those markings more distinct in males; dorsal surfaces of limbs pale brown with darker brown crossbands; dorsal and lateral surfaces of tail pale brown with darker brown crossbands; ventral scales pale brown to pale gray, many with dark brown flecks or small spots; scales of chin and throat pale brown, a few to many scales with darker brown flecking, flecking more distinct in males; proximal subcaudal surface pale brown, a few scales flecked with darker brown.

Hemipenis: unknown.

*Diagnosis/Similar Species. Norops wampuensis* is distinguished from all other Honduran species of *Norops*, except *N. quaggulus*, *N. tropidonotus*, and *N. uniformis*, by having a deep tubelike, scaleless axillary pocket. *Norops wampuensis* differs from *N. quaggulus* and *N. uniformis* in having the scales anterior to the ear opening much larger than those posterior to the ear opening (scales anterior to ear opening similar in size to those posterior to ear opening in *N. quaggulus* and *N. uniformis*). *Norops wampuensis* differs from *N. tropidonotus* in lacking a distinct dark central streak in the male dewlap and being a smaller species with males and females reaching 51 mm SVL (distinct dark central streak present in male

dewlap and males reaching 59 mm SVL and females 55 mm SVL in *N. tropidonotus*).

*Illustrations* (Figs. 71, 72). McCranie and Köhler, 2001 (adult, head scales, head and dewlap); Wilson and McCranie, 2004a (adult); McCranie et al., 2006 (adult).

*Remarks. Norops wampuensis* is included herein as incertae sedis in the *N. auratus* species group of Nicholson et al. (2012; see Remarks for *N. quaggulus*). Tissues of *N. wampuensis* were not available for the molecular studies of Nicholson et al. (2012). The species is extremely close to *N. tropidonotus*. Were it not for the extreme differences in habitat between the two nominal forms (see below), we would consider them conspecific.

*Natural History Comments. Norops wampuensis* is known from 95 to 110 m elevation in the Lowland Moist Forest formation. All specimens were collected inside undisturbed broadleaf rainforest in August and September 1992, with the exception of two collected in July 1994 at the edge of a cornfield recently carved from that primary forest. Lizards were active on the forest floor and in low leafy vegetation (to about 1.5 m above the ground). However, most were sleeping at night on leaves and stems of low vegetation (to about 1.5 m above the ground; JRM, pers. oberv.). On the other hand, we have never found the closely related *N. tropidonotus* sleeping on leaves or stems of vegetation at night, but have only found it sleeping in leaf litter on the ground. *Norops tropidonotus* has also never been found in undisturbed broadleaf rainforest in Honduras. Nothing has been reported on diet or reproduction in *N. wampuensis*.

*Etymology.* The specific name *wampuensis* is derived from the Río Wampú and the Latin suffix -*ensis* (denoting place or locality). This species is known only from a few localities near the Río Wampú, Olancho.

*Specimens Examined* (32 [0]; Map 36). OLANCHO: confluence of Quebrada Siksatara and Río Wampú, SMF 79246–47, 79906, USNM 539197; confluence of Ríos Aner and Wampú, SMF 79902–05, 79916–24, 79926, USNM 539193–95; confluence of

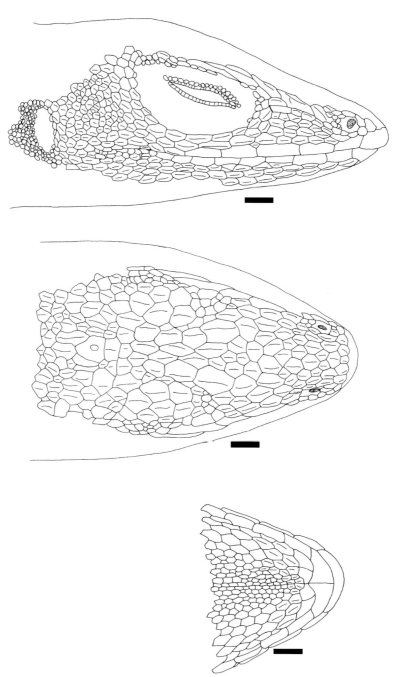

Figure 71.   *Norops wampuensis* head in lateral, dorsal, and ventral views. SMF 79902, adult male from confluence of Ríos Aner and Wampú, Gracias a Dios. Scale bar = 1.0 mm. Drawing by Gunther Köhler.

Figure 72. *Norops wampuensis.* (A) Adult male (USNM 539193) from confluence of Ríos Aner and Wampú, Olancho; (B) adult male dewlap (USNM 539197) from confluence of Quebrada Siksatara with Río Wampú, Gracias a Dios. Photographs by James R. McCranie.

Ríos Sausa and Wampú, USNM 539196; along Río Wampú between Ríos Aner and Sausa, SMF 79907–15, 79925.

Norops wellbornae (Ahl)
*Anolis sallaei*: Dunn and Emlen, 1932: 27 (in part).
*Anolis ustus wellbornae* Ahl, 1939: 246 (holotype ZMB 35710 [Stuart, 1955: 26; Köhler and Vesely, 2010: 215]; type locality: "El Salvador").

*Anolis sericeus sericeus*: Meyer, 1966: 175 (in part).
*Anolis sericeus*: Meyer, 1969: 231 (in part); Hahn, 1971: 111; Meyer and Wilson, 1973: 19 (in part); Lee, 1983: 340.1 (in part); Wilson et al., 1991: 70; Wilson and Townsend, 2007: 145 (in part); Townsend and Wilson, 2010b: 697 (in part).
*Norops sericeus*: Wilson and McCranie, 1998: 16 (in part); Wilson et al., 2001: 136 (in part); Lovich et al., 2006: 14.

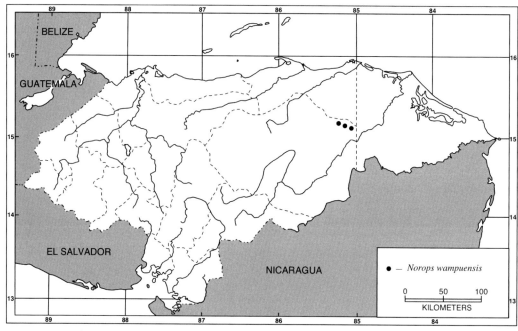

Map 36.   Localities for *Norops wampuensis*. Solid symbols denote specimens examined.

*Anolis wellbornae*: Köhler and Vesely, 2010: 215.

*Norops unilobatus*: Köhler and Vesely, 2010: 226 (in part).

*Norops wellbornae*: Köhler and Vesely, 2010: 227.

*Geographic Distribution. Norops wellbornae* occurs at low and moderate elevations on the Pacific versant from about Mazatenango, Guatemala, to northwestern Nicaragua.

*Description.* The following is based on 10 males (KU 194287; SDSNH 72751, 72753; SMF 78904; 87124; USNM 580294–95, 580298–99, 580705) and 10 females (CAS 152958; SDSNH 72752; SMF 79140, 87145, 91731–32; USNM 580296–97, 580723, 580730). *Norops wellbornae* is a moderately small anole (SVL 54 mm in largest Honduran male examined [USNM 580298], 49 mm in largest Honduran female examined [USNM 580297]); most dorsal head scales weakly to strongly keeled in internasal area, smooth or rugose in prefrontal, frontal

and parietal areas; deep frontal depression present; parietal depression absent; 3–6 (5.2 ± 1.0) postrostrals; anterior nasal usually single, contacting rostral and first supralabial; 4–7 (5.6 ± 0.9) internasals; canthal ridge sharply defined; scales comprising supraorbital semicircles smooth or weakly keeled, largest scale in semicircles usually larger than largest supraocular scale; supraorbital semicircles well defined, 1–3 (1.4 ± 0.6) scales separating supraorbital semicircles at narrowest point; 1–4 (2.5 ± 0.5) scales separating supraorbital semicircles and interparietal at narrowest point; interparietal well defined, irregular in outline, longer than wide, much larger than ear opening; 2–3 rows of about 3–8 (total number) enlarged keeled supraocular scales; 1–4 enlarged supraorbitals in contact with supraorbital semicircles; a single large elongated superciliary; usually 3 enlarged canthals; 7–10 (8.4 ± 0.8) scales between second canthals; 7–12 (8.9 ± 1.4) scales between posterior canthals; loreal region slightly concave, 27–41 (33.1 ± 3.8) mostly keeled loreal scales in maximum of 5–6 (5.2 ± 0.5) horizontal rows; 6–9 (7.2 ± 0.9) supralabials and 6–9 (7.2 ± 1.0) infralabials to level below center of eye; suboculars keeled, 2–4 suboculars broadly in contact with supralabials; ear opening very small, vertically oval or round; scales anterior to ear opening not granular, slightly larger than those posterior to ear opening; 6 postmentals, outer pair usually largest; gular scales faintly keeled; male dewlap moderately large, extending past level of axilla onto chest; male dewlap with about 5–9 horizontal gorgetal-sternal scale rows, about 5–10 scales per row (*n* = 7); 3 (modal number) anterior marginal pairs in male dewlap; female dewlap absent; low nuchal crest and dorsal ridge present in adult males; about 6–21 middorsal scale rows slightly enlarged, keeled, dorsal scales lateral to middorsal series grading into granular lateral scales; no enlarged scales scattered among laterals; 50–67 (58.2 ± 5.6) dorsal scales along vertebral midline between levels of axilla and groin in males, 46–70 (60.6 ± 7.0) in females; 28–37 (31.9 ± 3.5) dorsal scales along vertebral midline contained in 1 head length in males, 28–41 (34.2 ± 4.8) in females; ventral scales on midsection larger than largest dorsal scales; ventral body scales strongly keeled, mucronate, imbricate; 43–51 (47.3 ± 3.3) ventral scales along midventral line between levels of axilla and groin in males, 34–49 (43.0 ± 5.4) in females; 22–29 (25.9 ± 2.3) ventral scales contained in 1 head length in males, 18–29 (23.1 ± 3.2) in females; 109–134 (120.6 ± 7.8) scales around midbody in males, 98–116 (106.5 ± 6.3) in females; tubelike axillary pocket absent; precloacal scales smooth or weakly keeled; no enlarged postcloacal scales; tail nearly rounded to distinctly compressed, TH/TW 0.73–1.80 in 19; basal subcaudal scales keeled; lateral caudal scales keeled, homogeneous; dorsal medial caudal scale row slightly enlarged, keeled, not forming crest; most scales on anterior surface of antebrachium keeled, unicarinate; 20–28 (23.5 ± 2.0) subdigital lamellae on Phalanges II–IV of Toe IV of hind limbs; 5–8 (6.2 ± 0.8) subdigital scales on Phalanx I of Toe IV of hind limbs; SVL 43.0–54.0 (46.9 ± 4.4) mm in males, 31.0–49.2 (43.2 ± 5.1) mm in females; TAL/SVL 1.54–2.28 in seven males, 1.15–2.23 in eight females; HL/SVL 0.25–0.27 in males, 0.22–0.28 in females; SHL/SVL 0.20–0.24 in both males and females; SHL/HL 0.81–0.91 in males, 0.80–1.00 in females; longest toe of adpressed hind limb usually reaching between shoulder and ear opening, rarely beyond ear opening.

Color in life of an adult male (USNM 580294): dorsal surface of body Cinnamon (123A) with indistinct darker brown, small vertebral spots; dorsal surface of head Cinnamon with indistinct darker brown flecking; dorsal surfaces of limbs Cinnamon with dark brown mottling; belly pale yellowish brown; dewlap Trogon Yellow (153) with Ultramarine (270) central spot; iris brown with gold rim.

Color in alcohol: dorsal surfaces of head and body brown to grayish brown, without distinct pattern in males, many females with pale brown middorsal stripe, pale middorsal stripe extending well onto tail; lateral surface of head pale brown; dorsal surfaces of limbs brown with darker brown crossbands, at least on lower leg; dorsal surface of tail brown, usually with indistinct darker brown crossbars, at least proximally in those specimens in which pale middorsal stripe not extending onto tail; ventral surfaces of head and anterior third of body white or pale brown; ventral surface of midbody gray in males, gray laterally and dirty white medially in females; subcaudal surface with sparse to numerous brown flecks on scales at base, flecking becoming more prominent distally until subcaudal surface mostly brown for distal half; male dewlap with dark brown or gray central spot.

Hemipenis: the completely everted hemipenis of SMF 82667 (from Departamento Suchitepéquez, Guatemala) is a large bilobed organ; sulcus spermaticus bordered by well-developed sulcal lips, bifurcating at base of apex, branches opening into broad concave area distal to point of bifurcation on each lobe; truncus relatively long, longer than, or equal to, length of each lobe; asulcate surface of apex strongly calyculate, base of truncus without transverse folds; asulcate processus absent, although distinct ridge present on asulcate side.

*Diagnosis/Similar Species.* *Norops wellbornae* differs from all Honduran species of this genus *Norops*, except *N. unilobatus*, in the following combination of characters: male dewlap yellowish orange with large blue to purple blotch; short legged (longest toe of adpressed hind limb usually reaches to between shoulder and ear opening, rarely beyond ear opening); ear opening very small (less than one-quarter size of interparietal plate); a single conspicuously large and elongate superciliary; and ventral scales strongly keeled and mucronate. *Norops wellbornae* differs from its apparent nearest relative *N. unilobatus* by having bilobed hemipenes (hemipenes unilobed in *N. unilobatus*).

*Illustrations* (Figs. 73, 74, 106, 111). Stafford and Meyer, 1999 (head and dewlap only; as *N. sericeus*); Köhler and Acevedo, 2004 (adult, dewlap, head scales, male dewlap; as *N. sericeus* complex); Köhler et al., 2005b (adult, head scales; as *N. sericeus*); Köhler and Vesely, 2010 (head and dewlap, hemipenis; as *Anolis*); Köhler, 2014 (ventral scales; as *Anolis*).

*Remarks.* Mertens (1952b) placed *Anolis ustus wellbornae* in the synonymy of *Anolis* (= *Norops*) *sericeus* (Hallowell). Köhler and Vesely (2010) resurrected *N. wellbornae* from that synonymy. *Norops wellbornae* is a member of the *N. sericeus* subgroup of the *N. auratus* species group of Nicholson et al. (2012; see additional comments in the Remarks for *N. unilobatus*). Tissues of *N. wellbornae* were not utilized in the phylogenetic analyses of Nicholson et al. (2005, 2012).

*Natural History Comments.* *Norops wellbornae* occur from near sea level to 1,000 m elevation in the Lowland Dry Forest, Lowland Arid Forest, and Premontane Dry Forest formations. This species is active on the ground, on rock walls, in low leafy vegetation, and on tree trunks and branches up to at least 4 m above the ground. It sleeps at night on leaves, stems, and tree branches up to at least 4 m above the ground. *Norops wellbornae* was collected from March to July and during October and November and is likely active throughout the year. Nothing has been reported on its diet or reproduction.

*Etymology.* The specific name is a patronym for Margarete Vera Wellborn, who was affiliated with the ZMB and Ernest Ahl during the 1930s.

*Specimens Examined* (76 [16]; Map 37). CHOLUTECA: 1 km N of Cedeño, LSUMZ 33680, 36585; El Despoblado, CAS 152958; El Faro, USNM 580722–24; El Madreal, SMF 87124, 87145–46, 87149, 91731, USNM 580721; Finca Monterrey, SMF 91764; La Fortuna, SDSNH 72752–53; between San Lorenzo and Choluteca, AMNH 70404–06. EL PARAÍSO: El Rodeo, USNM 580702, 580706–08; near Mansaragua, AMNH 70909; 1 km S of Orealí, USNM 580725–30; Orealí, USNM 580703–05, 580709, 581192. FRANCISCO MORAZÁN: Cantarranas, ANSP 26080–81; El Zamorano, AMNH 70359–60, 70382–84, LACM 39769, MCZ R-48674, 49757, 49992; between San Juan de Flores and Talanga, UF 90205; near Tegucigalpa, LSUMZ 24141–42; Tegucigalpa, BYU 18195–96, LACM 39770, MCZ R-49932; Valle de Ángeles, TCWC 16192. INTIBUCÁ: Santa Lucía, SMF 78904, 79140. LEMPIRA: Erandique, CM 64604. VALLE: El Pacar, USNM 580710–17; Isla del Tigre, near communications tower, USNM 580294; Isla del Tigre, SDSNH 72751; Isla Garroba, USNM 580296–98; Isla Inglasera, USNM 580295; Isla Zacate Grande, KU 194287; Playona Isla Exposición, SMF 91732, USNM 580720; Punta El Molino, USNM 580299, 580718–19.

*Norops yoroensis* McCranie, Nicholson, and Köhler

*Norops yoroensis* McCranie, Nicholson, and Köhler, 2002: 466 (holotype, USNM 541012; type locality: "2.5 airline km

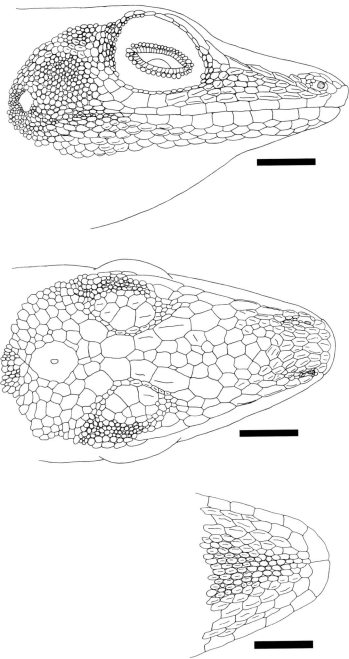

Figure 73. *Norops wellbornae* head in lateral, dorsal, and ventral views. SMF 82667, adult male from Finca San Julián, Suchitepéquez, Guatemala. Scale bar = 1.0 mm. Drawing by Gunther Köhler.

Figure 74.    *Norops wellbornae*. (A) Adult male dewlap (USNM 580294) from near communications tower, Isla del Tigre, Valle; (B) adult male dewlap (USNM 580294). Photographs by James R. McCranie.

NNE La Fortuna, 15°26′N, 87°18′W, 1600 m elevation, Cordillera Nombre de Dios, Departamento de Yoro, Honduras"); Köhler, 2003: 106, 2008: 114; Wilson and McCranie, 2003: 60, 2004b: 43; McCranie et al., 2006: 218; Wilson and Townsend, 2006: 105; McCranie and Solís, 2013: 242.

*Anolis yoroensis*: Köhler et al., 2007: 391; Townsend et al., 2007: 10, 2010: 12, 2012: 101, 2013: 197; McCranie and

Valdés Orellana, 2012: 304; Köhler, 2014: 210.

*Geographic Distribution. Norops yoroensis* occurs at moderate and intermediate elevations on the Atlantic versant from northwestern to north-central Honduras (but see Remarks).

*Description.* The following is based on 19 males (SMF 80765, 80769, 87163–67, 88675, 88696, 88698–701; USNM 541012–13, 541015, 541019, 563079, 563081) and

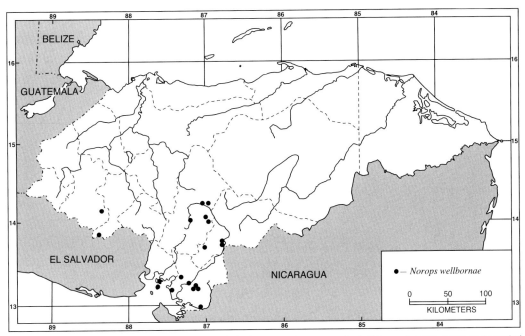

Map 37.   Localities for *Norops wellbornae*. Solid symbols denote specimens examined. This species also occurs on several of the smaller islands in the Golfo de Fonseca in extreme southern Honduras.

11 females (SMF 80766–68, 88697; USNM 541014, 541016–18, 541020, 563080, 565485). *Norops yoroensis* is a small anole (SVL 45 mm in largest male [SMF 88675], 47 mm in largest females [SMF 80766, 80768]); dorsal head scales weakly keeled in internasal region, rugose, weakly keeled, or tuberculate in prefrontal, frontal, and parietal areas; deep frontal depression present; parietal depression shallow; 4–9 (5.6 ± 1.3) postrostrals; anterior nasal usually entire, usually contacting rostral and first supralabial, occasionally only rostral; 5–10 (7.1 ± 1.3) internasals; canthal ridge sharply defined; scales comprising supraorbital semicircles weakly keeled, largest scale in semicircles about same size as largest supraocular scale; supraorbital semicircles well defined; 1–4 (2.7 ± 0.7) scales separating supraorbital semicircles at narrowest point; 2–4 (2.9 ± 0.6) scales separating supraorbital semicircles and interparietal at narrowest point; interparietal well defined, greatly enlarged relative to adjacent scales, surrounded by scales of moderate size, longer than wide, usually larger than ear opening; 2–3 rows of about 2–9 (total number) enlarged, keeled supraocular scales; 1–2 enlarged supraoculars in broad contact with supraorbital semicircles or completely separated by 1 row of small scales; 2 elongate superciliaries, posterior much shorter than anterior; usually 3 (occasionally 4–5) enlarged canthals; 5–14 (10.0 ± 1.8) scales between second canthals; 9–15 (11.9 ± 1.7) scales between posterior canthals; loreal region slightly concave, 30–63 (40.2 ± 7.5) mostly strongly keeled (some smooth or rugose) loreal scales in maximum of 5–8 (6.6 ± 0.8) horizontal rows; 6–9 (7.1 ± 0.8) supralabials and 5–9 (7.1 ± 1.1) infralabials to level below center of eye; suboculars weakly keeled, 2 suboculars usually separated from supralabials by 1 scale row or 1–2 suboculars in narrow contact with supralabials; ear opening vertically oval; scales anterior to ear opening granular, similar in size to those posterior to ear opening; 4–8 (6.0 ± 0.9) postmentals, outer

pair usually largest; keeled granular scales present on chin and throat; male dewlap moderately large, extending past level of axilla onto chest; male dewlap with 8–12 horizontal gorgetal-sternal scale rows, about 10–13 scales per row ($n = 2$); 2–4 (modal number) anterior marginal pairs in male dewlap; female dewlap rudimentary; no nuchal crest or dorsal ridge; 2 middorsal scale rows slightly enlarged, weakly keeled, dorsal scales lateral to middorsal series grading into granular lateral scales; no enlarged scales scattered among laterals; 55–77 (66.9 ± 6.7) dorsal scales along vertebral midline between levels of axilla and groin in males, 61–86 (75.1 ± 7.2) in females; 38–54 (46.4 ± 4.9) dorsal scales along vertebral midline contained in 1 head length in males, 28–56 (42.2 ± 9.1) in females; ventral scales on midsection larger than largest dorsal scales; ventral body scales weakly keeled, subimbricate; 44–61 (51.4 ± 5.2) ventral scales along midventral line between levels of axilla and groin in males, 46–60 (52.6 ± 5.4) in females; 30–44 (36.5 ± 4.4) ventral scales contained in 1 head length in males, 22–35 (27.9 ± 4.4) in females; 110–148 (123.8 ± 11.0) scales around midbody in males, 111–134 (124.2 ± 8.3) in females; tubelike axillary pocket absent; precloacal scales not keeled; no enlarged postcloacal scales in males; tail nearly rounded to slightly compressed, TH/TW 0.93–1.43 in 26; basal subcaudal scales keeled; lateral caudal scales keeled, homogeneous, although indistinct division in segments discernable; dorsal medial caudal scale row not enlarged, keeled, not forming crest; most scales on anterior surface of antebrachium keeled, unicarinate; 19–26 (23.1 ± 2.1) subdigital lamellae on Phalanges II–IV of Toe IV of hind limbs; 6–9 (7.3 ± 0.8) subdigital scales on Phalanx I of Toe IV of hind limbs; SVL 33.0–45.0 (39.5 ± 3.0) mm in males, 33.0–47.0 (42.0 ± 4.2) mm in females; TAL/SVL 1.69–2.00 in nine males, 1.49–1.88 in seven females; HL/SVL 0.25–0.29 in males, 0.22–0.27 in females; SHL/SVL 0.26–0.32 in males, 0.25–0.29 in females; SHL/HL 1.03–1.22 in males, 1.00–1.19 in females; longest toe of adpressed hind limb usually reaching between posterior border of eye and mideye.

Color in life of the male holotype (USNM 541012) was described by McCranie et al. (2002b: 469): "dorsum of head Dusky Brown (19) with a few Raw Umber (23) scales on snout; preocular bar Clay Color (26); supraocular bars Cinnamon (39); postocular stripe Clay Color (26), ground color above and below lateral stripe Brick Red (132A), mottled with dark brown; lateral stripe Buff-Yellow (53); dorsolateral stripe Light Drab (119C); middorsal area Dusky Brown (19), mottled with Buff-Yellow (53) and Light Drab (119C); arms Fuscous (21) with Buff-Yellow (53) bars and spots, legs Brick Red (132A) with Orange-Rufous (132C) crossbars; tail banded brown, dark brown, and pale brown, underside with Mahogany Red (132B) scales; upper lip, chin, and belly Buff-Yellow (53); dewlap Orange-Rufous (132C) with gray-flecked white scales." Color in life of another adult male (SMF 88699): dorsal surface of head brown with darker brown interocular bar, bar preceded and followed by orange-brown bar; dorsal surface of body brown with darker brown row of middorsal spots; dark brown stripe extending from ear opening to above forelimb; dorsal surfaces of shanks with pale, narrow cross stripes; dorsal surface of tail banded pale brown and dark brown; dewlap Spectrum Orange (17). Color in life of a third adult male (SMF 88944): dorsal surface of head with dark brown snout, pale olive yellow band anterior to eyes, dark brown interocular bar, narrow pale olive yellow band posterior to interocular bar, and dark brown blotch on parietals; middorsal area olive yellow with series of 3–4 dark brown middorsal spots and dark brown smudging elsewhere in this field; dark brown band beginning on snout, passing around eye onto neck, thence onto body where it begins to break up to disappear at about midbody, this band bordered below by narrow pale olive yellow stripe beginning at axilla and fading out at about midbody; dorsal surfaces of forelimbs olive yellow with dark brown smudging;

dorsal surfaces of hind limbs olive yellow with pale olive yellow and olive brown crossbars; tail olive yellow with dark streaks at base and darker crossbands distally; supralabials pale olive yellow; chin pinkish cream; belly olive cream; dewlap pale chocolate brown with pale yellow scales; iris copper with gold rim around pupil. Color in life of a fourth adult male (SMF 88939): dorsal surface of head dark brown, grading to olive brown on back; pair of pale olive stripes extending posteriorly from temporal region to point above shoulder; series of posteriorly pointing chevrons extending down back; lateral surfaces of body olive brown with yellow stripe extending from axilla to about level of dewlap; dorsal surfaces of limbs brown with tan crossbars; tail tan with dark brown bands; upper lip cream; chin cream with brown spotting; dewlap pale dirty orange with white scales and small brown spots on skin. Color in life of dewlap of another adult male (SMF 88698) was similar to that of SMF 88699 (described above), whereas another adult male (SMF 88700) had a Cinnamon Rufous (40) dewlap with dirty white scales (those scales dark brown when specimen first collected). Dewlap color of a subadult male (SMF 88946) was Cinnamon (39). Color in life of an adult female (SMF 88942): dorsal surface of head olive green with rust patina; dorsal surface of body pale olive yellow with tan middorsal band outlined at intervals by short dark brown streaks; dorsal surfaces of forelimbs pale olive yellow with rust patina; dorsal surfaces of hind limbs rust red with golden yellow crossbars; tail pale olive yellow with rust patina, faint evidence of middorsal band continuing onto base, band gradually fading posteriorly, distal portion of tail with rust brown crossbands; belly cream with brown smudging; chin cream with dark brown punctations; dewlap white with dark brown scales; iris rust red. Color in life of an adult female paratype (USNM 541014) was also described by McCranie et al. (2002b: 470–471): "dorsum dull olive green, with coppery sheen on middorsum, and ca. 5 dark

brown H-shaped middorsal blotches; lateral surface of body with similar dark olive green markings; dorsal surface of head dull olive green with dark brown to very dark brown markings; front limbs olive green with indistinct dark brown transverse banding; hind limbs dull olive green with dark brown and pale olive green transverse banding; iris coppery red with thin bronze ring around pupil; side of head same as ground color for middorsum; tail olive green with dark brown and cream bands; throat cream with brown flecking; venter pale yellow with scattered dusting of rust red; subcaudal surface rust red." Color in life of another adult female (SMF 80766): dorsal surface of head Greenish Olive (49) with Citrine (51) preocular bar; dorsal surface of body Citrine with Dark Grayish Brown (20) middorsal markings; dorsal surfaces of forelimbs Olive-Yellow (52) with Greenish-Olive crossbands; dorsal surfaces of hind limbs Olive-Yellow with rust-red patina; chin gray with olive markings; ventral surface of body Sulphur Yellow (57) with olive markings; dewlap white with gray scales; iris copper.

Color in alcohol: dorsal surfaces of head and body dark brown; lateral surface of body brown, with small dark brown middorsal blotches present in some (both sexes); occasional females have a pale brown middorsal stripe that is outlined by a nearly straight dark brown stripe on each side, also occasional females have a dark brown middorsum flanked by a thin, incomplete slightly incurved pale brown line, areas inside pale line filled with dark brown pigment; lateral surface of head brown, except suboculars, row of scales separating suboculars from supralabials, lower row of loreals, and supralabials usually mostly white; dorsal surfaces of limbs brown with indistinct paler brown crossbars; dorsal surface of tail brown, without distinct markings; scales of ventral surfaces of head and body white, flecked with brown, lightly to moderately flecked anteriorly, becoming progressively more heavily flecked posteriorly; scales of subcaudal surface white, flecked with brown, lightly flecked proximally, becoming heavily flecked for distal two-thirds.

Hemipenis: the completely everted hemipenis of SMF 87167 is a medium-sized, bilobed organ; sulcus spermaticus bordered by well-developed sulcal lips, bifurcating at base of apex, branches opening into broad concave areas distal to point of bifurcation on each lobe; no asulcate processus; lobes strongly calyculate; truncus with transverse folds.

*Diagnosis/Similar Species.* *Norops yoroensis* is distinguished from all other Honduran *Norops*, except *N. carpenteri*, *N. limifrons*, *N. ocelloscapularis*, *N. rodriguezii*, and *N. zeus*, by the combination of having a single elongated prenasal scale, weakly keeled ventral scales, and slender habitus. *Norops yoroensis* differs from *N. carpenteri* by having a pale lateral stripe and brown dorsal surfaces in life (no pale lateral stripe and greenish brown dorsal surfaces with pale spots in life in *N. carpenteri*). *Norops yoroensis* differs from *N. rodriguezii* by having weakly keeled, subimbricate ventral scales (ventrals smooth, nonimbricate in *N. rodriguezii*). *Norops yoroensis* differs from *N. limifrons* and *N. zeus* by having weakly keeled, subimbricate ventral scales, shorter hind legs with longest toe of adpressed hind limb usually reaching between posterior border of eye and mideye, and by having a predominantly orange male dewlap in life (ventrals smooth, nonimbricate, adpressed hind limb reaching between anterior border of eye and tip of snout, dewlap dirty white with or without basal orange-yellow spot in life in *N. limifrons* and *N. zeus*). *Norops yoroensis* differs from *N. ocelloscapularis* by lacking an ocellated shoulder spot (such a spot usually present in *N. ocelloscapularis*).

*Illustrations* (Figs. 75, 76, 84). McCranie et al., 2002b (adult, head scales); Köhler, 2003 (head scales), 2008 (head scales), 2014 (ventral scales; as *Anolis*); Townsend et al., 2012 (adult).

*Remarks.* McCranie et al. (2002b) considered *N. yoroensis* to be a member of the *N. cupreus* species group. However, unlike

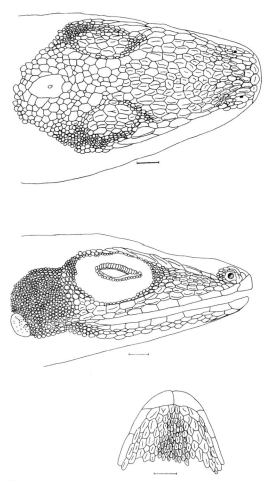

Figure 75. *Norops yoroensis* head in dorsal, lateral, and ventral views. SMF 80769, adult male from Cerro de Pajarillos, Yoro. Scale bar = 1.0 mm. Drawing by Lara Czupalla.

*N. cupreus*, *N. yoroensis* usually has an entire anterior nasal scale (divided in *N. cupreus*). An entire anterior nasal scale and smooth to weakly keeled ventral scales are characteristic of the species assigned to the *N. fuscoauratus* species group by Savage and Guyer (1989). However, the monophyly of that group was rejected by both molecular (Nicholson, 2002; Nicholson et al., 2005, 2012) and morphological and molecular (Poe, 2004) data. In its external morphology, *N. yoroensis* is very similar to *N. carpenteri*, *N. ocelloscapularis*, and *N. rodriguezii*. Therefore, we include *N. yoroensis*, along with the three just mentioned

Figure 76.  *Norops yoroensis*. (A) Adult male (in UNAH collection); (B) adult male dewlap (in UNAH collection), both from Quebrada Las Cuevas, Santa Bárbara. Photographs by James R. McCranie.

species plus *N. limifrons* and *N. zeus* in the *N. fuscoauratus* species subgroup of the *N. auratus* species group of Nicholson et al. (2012; also see Remarks for *N. ocelloscapularis*). Tissues of *N. yoroensis* were not available for the phylogenetic analyses of Nicholson et al. (2005, 2012).

Although not described until 2002, it was recently discovered that *N. yoroensis* was first collected in Honduras during 1923 by Schmidt. Also, Meyer and Wilson collected a few specimens of this species during 1967. McCranie has an illustration of a specimen of *N. yoroensis* that was drawn by Meyer, who thought the specimen represented an undescribed species. However, Meyer never prepared a manuscript describing this species, nor did Meyer and Wilson (1973) list those specimens in their review of the lizards of Honduras.

*Norops yoroensis* has yet to be reported from Guatemala, but likely occurs in the

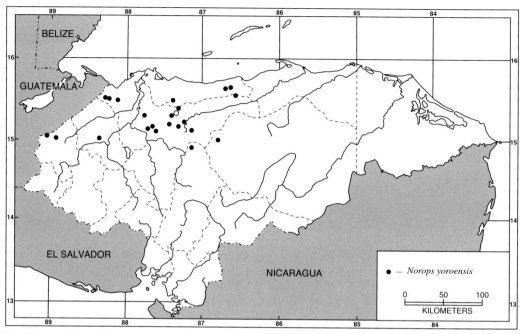

Map 38.   Localities for *Norops yoroensis*. Solid symbols denote specimens examined and the open symbol an accepted record.

eastern portion of that country near the border with Honduras.

*Natural History Comments. Norops yoroensis* is known from 650 to 1,600 m elevation in the Premontane Wet Forest and Lower Montane Wet Forest formations and peripherally in the Premontane Moist Forest formation. It is most common inside deeply shaded pristine broadleaf forest, although it can be found in somewhat open areas in disturbed broadleaf forest. Individuals are active on leafy vegetation and tree trunks up to at least 3 m above the ground. It is also active on the ground and attempts to escape by climbing the nearest tree trunk. Seven specimens of the Cerro de Pajarillos, Yoro, series were collected during a single hour on an afternoon following a heavy midday rain. It sleeps at night on stems and leaves about 1–2 m above the ground. The species was collected during January, February, from April to July, and during September and November; thus, is active throughout the

year. Nothing has been published on its diet or reproduction.

*Etymology.* The specific name *yoroensis* is formed from Yoro (a Honduran department) and the Latin suffix *-ensis* (denoting place, locality), in reference to the Honduran department in which the type specimens were collected.

*Specimens Examined* (274 [0] + 3 skeletons; Map 38). ATLÁNTIDA: S slope of Cerro Búfalo, SMF 88948; Cerro El Chino, USNM 578810; La Liberación, USNM 578779–809, 578820; near Los Planes, UTA R-41232; Quebrada de Oro, SMF 88696–701, 88939–47, 88949–52, UTA R-41237. COPÁN: Quebrada las Piedras, SMF 93369; San Isidro, SMF 91311. CORTÉS: Banaderos, SMF 87156–57; 0.5 km N of Buenos Aires, SMF 87158–61; Buenas Aires, SMF 87162; Finca Naranjito, SMF 91309–10; mountains W of San Pedro Sula, FMNH 5278–79. FRANCISCO MORAZÁN: Los Planes, SMF 87163–67. OLANCHO: Río de Enmedio, USNM 563079–80. SANTA

BÁRBARA: Nuevo Joconales, SMF 93353; Quebrada de las Minas, SMF 93354–58; Quebrada Las Cuevas, SMF 93359–61, USNM 580308–19. YORO: Cerro de Pajarillos, SMF 80765–69, USNM 541015–20; El Panál, UF 166309–24; near El Porvenir de Morazán, UF 166325–32; 2.5 airline km NNE La Fortuna, USNM 541012–14, 563081, 565485; La Libertad, SMF 88675; Montaña La Ruidosa, USNM 563082; Pino Alto, UF 166333–35; Portillo Grande, FMNH 21876 (25), MCZ R-48703 (skeleton), 48704–09, 175282–83 (both skeletons), 175284–337, UMMZ 94043 (6), 131486 (2), 131487 (8); Subirana Valley, FMNH 21846 (5), MCZ R-48710–14, UMMZ 94039 (5), USNM 121125–30; 10 km W of Yoro, ANSP 30691–92; 32.0 km W of Yoro, USNM 217594; near Yuqüela, UTA R-53987; Yuqüela, UTA R-53980–86.

*Other Records* (Map 38). YORO: 1.5 km N of San Francisco, (photographs, UNAH).

*Norops zeus* Köhler and McCranie

*Anolis rodriguezii*: Barbour and Loveridge, 1929a: 140; Townsend and Wilson, 2010b: 697 (in part).

*Anolis limifrons*: Dunn and Emlen, 1932: 27; Meyer, 1969: 225 (in part); Meyer and Wilson, 1973: 18 (in part); Wilson et al., 1979a: 25.

*Norops limifrons*: Wilson et al., 2001: 136 (in part).

*Norops zeus* Köhler and McCranie, 2001: 236 (holotype, SMF 77196; type locality: "Liberia, Parque Nacional Pico Bonito, 90 m elevation, Departamento de Atlántida, Honduras"); Köhler, 2003: 106, 2008: 114; Wilson and McCranie, 2003: 60; McCranie and Castañeda, 2005: 14; McCranie et al., 2006: 218; Wilson and Townsend, 2006: 105; Diener, 2008: 17; Nicholson et al., 2012: 12; McCranie and Solís, 2013: 242

*Anolis zeus*: Nicholson et al., 2005: 933; Köhler et al., 2007: 391; Townsend et al., 2012: 102; Pyron et al., 2013: fig. 19.

*Geographic Distribution. Norops zeus* occurs at low and moderate elevations of the Atlantic versant from the northern slopes and foothills of the Cordillera Nombre de Dios to the western slopes of Cerro Azul Meámbar and adjacent lowlands in northern Honduras.

*Description.* The following is based on 10 males (SMF 77194, 77196, 80698–99, 80702–03, 87002, 87006–07; USNM 565488) and 12 females (SMF 77193, 77195, 80700–01, 80703–05, 87008–09; USNM 541022–24). *Norops zeus* is a small anole (SVL 43 mm in largest male [SMF 87007], 44 mm in largest female [SMF 77193]); dorsal head scales weakly keeled in internasal and prefrontal regions, rugose to tuberculate in frontal and parietal areas; deep frontal depression present; parietal depression absent; 4–8 (5.0 ± 0.7) postrostrals; anterior nasal entire, usually contacting rostral and first supralabial, occasionally only rostral; 5–10 (7.9 ± 1.2) internasals; canthal ridge sharply defined; scales comprising supraorbital semicircles weakly keeled, largest scale in semicircles about same size as largest supraocular scale; supraorbital semicircles well defined; 1–3 (2.1 ± 0.7) scales separating supraorbital semicircles at narrowest point; 2–4 (3.0 ± 0.6) scales separating supraorbital semicircles and interparietal at narrowest point; interparietal well defined, greatly enlarged relative to adjacent scales, irregular in outline, longer than wide, surrounded by scales of moderate size, about equal in size to ear opening; 2–3 rows of about 4–8 (total number) enlarged, keeled supraocular scales; enlarged supraoculars varying from completely separated to almost separated from supraorbital semicircles by 1 row of small scales; 2 elongate superciliaries, posterior much shorter than anterior; 3 enlarged canthals; 7–13 (10.3 ± 1.6) scales between second canthals; 9–18 (12.3 ± 2.0) scales between posterior canthals; loreal region slightly concave, 30–65 (43.9 ± 9.7) mostly strongly keeled (some smooth or rugose) loreal scales in maximum of 5–7 (6.0 ± 0.8) horizontal rows; 6–9 (7.3 ± 0.8) supralabials and 6–8 (7.4 ± 0.6) infralabials to level below center of eye; suboculars weakly keeled, 1–3 suboculars in broad contact with supralabials; scales anterior to ear opening granular, similar in size to those posterior to ear opening; 6–7 (6.1 ± 0.3) postmentals, outer pair usually largest; keeled granular scales present on chin and throat; male dewlap moderately large, extending to level of axilla; male dewlap with 10–14 horizontal gorgetal-sternal scale rows, about 16–19 scales per row ($n$ = 4); 2 (modal number) anterior marginal pairs in male dewlap; female dewlap absent; no nuchal crest or dorsal ridge; 2 middorsal scale rows slightly enlarged, smooth to rugose, dorsal scales lateral to middorsal series grading into granular lateral scales; no enlarged scales scattered among laterals; 75–114 (91.5 ± 14.6) dorsal scales along vertebral midline between levels of axilla and groin in males, 96–118 (106.9 ± 8.2) in females; 47–72 (59.9 ± 7.7) dorsal scales along vertebral midline contained in 1 head length in males, 52–68 (58.3 ± 5.4) in females; ventral scales on midsection about same size as largest dorsal scales; ventral body scales smooth, slightly swollen with rounded posterior edges, sub-imbricate; 56–75 (64.7 ± 6.7) ventral scales along midventral line between levels of axilla and groin in males, 59–82 (71.2 ± 7.9) in females; 38–59 (46.2 ± 6.9) ventral scales contained in 1 head length in males, 38–56 (46.2 ± 6.2) in females; 112–152 (139.2 ± 11.6) scales around midbody in males, 116–144 (134.0 ± 10.3) in females; tubelike axillary pocket absent; precloacal scales not keeled; usually pair of slightly enlarged postcloacal scales in males; tail nearly rounded to distinctly compressed, TH/TW 1.07–1.50; basal subcaudal scales smooth; lateral caudal scales faintly keeled, homogeneous, although indistinct division in segments discernable; dorsal medial caudal scale row not enlarged, keeled, not forming crest; most scales on anterior surface of antebrachium rugose to unicarinate; 22–25 (23.2 ± 0.8) subdigital lamellae on Phalanges II–IV of Toe IV of hind limbs; 6–8 (6.9 ± 0.8) subdigital scales on Phalanx I of Toe IV of hind limbs; SVL 33.0–43.0 (38.8 ± 2.7) mm in males, 27.0–44.0 (38.7 ± 5.1) mm in females; TAL/SVL 2.11–2.32 in eight males, 1.69–2.26 in eight females; HL/SVL 0.24–0.28 in males, 0.25–0.31 in females; SHL/SVL 0.26–0.32 in males, 0.25–0.33 in females; SHL/HL 1.05–1.20 in males, 1.01–1.13 in 10 females; longest toe of adpressed hind limb usually reaching between anterior border of eye and tip of snout.

Color in life of an adult male (SMF 87002): dorsal surface of head Olive-Brown (28); dorsal surface of body Olive-Brown with Burnt Umber (22) middorsal blotches; dorsal surface of tail with alternating Burnt Umber and pale brown crossbands; belly pale brown; dewlap uniformly dirty white with brown scales; iris Burnt Umber with pale gold rim. Two other adult males (SMF 77196, 80703) had uniformly dirty white dewlaps (including the gorgetal scales).

Color in alcohol: dorsal surfaces of head and body brown to grayish brown; middorsal pattern of small dark brown spots or blotches present in some, some females have a pale brown middorsal stripe that is bounded on both sides by thin, dark brown border, pale middorsal stripe extending well onto tail; lateral surface of head pale brown; dorsal surfaces of limbs brown with paler brown crossbands; dorsal surface of tail brown with indistinct darker brown crossbars in those specimens in which pale middorsal stripe not extending onto tail; ventral surface of head white with brown flecks on many scales; ventral surface of body white with sparse to numerous brown flecks on scales of chest region, brown flecking on scales becoming more prominent posteriorly; subcaudal surface with sparse to numerous brown flecks on scales proximally, flecking becoming more prominent distally until subcaudal surface mostly brown for distal half; male dewlap colorless.

Hemipenis: the completely everted hemipenis of SMF 77194 is a small unilobed organ; sulcus spermaticus bordered by well-developed sulcal lips, opening at base of apex; no discernable surface structure on truncus or lobe; no asulcate processus.

*Diagnosis/Similar Species.* *Norops zeus* is distinguished from all other Honduran *Norops*, except *N. carpenteri, N. limifrons,*

*N. ocelloscapularis*, *N. rodriguezii*, and *N. yoroensis*, by the combination of having a single elongated prenasal scale, smooth and subimbricate ventral scales, and slender habitus. *Norops zeus* differs from *N. carpenteri*, *N. ocelloscapularis*, *N. rodriguezii*, and *N. yoroensis* by having long hind legs (longest toe of adpressed hind limb usually reaching between anterior border of eye and tip of snout in *Norops zeus* versus not reaching beyond anterior border of eye in *N. carpenteri*, *N. ocelloscapularis*, *N. rodriguezii*, and *N. yoroensis*). *Norops zeus* also differs from those four species in having an uniformly dirty white male dewlap in life (dewlap orange in life in *N. carpenteri*, *N. ocelloscapularis*, *N. rodriguezii*, and *N. yoroensis*). *Norops zeus* differs further from *N. carpenteri* in having brown dorsal surfaces without paler spots in life (dorsal surfaces greenish brown with paler spots in life in *N. carpenteri*). *Norops zeus* differs from *N. limifrons* by having a uniformly dirty white male dewlap, without a basal orange-yellow spot in life (basal orange-yellow spot present in life in dewlap of *N. limifrons*).

*Illustrations* (Figs. 77, 78). Köhler, 2000 (head and dewlap; as *N.* cf. *limifrons*), 2003 (head and dewlap), 2008 (head and dewlap); Köhler and McCranie, 2001 (adult, head scales, head and dewlap); Diener, 2008 (adult); Townsend et al., 2012 (adult).

*Remarks. Norops zeus* appears to be most closely related to *N. limifrons* based on their similar external morphology. The latter species was included in the *N. fuscoauratus* species group by Köhler et al. (2001) and that relationship is retained herein, but as the *N. fuscoauratus* species subgroup (also see Remarks for *N. limifrons* and *N. ocelloscapularis*).

*Norops zeus* was not recognized as a distinct species until 2001, despite museum specimens existing sometime before 1929.

Townsend et al. (2012) stated that morphological and molecular data suggests that two species are included in the presently understood concept of *N. zeus*.

*Natural History Comments. Norops zeus* is known from 5 to 900 m elevation in the Lowland Moist Forest and Premontane Wet Forest formations and peripherally in the Lowland Dry Forest formation. Specimens are active on stems of low vegetation and low on tree trunks in primary and secondary forest. It sleeps at night on leaves, stems, and at the ends of branches in low vegetation. *Norops zeus* has been collected from January to August and during October and December, thus is active throughout the year. Nothing has been reported on its diet or reproduction.

*Etymology.* The name *zeus* is derived from Zeus, the chief of the Greek gods. The name refers to the Cordillera Nombre de Dios (= in the name of God mountain range), where the type locality of this species is located.

*Specimens Examined* (83 [10] + 2 C&S; Map 39). ATLÁNTIDA: Cascada de Río Corinto, SMF 91355; N slope of Cerro Búfalo, SMF 80698, 80700; mountains S of Corozal, LACM 47268, 47275; Estación Forestal CURLA, SMF 80699, 80701–02, 87002–07, 88712, USNM 541023–25, 565486–87; Guaymas District, MCZ R-20635, UMMZ 58404–07; 7.4 km SE of La Ceiba, USNM 541022; 8 km E of La Ceiba, KU 101401–03; 12 km E of La Ceiba, KU 101404; 14.5 km E of La Ceiba, LACM 47274; La Ceiba, LSUMZ 21380 (C&S); La Liberación, USNM 578818; Lancetilla, AMNH 70430, MCZ R-29822–23, 32211–13, USNM 578817; Liberia, SMF 77193, 77196; Mezapita, USNM 580320; Quebrada La Muralla, SMF 77194–95; Salado Barra, MCZ R-191085, 191088; San Marcos, SMF 87008–09; 32.2 km E Tela, LACM 47272–73; Tela, UMMZ 58412–13, 70319 (5). COLÓN: Cerro Calentura, LSUMZ 22515–16, 27735 (C&S), SMF 80703. CORTÉS: NE end of Lago de Yojoa, KU 194867, 194890; Los Pinos, SMF 91308, 92840, UF 166180–83, 166231–37, USNM 565488; near Peña Blanca, UF 166213, 166238; Potrerillos, MCZ R-29816; 7.2 km ENE of Villanueva, LACM 47271. YORO: San José de Texíguat, USNM 580321; Subirana Valley, MCZ R-38836; near Yuquela, UTA R-53989.

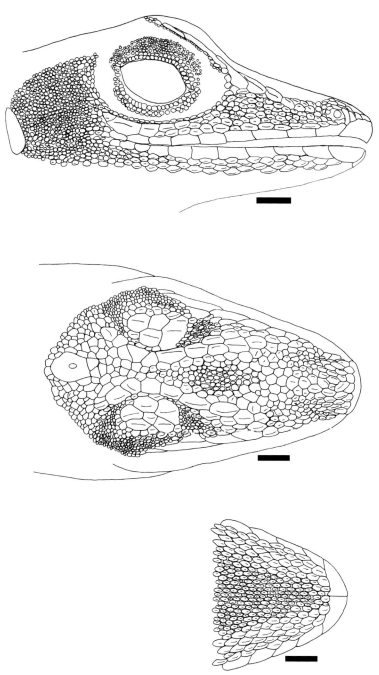

Figure 77.   *Norops zeus* head in lateral, dorsal, and ventral views. SMF 77196, adult male from Liberia, Atlántida. Scale bar = 1.0 mm. Drawing by Gunther Köhler.

Figure 78.   *Norops zeus.* (A) Adult male (in UNAH collection) from Estación Forestal CURLA, Atlántida; (B) adult male dewlap (from series SMF 77193–96) from Liberia, Atlántida. Photographs by James R. McCranie (A) and Gunther Köhler (B).

## KEY TO THE HONDURAN SPECIES OF THE GENERA *ANOLIS* AND *NOROPS*

1 A.  All autotomic caudal vertebrae without anterolaterally directed transverse processes (Fig. 79); ear opening horizontally elongate, with posterior margin forming long depression (Fig. 1); adult heads conspicuously long (Fig. 1); >35 lamellae under fourth toe ...............
................. *Anolis allisoni* (p. 11)

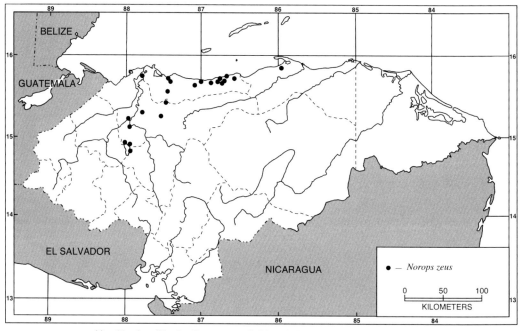

Map 39.   Localities for *Norops zeus*. Solid symbols denote specimens examined.

Figure 79.   (A) Automic caudal vertebrae without anterolaterally directed transverse processes (*alpha* anoles) in *Anolis allisoni* (SMF 78320) from Isla de Guanaja, Islas de la Bahía; (B) autotomic caudal vertebrae with anterolaterally directed transverse processes (*beta* anoles) in *Norops subocularis* (Davis) (USNM 133731) from Guerrero, Mexico. Photographs by Linda Acker.

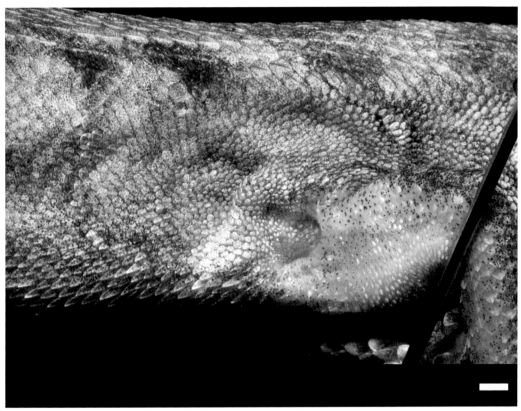

Figure 80.   Scaleless tubelike axillary pocket in *Norops tropidonotus* (SMF 79058) from Cerro Uyuca, Francisco Morazán. Scale bar = 1.0 mm. Photograph by Gunther Köhler.

B. At least some autotomic caudal vertebrae with anterolaterally directed transverse processes (Fig. 79); ear opening vertically oval (Fig. 3); adult heads not conspicuously elongate (Fig. 3); <30 lamellae under fourth toe ...   2

2 A. A deep, scaleless, tubelike axillary pocket present (Fig. 80) ....   3

B. No deep, tubelike axillary pocket present, at most a shallow, scaled axillary depression .......   6

3 A. Scales anterior to ear opening flat, keeled, about twice as large as scales posterior to ear opening (the latter usually granular; Fig. 63) .................   4

B. Scales anterior to ear opening about same size as scales posterior to ear opening (all granular; Fig. 51) ...................   5

4 A. Male dewlap orange-red in life, with yellow margin and distinct central dark streak (Fig. 64); adult males to 59 mm snout–vent length (SVL), adult females to 55 mm SVL ..............
...... *Norops tropidonotus* (p. 177)

B. Male dewlap orange-red in life with yellow margin, but without central dark streak (Fig. 72); adult males and adult females to 51 mm SVL.........................
...... *Norops wampuensis* (p. 201)

5 A. Flanks usually with 1–3 pale vertical lines (lines occasionally broken into vertical dashes); male dewlap rose with large

Figure 81.   Laterally compressed tail with middorsal crest of enlarged scales in a male *Norops sagrei* (SMF 77743) from San Pedro Sula, Cortés. Scale bar = 5.0 mm. Photograph by Gunther Köhler.

purple spot in life (Fig. 66); 23–32 (rarely up to 37; mean 28.5) middorsal scales between levels of axilla and groin....................
.......... *Norops uniformis* (p. 185)

B. Flanks without pale vertical lines or dashes; male dewlap reddish orange with yellow margin in life (Fig. 52); 30–44 (mean 37.0) middorsal scales between levels of axilla and groin........
........ *Norops quaggulus* (p. 146)

6 A. Outer postmental scale on each side greatly enlarged, length greater than that of mental scale (Fig. 59); tail of adult males distinctly compressed, with middorsal crest of enlarged scales (Fig. 81); maximum known SVL 76 mm ....... 7

B. Outer postmental scale on each side not, or only moderately, enlarged, length less than that

of mental scale (Fig. 75); tail of adult males usually not, or only slightly, compressed (Fig. 82), or if noticeably compressed, adults large (adult SVL > 76 mm); tail with or without crest of enlarged scales .......... 8

7 A. Usually 4 (mean 4.3) postmental scales; male dewlap orange or orange red in life (Fig. 60); occurs on islas de Roatán and Utila and extreme northern mainland .. *Norops sagrei* (p. 166)

B. Usually 5 or more (rarely 4; mean 5.8) postmentals; male dewlap dark brown to gray in life (Fig. 40); occurs only on Islas del Cisne (Swan Islands) .... *Norops nelsoni* (p. 114)

8 A. Usually single elongate prenasal scale (Fig. 31); ventral scales smooth (Fig. 83) to weakly keeled (Fig. 84); habitus slender ........... 9

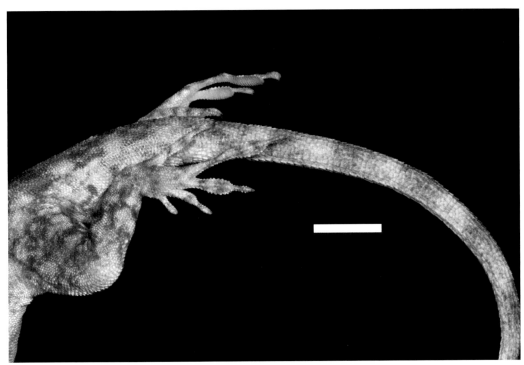

Figure 82.    Rounded tail of *Norops utilensis* (SMF 79866) from Isla de Utila, Islas de la Bahía. Scale bar = 5.0 mm. Photograph by Gunther Köhler.

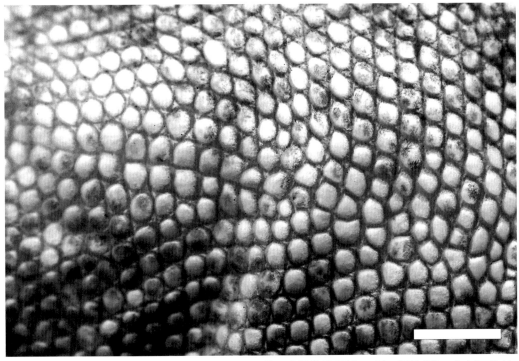

Figure 83.    Smooth, nonimbricate ventral scales of *Norops rodriguezii* (SMF 79087) from 1 km SSE of Tegucigalpita, Cortés. Scale bar = 1.0 mm. Photograph by Gunther Köhler.

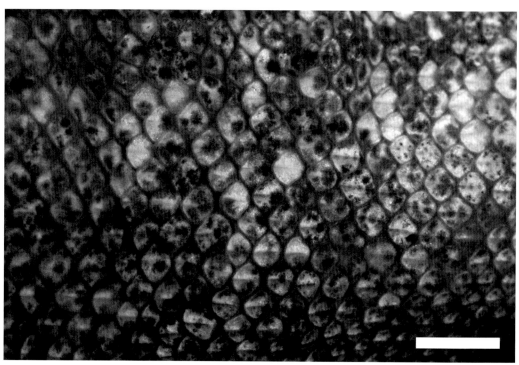

Figure 84.   Weakly keeled and imbricate ventrals in *Norops yoroensis* (SMF 80768) from Cerro de Pajarillos, Yoro. Scale bar = 1.0 mm. Photograph by Gunther Köhler.

B. Usually two rounded prenasal scales (Fig. 11); ventral scales variable; habitus usually not slender, if so, ventral scales strongly keeled (Fig. 85) and/ or male dewlap yellowish orange with large blue blotch in life .................................. 14

9 A. Long-legged (fourth toe of adpressed hind limb reaches at least posterior border of eye, usually well beyond eye); male dewlap dirty white, with or without basal orange-yellow spot in life .......................... 10

B. Shorter legged (fourth toe of adpressed hind limb usually not reaching past posterior border of eye, reaching anterior margin of eye in one species [*N. ocelloscapularis*]); male dewlap orange or yellowish orange in life ........................ 11

10 A. Male dewlap dirty white with basal orange-yellow spot in life (Fig. 32) ... *Norops limifrons* (p. 96)

B. Male dewlap uniformly dirty white, no basal orange-yellow spot in life (Fig. 78) .............. ................. *Norops zeus* (p. 217)

11 A. Usually ocellated shoulder spot present (Fig. 42); longer legged (fourth toe of adpressed hind limb reaches at least between posterior and anterior borders of eye); pale lateral stripe usually present in life................... .. *Norops ocelloscapularis* (p. 121)

B. Ocellated shoulder spot absent; shorter legged (fourth toe of adpressed hind limb reaches between ear opening and mid-eye); pale lateral stripe present or absent in life .................. 12

12 A. Dorsum greenish brown with pale spots in life (Fig. 14); no

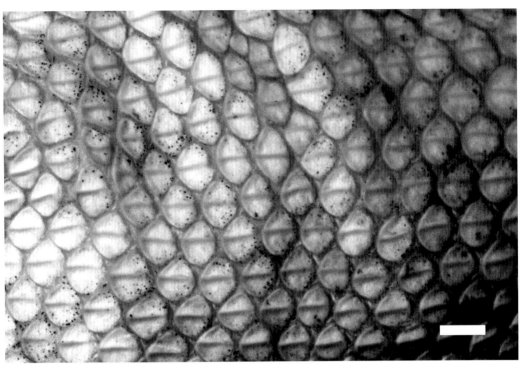

Figure 85. Strongly keeled ventrals in *Norops capito* (SMF 91252) from La Cafetalera, Santa Bárbara. Scale bar = 1.0 mm. Photograph by Gunther Köhler.

pale lateral stripe .................
......... *Norops carpenteri* (p. 45)

B. Dorsum brown without pale spots in life; pale lateral stripe present or absent .................  13

13 A. Midventral scales smooth and nonimbricate (Fig. 83); 1–4 suboculars in broad contact with supralabials ...................
...... *Norops rodriguezii* (p. 156)

B. Ventral scales weakly keeled, imbricate (Fig. 84); suboculars usually separated from supralabials by 1 scale row, or 1–2 suboculars in narrow contact with supralabials ...................
......... *Norops yoroensis* (p. 208)

14 A. Head conspicuously broad and stout, snout rather blunt in lateral aspect (Fig. 11); SHL/HL ≥ 1.1; usually with pale chin crossband, band most distinct in juveniles; ventral

scales strongly keeled (Fig. 85); dorsal scales smooth, juxtaposed, most pentagonal or hexagonal (Fig. 86); adult SVL 66–100 mm; male dewlap uniformly greenish yellow, brown, or brownish yellow in life, small (Fig. 12), not extending posterior to level of axilla ......
............... *Norops capito* (p. 39)

B. Head not conspicuously broad and stout (Fig. 7); SHL/HL usually <1.0, or if >1.0, dorsal scales keeled (Fig. 87); no pale chin crossband; combination of other characters different from above .................................  15

15 A. Large adult size (adult SVL 70 to slightly >100 mm); 9–13 supralabials to level below center of eye; males with or without enlarged postcloacal scales ....  16

Figure 86.   Dorsal scales smooth, juxtaposed, most pentagonal or hexagonal in *Norops capito* (SMF 91252) from La Cafetalera, Santa Bárbara. Scale bar = 1.0 mm. Photograph by Gunther Köhler.

B.  Medium-small to medium-large species (maximum SVL ≤ 70 mm, except *N. lemurinus*, which reaches 79 mm SVL [to 73 mm SVL in Honduran specimens] and *N. bicaorum*, which reaches 86 mm SVL); combination of other characters different from above ...... 18

16 A.  Total loreal scales >60; usually >130 scales around midbody; longest toe of adpressed hind limb usually reaching between posterior and anterior borders of eye; tail compressed (Fig. 88); male dewlap orange with purple streaks in life (Fig. 34) ........ *Norops loveridgei* (p. 101)

B.  Total loreal scales <62; usually fewer than 130 scales around midbody; longest toe of adpressed hind limb usually reaching ear opening; tail

rounded to slightly compressed; male dewlap pinkish brown or with large blue basal spot in life .......................... 17

17 A.  Anterior dorsal head scales rugose or weakly keeled (Fig. 45); midventral scales weakly keeled (Fig. 89) with rounded or truncated posterior margins, 62–86 (68–86 in Honduran specimens) scales between levels of axilla and groin; 116–140 (116–133 in Honduran specimens) scales around midbody; male dewlap pinkish brown with pale yellow margin in life (Fig. 46); dorsal ground color mottled olive yellow and brownish olive in life (Fig. 46) ............. *Norops petersii* (p. 131)

B.  Anterior dorsal head scales strongly keeled (Fig. 9); midventral scales strongly keeled,

Figure 87.   Keeled dorsal scales in *Norops lemurinus* (SMF 85885) from Raudal Kiplatara, Gracias a Dios. Scale bar = 1.0 mm. Photograph by Gunther Köhler.

mucronate and/or imbricate (Fig. 90), 34–56 between levels of axilla and groin; 76–120 scales around midbody; male dewlap with large blue basal spot and pinkish to dark orange outer border in life (Fig. 10); dorsal ground color green in life when not stressed (Fig. 10) .....
............ *Norops biporcatus* (p. 33)

18 A. Lateral body scales heterogeneous (at least in majority of individuals; Fig. 91), solitary, enlarged, keeled, or elevated scales (sometimes slightly paler color than other laterals) frequently scattered among laterals, or if lateral body scales homogeneous, male dewlap dirty white, pale yellow, or pale gray in life and ventral scales distinctly keeled; males with enlarged postcloacal

scales (Fig. 92); usually at least 1 subocular in broad contact with a supralabial (Fig. 3) .....     19

B. Lateral body scales homogeneous; combination of other characters different from above ................     28

19 A. Enlarged dorsal scales abruptly different in size from much smaller lateral body scales (Fig. 93); midventral scales smooth (Fig. 94); male dewlap orange-yellow in life (Fig. 4) ......
..... *Norops amplisquamosus* (p. 18)

B. Enlarged dorsal scales usually grading into smaller lateral body scales (Fig. 95); midventral scales smooth to strongly keeled; male dewlap not orange-yellow in life ............     20

20 A. Male dewlap dirty white, pale yellow, or pale gray in life; 0–5 slightly enlarged middorsal scale rows ............................     21

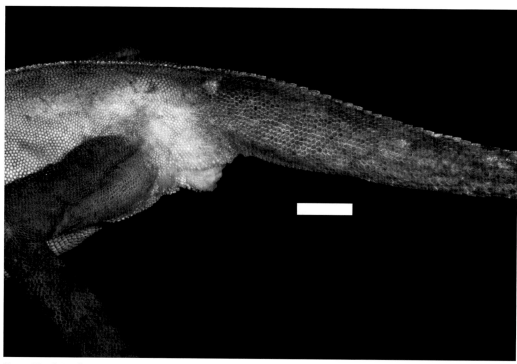

Figure 88.   Laterally compressed tail in a male *Norops loveridgei* (SMF 86951) from Quebrada de Oro, Atlántida. Scale bar = 5.0 mm. Photograph by Gunther Köhler.

   B.  Male dewlap red, bright orange, or orange-red in life; at least 6 distinctly enlarged middorsal scale rows ........................... 23

21 A.  Male dewlap extending to about level of axilla (Fig. 28)..........
........ *Norops laeviventris* (p. 82)

   B.  Male dewlap extending well onto chest posterior to level of axilla (Fig. 26) .................. 22

22 A.  Male dewlap pale yellow with purple gorgetal scales in life (Fig. 26); occurs in western portion of Cordillera Nombre de Dios in north-central Honduras ....... *Norops kreutzi* (p. 79)

   B.  Male dewlap dirty white, pale yellow, or pale gray, with gorgetal scales of same color (Fig. 20); occurs in Parque Nacional Cusuco and Parque Nacional Cerro Azul in north-western Honduras................
.............. *Norops cusuco* (p. 65)

23 A.  Midventral scales strongly keeled (Fig. 96) .................................. 24

   B.  Midventral scales smooth or only weakly keeled ............... 26

24 A.  Hemipenis with divided asulcate processus (Fig. 97)...............
.......... *Norops morazani* (p. 106)

   B.  Hemipenis with undivided asulcate processus (Fig. 98) ........ 25

25 A.  Usually about 8–11 rows of enlarged dorsal scales; female dewlap small, white or orange-red in life; occurs on Montaña Santa Bárbara......................
... *Norops rubribarbaris* (p. 161)

   B.  Usually about 14–18 rows of enlarged dorsal scales; female dewlap well-developed, but smaller than male dewlap, bright orange or orange red

Figure 89.   Midventrals weakly keeled in *Norops petersii* (SMF 86943) from Baja Verapaz, Guatemala. Scale bar = 1.0 mm. Photograph by Gunther Köhler.

in life; occurs in mountains of SW and east-central Honduras ......... *Norops crassulus* (p. 49)

26 A. Midventral scales weakly keeled (Fig. 99) ... *Norops sminthus* (p. 172)

B. Midventral scales smooth (Fig. 100) ..................................... 27

27 A. Medial dorsal scales uniform in size, without interspersed small scales (Fig. 101) ........... . *Norops heteropholidotus* (p. 70)

B. Small scales irregularly interspersed among enlarged medial dorsal scales (Fig. 102) ...... ............ *Norops muralla* (p. 110)

28 A. Dorsal scales flat, juxtaposed, hexagonal (Fig. 103); male dewlap orange-yellow in life (Fig. 44); semiaquatic along streams in NE Honduras....... ......... *Norops oxylophus* (p. 126)

B. Dorsal scales weakly keeled and subimbricate (Fig. 104) to dis-

tinctly keeled and imbricate (Fig. 105), not hexagonal; male dewlap not orange-yellow in life ...................................... 29

29 A. Midventral scales weakly keeled or smooth, not mucronate (Figs. 99–100) ..................... 30

B. Midventral scales distinctly to strongly keeled (Fig. 106), often mucronate, or if midventrals weakly keeled, male dewlap brown in life, usually with large darker brown spot ........ 34

30 A. Ventrals granular or conical, subimbricate or nonimbricate (Fig. 107); short-legged (fourth toe of adpressed hind limb reaches not much further than shoulder); usually 4 (rarely 3 or 5) rows of loreals; supraoculars smooth to weakly keeled (Fig. 5) ..................................... 31

Figure 90.   Midventrals strongly keeled in *Norops biporcatus* (SMF 79147) from Laguna del Cerro, Copán. Scale bar = 1.0 mm. Photograph by Gunther Köhler.

B. Ventrals flat, distinctly imbricate (Fig. 108); longer-legged (fourth toe of adpressed hind limb reaches beyond ear opening); 5 or more (rarely 4) rows of loreals; supraoculars distinctly keeled (Fig. 23) .......... 32

31 A. Five supracaudal scales present per caudal segment (Fig. 109); occurs only on Isla de Utila ........ *Norops utilensis* (p. 196)

B. Four supracaudal scales present per caudal segment (Fig. 110); occurs across northern mainland ........ *Norops beckeri* (p. 22)

32 A. Male dewlap orange-red with large central blue blotch in life (Fig. 24); female dewlap well developed, yellow with large central blue blotch in life (Fig. 24); SVL to 73 mm in males, 68 mm in females....... ......... *Norops johnmeyeri* (p. 74)

B. Male dewlap rose with purple spot or uniform purple in life; female dewlap small, usually similar color to that of male; SVL to 59 mm in males, 60 mm in females ............................. 33

33 A. Male dewlap purple in life (Fig. 50); 6–9 (mean 7.7) scales between second canthals; snout scales multicarinate (Fig. 49); 6–10 (mean 8.5) scales between posterior canthals .................... .... *Norops purpurgularis* (p. 141)

B. Male dewlap rose with purple central spot in life (Fig. 48); 6–12 (mean 8.6) scales between second canthals; snout scales unicarinate (Fig. 47); 7–15 (mean 9.9) scales between

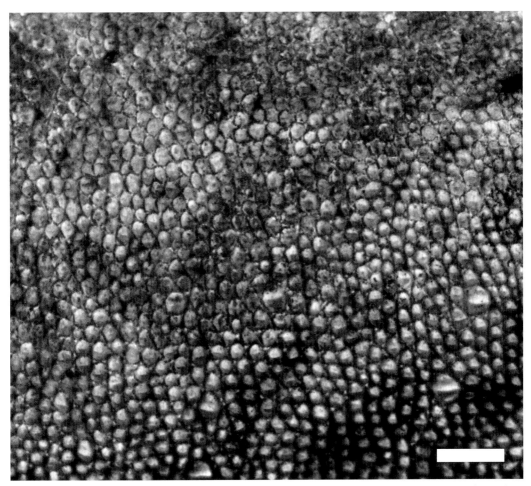

Figure 91.   Lateral heterogeneous body scales in *Norops heteropholidotus* (SMF 78027) from Quebrada La Quebradona, Ocotepeque. Scale bar = 1.0 mm. Photograph by Gunther Köhler.

posterior canthals..................
.......... *Norops pijolense* (p. 137)

34 A. Male dewlap yellowish orange with large blue blotch in life (Fig. 74); short-legged (fourth toe of adpressed hind limb reaches between shoulder and ear opening, rarely beyond ear opening) ............................. 35

B. Male dewlap brown or some shade of red in life; long-legged (fourth toe of adpressed hind limb reaches between ear opening and anterior margin of eye, usually beyond posterior margin of eye) ...................... 36

35 A. Hemipenis bilobed (Fig. 111); occurs on Pacific versant in southern Honduras ..............
....... *Norops wellbornae* (p. 205)

B. Hemipenis unilobed (Fig. 112); occurs on Caribbean versant, including several interior valleys in south-central Honduras ........ *Norops unilobatus* (p. 189)

36 A. Male dewlap brown in life, usually with large darker brown spot (Fig. 18); no con-

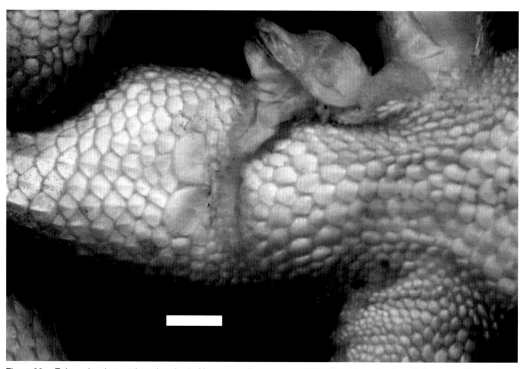

Figure 92.    Enlarged male postcloacal scales in *Norops amplisquamosus* (SMF 77750) from Sendero El Danto, Cortés. Scale bar = 1.0 mm. Photograph by Gunther Köhler.

spicuous dark brown lines radiating outward from eye; maximum SVL about 53 mm; ventral scales not mucronate and varying from subimbricate to imbricate (Fig. 113) .............. .............. *Norops cupreus* (p. 55)

B. Male dewlap some shade of red in life; conspicuous dark brown lines radiating outward from eye (Fig. 114); maximum SVL 62–86 mm; ventral scales mucronate and imbricate (Fig. 115) ................................... 37

37 A. Male dewlap without suffusion of black pigment centrally, often with black to dark brown edged gorgetal scales (Fig. 30); males average about 56 mm SVL; 5–8 (rarely 9) horizontal loreal rows; occurs on mainland and Cayos Cochinos ...... .......... *Norops lemurinus* (p. 87)

B. Male dewlap with suffusion of black pigment centrally and mostly white gorgetal scales (Fig. 54); 8–10 (rarely 6 or 7) horizontal loreal rows; occurs on Isla de Roatán or on Isla de Utila ................................... 38

38 A. Males average about 64 mm SVL, females about 66 mm; sulcal branches of hemipenis opening into broad concave area distal to point of bifurcation on each lobe (Fig. 116); low asulcate processus present; endemic to Isla de Utila........ .......... *Norops bicaorum* (p. 27)

B. Males average about 56 mm SVL, females about 58 mm; sulcal branches of hemipenis extending

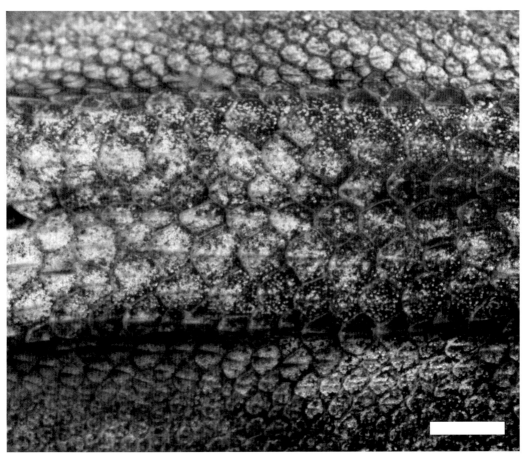

Figure 93.   Dorsal scales abruptly enlarged in *Norops amplisquamosus* (SMF 77747) from Sendero El Danto, Cortés. Scale bar = 1.0 mm. Photograph by Gunther Köhler.

to tip of each lobe (Fig. 117); asulcate processus absent .........
......... *Norops roatanensis* (p. 151)

CLAVE PARA LAS ESPECIES HONDUREÑAS DE LOS GÉNEROS *ANOLIS* Y *NOROPS*

1 A.  Todas de las vertebras caudales autotómicas sin el proceso transversal dirigido anterolateral (Fig. 79); abertura ótica horizontalmente elongada y con el margen posterior formando una depresión alargada (Fig. 1); cabeza en adultos conspicuamente alargadas (Fig. 1); >35 lamelas abajo del cuarto dedo del pie ...........
.................. *Anolis allisoni* (p. 11)

B.  Al meños algunas de las vertebras caudales autotómicas con procesos transversales dirigidos anterolateralmente (Fig. 79); abertura ótica verticalmente oval (Fig. 3); cabeza en adultos no conspicuamente alargadas (Fig. 3); <30 lamelas abajo del cuarto dedo del pie .................   **2**

2 A.  Un bolso axilar en forma de tubo y sin escamas presente (Fig. 80) .............................................   **3**

Figure 94.   Smooth midventral scales in *Norops amplisquamosus* (SMF 77747) from Sendero El Danto, Cortés. Scale bar = 1.0 mm. Photograph by Gunther Köhler.

B. No hay un bolso axilar en forma de tubo y sin escamas, o no existe más que una pequeña depresión axilar con escamas ...   **6**

3 A. Las escamas anteriores al la abertura ótica planas y quilladas, aproximadamente del doble de tamaño de las que están posteriormente a ésta (éstas últimas usualmentre granulares; Fig. 63) ........................   **4**

B. Las escamas anteriores al la abertura ótica aproximadamente del mismo tamaño que las que están posteriormente a ésta (todas granulares; Fig. 51) ...................   **5**

4 A. El abanico gular en machos de color naranja-rojo en vida con margen amarillo y con una banda central oscura (Fig. 64); los machos adultos con una longitud hocico-cloaca (LHC) hasta de 59 mm, hembras adul-

tas hasta de 55 mm LHC .........
....... *Norops tropidonotus* (p. 177)

B. El abanico gular en machos de color naranja-rojo en vida, con margen amarillo, pero sin una banda central oscura (Fig. 72); machos y hembras con una LHC hasta de 51 mm ...........
..... *Norops wampuensis* (p. 201)

5 A. Los flancos de cuerpo con 1–3 líneas verticales pálidas (las líneas pueden estar interrumpidas); abanico gular en machos de color rosa con un punto grande de color púrpura en vida (Fig. 66); usualmente entre 23–32 (en promedio 28.5) escamas medio-dorsales entre los bordes de la axila y la ingle, raramente hasta 37 ......
......... *Norops uniformis* (p. 185)

B. Los flancos de cuerpo sin líneas verticales pálidas (completas o

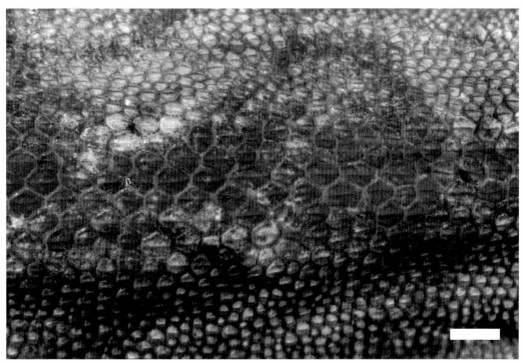

Figure 95.   Enlarged dorsal scales grading into smaller laterals in *Norops crassulus* (SMF 78799) from Pico La Picucha, Olancho. Scale bar = 1.0 mm. Photograph by Gunther Köhler.

interrumpidas); abanico gular en machos de color rojizo anaranjado con margen amarillo en vida (Fig. 52); usualmente entre 30–44 (en promedio 37.0) escamas mediodorsales entre los bordes de la axila y la ingle ...................
........ *Norops quaggulus* (p. 146)

6 A. Las escamas postmentales más externas, en cada lado, muy agrandada, su longitud mayor a la de la escama mental (Fig. 59); la cola de los machos adultos distintivamente comprimida lateralmente, y con una cresta medio-dorsal de escamas alargadas (Fig. 81); LHC máxima en adultos 76 mm ................................. **7**

B. Las escamas postmentales más externas, en cada lado, no muy largo, su longitud menor que la

longitud de la escama mental (Fig. 75); la cola de los machos adultos no o ligeramente comprimida (Fig. 82), o si está notoriamente comprimida, los adultos son de tamaño grande (LHC 76 mm); cola con o sin una cresta medio-dorsal de escamas alargadas .................. **8**

7 A. Usualmente cuatro (en promedio 4.3) escamas postmentales; abanico gular en machos de color anaranjado o naranja-rojo en vida (Fig. 60); se encuentran en las islas de Roatán y Utila y el extremo norte de la parte continental del país ......
.............. *Norops sagrei* (p. 166)

B. Usualmente cinco o más (raramente 4; en promedio 5.8) escamas postmentales; abanico gular en machos de color café oscuro a gris en vida (Fig. 40);

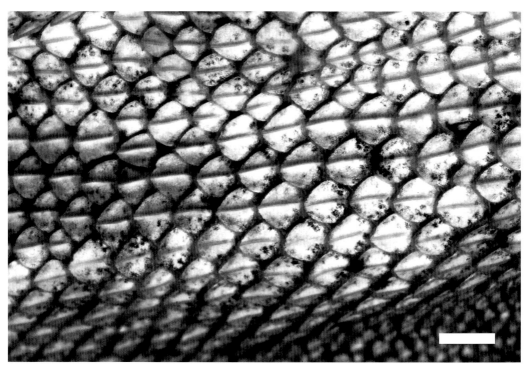

Figure 96.   Midventral scales strongly keeled in *Norops crassulus* (SMF 78799) from Pico La Picucha, Olancho. Scale bar = 1.0 mm. Photograph by Gunther Köhler.

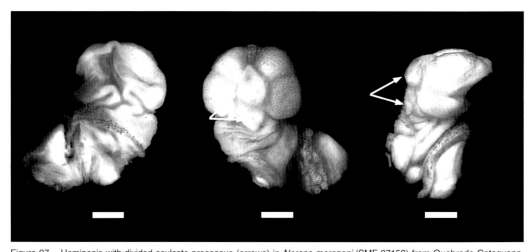

Figure 97.   Hemipenis with divided asulcate processus (arrows) in *Norops morazani* (SMF 87153) from Quebrada Cataguana, Francisco Morazán. Scale bar = 1.0 mm. Photograph by Gunther Köhler.

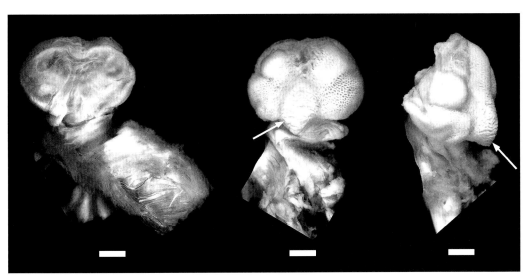

Figure 98. Hemipenis with undivided asulcate processus (arrows) in *Norops crassulus* (SMF 78104) from Sonsonate, El Salvador. Scale bar = 1.0 mm. Photograph by Gunther Köhler.

Figure 99. Weakly keeled midventral scales in *Norops sminthus* (SMF 77181) from Cerro La Tigra, Francisco Morazán. Scale bar = 1.0 mm. Photograph by Gunther Köhler.

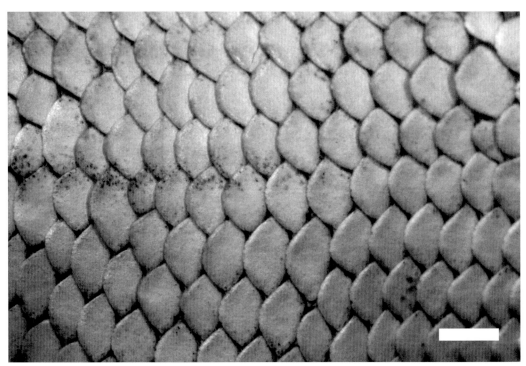

Figure 100.  Smooth midventral scales in *Norops heteropholidotus* (SMF 78030) from Quebrada La Quebradona, Ocotepeque. Scale bar = 1.0 mm. Photograph by Gunther Köhler.

Figure 101.  Median dorsal scales uniform in size in *Norops heteropholidotus* (SMF 78027) from Quebrada La Quebradona, Ocotepeque. Scale bar = 1.0 mm. Photograph by Gunther Köhler.

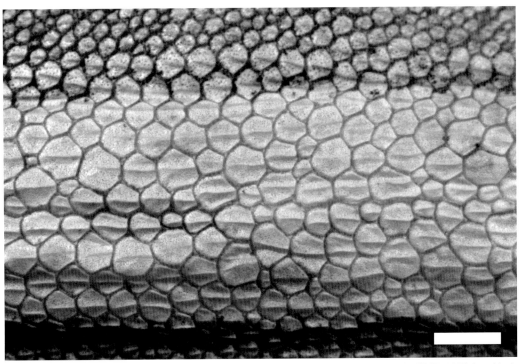

Figure 102. Small scales irregularly interspersed among enlarged dorsal scales in *Norops muralla* (SMF 78378) from Monte Escondido, Olancho. Scale bar = 1.0 mm. Photograph by Gunther Köhler.

se encuentran en las Islas del Cisne .... *Norops nelsoni* (p. 114)

8 A. Usualmente sólo una escama prenasal elongada (Fig. 31); escamas ventrales lisas (Fig. 83) a ligeramente quilladas (Fig. 84); forma corporal alargada ........................................ **9**

B. Usualmente dos escamas prenasales redondeadas (Fig. 11); escamas ventrales variables; forma corporal usualmente no alargada, o si lo está, las escamas ventrales son fuertemente quilladas (Fig. 85) y/o abanico gular en machos de color amarillo-naranja con una mancha grande de color azul en vida ................................ **14**

9 A. Piernas largas (punta del cuatro dedo del pie cuando las piernas se doblan hacia la parte anterior del cuerpo, llega al borde posterior del ojo, usualmente más allá del ojo); abanico gular en machos de color blancuzco, con o sin una mancha naranja-amarillenta en la base, en vida ........................ **10**

B. Piernas cortas (punta del cuatro dedo del pie cuando las piernas se doblan hacia la parte anterior del cuerpo, usualmente no rebasa más allá del borde posterior del ojo, usualmente más allá del ojo, llegando al borde anterior del ojo en una especie [*N. ocelloscapularis*]); abanico gular en machos de color naranja o amarillo-naranja en vida ........ **11**

10 A. Abanico gular en machos de color blancuzco, con una mancha naranja-amarillenta en la base, en vida (Fig. 32)...........
........... *Norops limifrons* (p. 96)

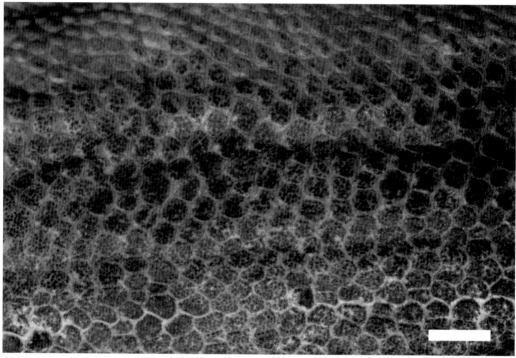

Figure 103.    Dorsal scales flat, juxtaposed, hexagonal in *Norops oxylophus* (SMF 88677) from Matamoros, Olancho. Scale bar = 1.0 mm. Photograph by Gunther Köhler.

B. Abanico gular en machos de color blancuzco uniforme, sin una mancha naranja-amarillenta en la base, en vida (Fig. 78) .... *Norops zeus* (p. 217)

11 A. Usualmente con ocelo sobre el hombro (Fig. 42); piernas relativamente largas (punta del cuatro dedo del pie cuando las piernas se doblan hacia la parte anterior del cuerpo, llega entre el borde posterior y anterior del ojo); usualmente una banda lateral clara sobre el cuerpo presente en vida ....................................
... *Norops ocelloscapularis* (p. 121)

B. Sin ocelo sobre el hombro; piernas relativamente cortas (punta del cuatro dedo del pie cuando las piernas se doblan hacia la parte anterior del cuerpo, llega entre la abertura ótica y la mitad del ojo); con o sin una banda lateral clara sobre el cuerpo presente en vida ................... **12**

12 A. Dorso de color café-verdoso con puntas claros en vida (Fig. 14); sin una banda lateral clara en vida .... *Norops carpenteri* (p. 45)

B. Dorso de color café, sin puntas claros en vida; con o sin una banda lateral clara sobre el cuerpo presente en vida ........ **13**

13 A. Escamas medio-ventrales lisas y no imbricadas (Fig. 83); 1–4 suboculares en amplio contacto con las supralabiales .....
....... *Norops rodriguezii* (p. 156)

B. Escamas medio-ventrales ligeramente quilladas e imbricadas (Fig. 84); suboculares usualmente separadas de las supralabiales por una hilera de escamas o 1–2 suboculares en

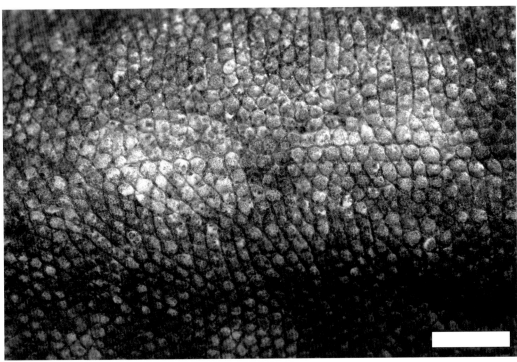

Figure 104.   Dorsal scales weakly keeled, subimbricate in *Norops utilensis* (SMF 79866) from Isla de Utila, Islas de la Bahía. Scale bar = 1.0 mm. Photograph by Gunther Köhler.

ligero contacto con supralabiales ...... *Norops yoroensis* (p. 208)

14 A. La cabeza es conspicuamente ancha y robusta, la punta del hocico es roma en vista lateral (Fig. 11);la proporción longitud de la tibia/longitud de la cabeza (LT/LC) ≥1.1; usualmente una banda clara cruzando la barbilla, banda más evidente en juveniles; escamas ventrales fuertemente quilladas (Fig. 85); escamas dorsales lisas, yuxtapuestas, la mayoría pentagonales o hexagonales (Fig. 86); LHC en adultos 66–100 mm; abanico gular en machos uniformemente verdoso-amarillo, café o cafesoso-amarillo en vida, pequeño (Fig. 12), no se extiende más allá del nivel de la axila.........
............... *Norops capito* (p. 39)

B. La cabeza no es conspicuamente ancha y robusta (Fig. 7); LT/LC usualmente <1.0, o si es >1.0, escamas dorsales quilladas (Fig. 87); no hay una banda clara cruzando la barbilla; combinación de otros caracteres no como lo descrito en la opción 14 A ............... **15**

15 A. Adultos de tamaño grande (LHC 70 a ligeramente mayores de 100 mm); 9–13 supralabiales a el nivel de la parte medio del ojo; machos con o sin escamas postcloacales alargadas ..................... **16**

B. Especies de tamaño mediano-pequeño hasta mediano largo (LHC máxima ≤70 mm, excepto en *N. lemurinus*, el cual puede llegar a 79 mm LHC [hasta 73 mm LHC en especímenes Hondureñas] y *N. bi-*

Figure 105.    Dorsal scales distinctly keeled, imbricate in *Norops lemurinus* (SMF 85885) from Raudal Kiplatara, Gracias a Dios. Scale bar = 1.0 mm. Photograph by Gunther Köhler.

*caorum*, el cual puede llegar a medir de LHC 86 mm); combinación de otros caracteres no como lo descrito en la opición 15 A ....................................... **18**

16 A.  Escamas loreales >60; usualmente >130 escamas alrededor de la parte media del cuerpo; punta del cuatro dedo del pie cuando las piernas se doblan hacia la parte anterior del cuerpo, usualmente llega al menos hasta el borde posterior del ojo; cola lateralmente comprimida (Fig. 88); abanico gular en machos de color naranja con lineas de púrpura en vida (Fig. 34) ..............................
........ *Norops loveridgei* (p. 101)

B.  Escamas loreales <62; usualmente menos de 130 escamas alrededor de la parte media del cuerpo; punta del cuatro

dedo del pie cuando las piernas se doblan hacia la parte anterior del cuerpo, usualmente llega solamente hasta el tímpano; cola redondeado hasta ligeramente comprimida; abanico gular en machos rosáceo café o con un punto azul grande basal en vida ............. **17**

17 A.  Escamas dorsales anteriores de la cabeza rugosas o ligeramente quilladas (Fig. 45); escamas medio-ventrales ligeramente quilladas (Fig. 89) con márgenes posteriores redondeados o truncados, de 62–86 (68–86 en especímenes Hondureñas) entre la axila y la ingle; 116–140 (116–133 en especímenes Hondureñas) escamas alrededor de la parte media del cuerpo; abanico gular en machos rosáceo café con margen amarillo

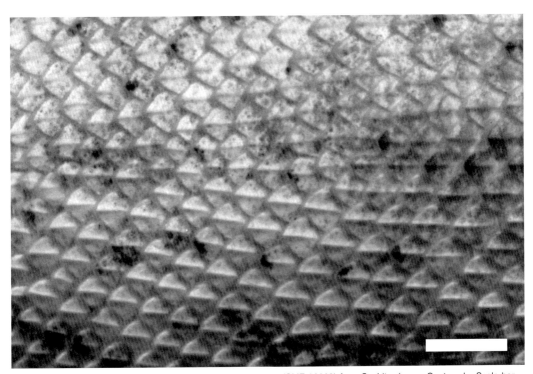

Figure 106.   Strongly keeled ventral scales in *Norops wellbornae* (SMF 82668) from Suchitepéquez, Guatemala. Scale bar = 1.0 mm. Photograph by Gunther Köhler.

claro en vida (Fig. 46); color del cuerpo moteado de olivo amarillo y cafesoso olivo en vida (Fig. 46) .. *Norops petersii* (p. 131)

B. Escamas dorsales anteriores de la cabeza fuertemente quilladas (Fig. 9); escamas medio-ventrales fuertemente quilladas (Fig. 90) y mucronadas y/o imbricadas, de 34–56 entre la axila y la ingle; 76–120 escamas alrededor de la parte media del cuerpo; abanico gular en machos con un punto basal de color azul de gran tamaño y bordeado de rosáceo a naranja oscuro en vida (Fig. 10); coloración dorsal verde en vida, al no estar estresados (Fig. 10)....
........... *Norops biporcatus* (p. 33)

18 A. Escamas laterales del cuerpo heterogéneas (al menos en

algunos individuos; Fig. 91), algunas escamas alargadas, quilladas o elevadas (algunas veces ligeramente de color más pálido que otras escamas laterales) frecuentemente esparcidas entre otras escamas laterales, o si las escamas laterales del cuerpo son homogéneas, el color del abanico gular en machos es blancuzco, amarillo pálido o gris pálido en vida y las escamas ventrales están distintivamente quilladas; los machos poseen escamas postcloacales alargadas (Fig. 92); usualmente por lo menos una escama subocular en amplio contacto con una supralabial (Fig. 3) ...............................   **19**

B. Escamas laterales del cuerpo homogéneas; combinación de

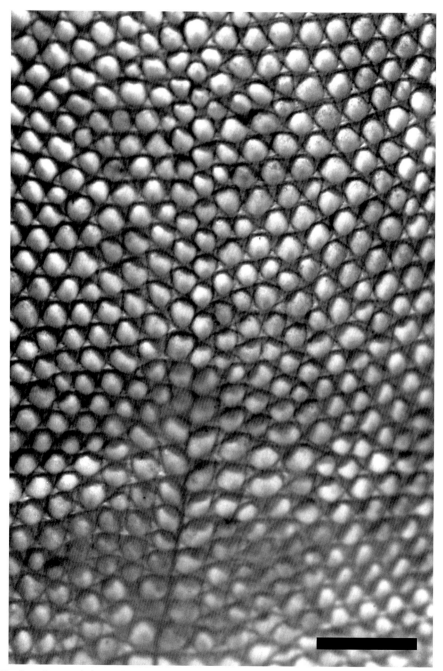

Figure 107.    Granular or conical ventral scales in *Norops utilensis* (SMF 79866) from Isla de Utila, Islas de la Bahía. Scale bar = 1.0 mm. Photograph by Gunther Köhler.

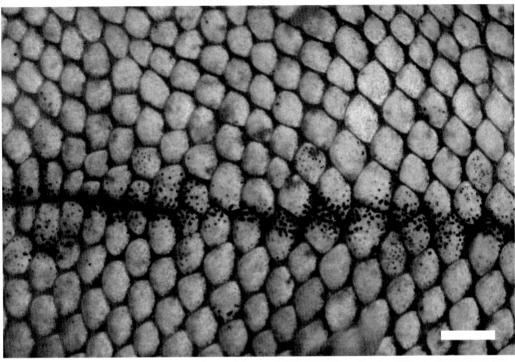

Figure 108.   Flat and imbricate ventral scales in *Norops johnmeyeri* (SMF 77756) from Sendero El Danto, Cortés. Scale bar = 1.0 mm. Photograph by Gunther Köhler.

otros caracteres no como lo descrito en la opición 18 A ....  **28**

19 A.  Escamas dorsales del cuerpo abruptamente más grandes en tamaño que las escamas later-ales del cuerpo (Fig. 93); esca-mas medio-ventrales lisas (Fig. 94); abanico gular en machos de color naranja amar-illento en vida (Fig. 4)..........
. *Norops amplisquamosus* (p. 18)

B.  Escamas dorsales del cuerpo usualmente disminoyendo gra-dualmente de tamaño de las pequeñas escamas laterales (Fig. 95); escamas medio-ven-trales fuertemente quilladas; abanico gular en machos no es de color naranja amarillento en vida ................................  **20**

20 A.  Abanico gular en machos blan-cuzco, amarillo pálido o gris pálido en vida; de 0–5 escamas medio-dorsales ligeramente agrandadas ..........................  **21**

B.  Abanico gular en machos color rojo, anaranjado brillante o rojo-naranja en vida; por los menos 6 escamas medio-dor-sales distintamente agrandadas ...........................................  **23**

21 A.  Abanico gular en machos exten-diéndose hasta el nivel de la axila (Fig. 28)......................
........ *Norops laeviventris* (p. 82)

B.  Abanico gular en machos exten-diéndose hasta el nivel del pecho más allá del nivel de la axila (Fig. 26) ......................  **22**

22 A.  Abanico gular en machos amarillo pálido, con escamas gorgetales púrpura en vida (Fig. 26); se encuentran en la porción oeste de la Cordillera Nombre de Dios,

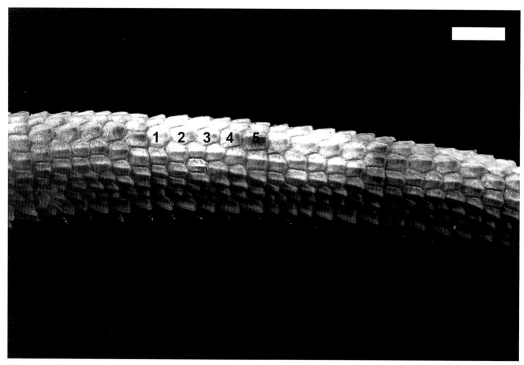

Figure 109.   Five supracaudal scales per caudal segment in *Norops utilensis* (SMF 77055) from Isla de Utila, Islas de la Bahía. Scale bar = 1.0 mm. Photograph by Gunther Köhler.

en el centro-norte de Hondu-ras ............. *Norops kreutzi* (p. 79)

B. Abanico gular en machos blan-cuzco, amarillo pálido o gris pálido con escamas gorgetales del mismo color (Fig. 20); se encuentran en Parque Nacional Cusuco and Parque Nacional Cerro Azul en el noroeste de Honduras ... *Norops cusuco* (p. 65)

23 A. Escamas medio-ventrales fuer-temente quilladas (Fig. 96) ... **24**

B. Escamas medio-ventrales lisas o ligeramente quilladas ............ **26**

24 A. Hemipenes con el proceso asul-cado dividido (Fig. 97)..........
.......... *Norops morazani* (p. 106)

B. Hemipenes con el proceso asul-cado no dividido (Fig. 98) ..... **25**

25 A. Usualmente entre 8–11 escamas dorsales alargadas; abanico gu-lar en hembras pequeño, de color blanco o naranja-rojo en vida; se encuentran en la Montaña Santa Bárbara.........
... *Norops rubribarbaris* (p. 161)

B. Usualmente entre 14–18 esca-mas dorsales alargadas; aba-nico gular en hembras bien desarrollado, pero más pe-queño que el de los machos, de color anaranjado brillante o naranja-rojo en vida; se en-cuentran en las montañas del suroeste y centro-este de Hon-duras .... *Norops crassulus* (p. 49)

26 A. Escamas medio-ventrales ligera-mente quilladas (Fig. 99).......
.......... *Norops sminthus* (p. 172)

B. Escamas medio-ventrales lisas (Fig. 100) ............................. **27**

27 A. Escamas medio-dorsales de ta-maño uniforme, sin escamas pequeñas intercaladas (Fig.

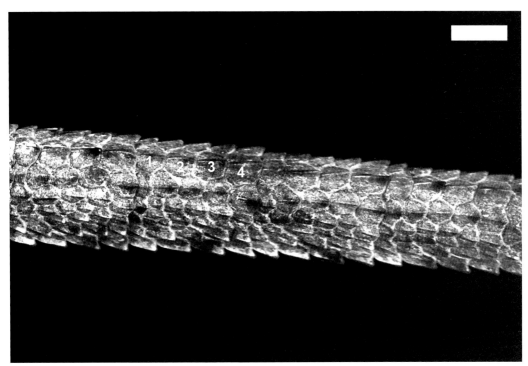

Figure 110.   Four supracaudal scales per caudal segment in *Norops beckeri* (SMF 91288) from San José de Colinas, Santa Bárbara. Scale bar = 1.0 mm. Photograph by Gunther Köhler.

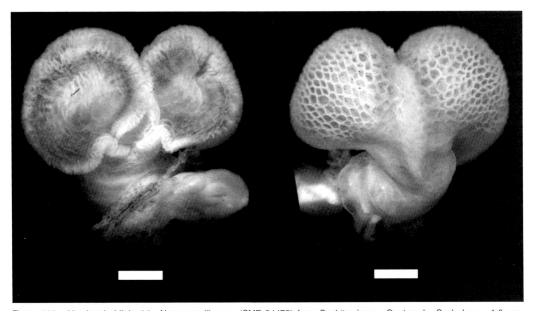

Figure 111.   Hemipenis bilobed in *Norops wellbornae* (SMF 84473) from Suchitepéquez, Guatemala. Scale bar = 1.0 mm. Photograph by Gunther Köhler.

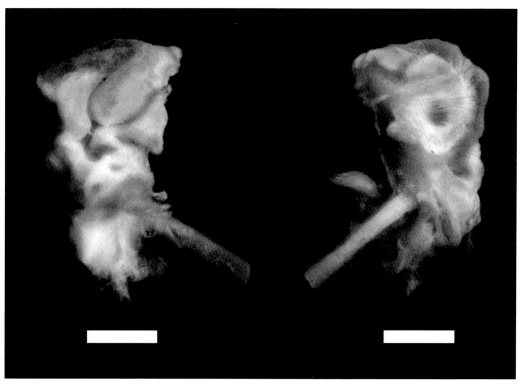

Figure 112.　Hemipenis unilobed in *Norops unilobatus* (SMF 79366) from Isla de Utila, Islas de la Bahía. Scale bar = 1.0 mm. Photograph by Gunther Köhler.

101) ..........................................
.... *Norops heteropholidotus* (p. 70)

B. Pequeñas escamas irregularmente intercaladas entre las medio-dorsales que son más grandes (Fig. 102) ................
............ *Norops muralla* (p. 110)

28 A. Escamas dorsales aplanadas, yuxtapuestas, hexagonales (Fig. 103); abanico gular en machos naranja-amarillento en vida (Fig. 44); especie semiacuática, vive a lo largo de arroyos en el noreste de Honduras ... *Norops oxylophus* (p. 126)

B. Escamas dorsales ligeramente quilladas y subimbricadas (Fig. 104) hasta distintivamente quilladas y imbricadas (Fig. 105), no hexagonales; abanico gular en machos no

de color naranja-amarillento en vida ................................ **29**

29 A. Escamas medio-ventrales ligeramente quilladas o lisas, no mucronadas (Figs. 99–100) .... **30**

B. Escamas medio-ventrales de distintivamente a fuertemente quilladas (Fig. 106), frecuentemente mucronadas, o si las medio-ventrales están ligeramente quilladas, el abanico gular de los machos es café en vida, usualmente con un punto grande de color café oscuro .... **34**

30 A. Ventrales granulares o cónicas, subimbricadas o no (Fig. 107); piernas relativamente cortas (punta del cuatro dedo del pie cuando las piernas se doblan hacia la parte anterior del cuerpo, no llega más allá del

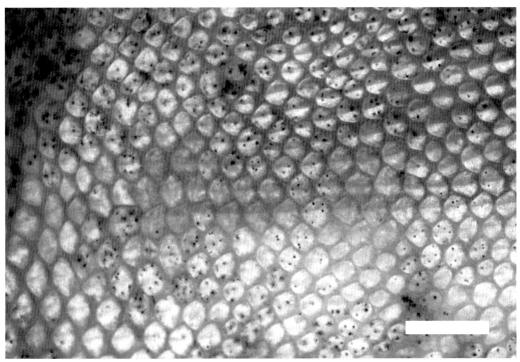

Figure 113.   Subimbricate to imbricate ventral scales in *Norops cupreus* (SMF 91254) from Bachi Kiamp, Gracias a Dios. Scale bar = 1.0 mm. Photograph by Gunther Köhler.

hombro); usualmente 4 (raramente 3 o 5) hileras de loreales; supraoculares lisas o débilmente quilladas (Fig. 5) .......... **31**

B. Ventrales planas y distintivamente imbricadas (Fig. 108); piernas relativamente largas (punta del cuatro dedo del pie cuando las piernas se doblan hacia la parte anterior del cuerpo, no llega más allá de la abertura ótica); 5 o más (raramente 4) hileras de loreales; supraoculares distintivamente quillidas (Fig. 23) ....... **32**

31 A. Cinco escamas supracaudales presente por cada segmento caudal (Fig. 109); se encuentran solamente en la Isla de Utila ... *Norops utilensis* (p. 196)

B. Cuatro escamas supracaudales presente por cada segmento

caudal (Fig. 110); se encuentran en el norte de Honduras ........... *Norops beckeri* (p. 22)

32 A. Abanico gular en machos naranja-rojo con un punto grande central de color azul en vida (Fig. 24); abanico gular en hembras bien desarrolado, amarillo con un punto central grande de la color azul en vida (Fig. 24); LHC hasta 73 mm en machos, 68 mm en hembras ... *Norops johnmeyeri* (p. 74)

B. Abanico gular en machos rosado con un punto púrpura o de color uniformemente púrpura en vida; abanico gular en hembras bien dessarollado, pero más pequeño que el abanico gular de machos, y usualmente de color similar al del macho; LHC hasta 59 mm en machos, 60 mm en hembras .............. **33**

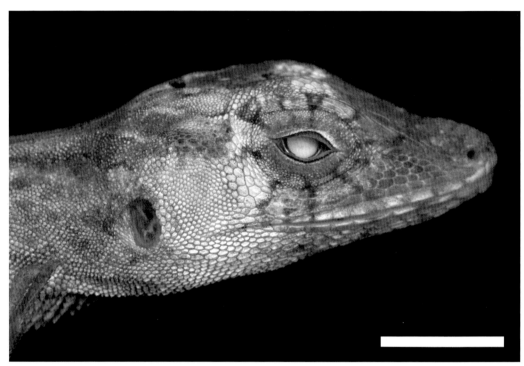

Figure 114.   Conspicuous lines radiating outward from eye in *Norops lemurinus* (SMF 85887) from Raudal Kiplatara, Gracias a Dios. Scale bar = 5.0 mm. Photograph by Gunther Köhler.

Figure 115.   Ventral scales mucronate and imbricate in *Norops lemurinus* (SMF 85885) from Raudal Kiplatara, Gracias a Dios. Scale bar = 1.0 mm. Photograph by Gunther Köhler.

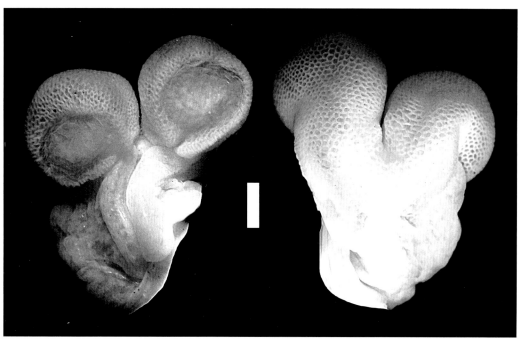

Figure 116.   Sulcal branches of hemipenis opening into broad concave area distal to bifurcation on each lobe in *Norops bicaorum* (SMF 81127) from Isla de Utila, Islas de la Bahía, Honduras. Scale bar = 1.0 mm. Photograph by Gunther Köhler.

33 A.  Abanico gular en machos púrpura en vida (Fig. 50); 6–9 (en
promedio 7.7) escamas entre
las cantales secundarias; escamas de la hocico multicarina-
das (Fig. 49); 6–10 (en promedio 8.5) escamas entre las
cantales posteriores..............
... *Norops purpurgularis* (p. 141)

B.  Abanico gular en machos rosa
con un punto central púrpura
en vida (Fig. 48); 6–12 (en
promedio 8.6) escamas entre
las cantales secundarias; escamas de la hocico unicarinadas
(Fig. 47); 7–15 (en promedio
9.9) escamas entre las cantales
posteriores..........................
.......... *Norops pijolense* (p. 137)

34 A.  Abanico gular en machos amar-
illento naranja con un punto
grande de color azul en vida
(Fig. 74); piernas cortas (punta
del cuatro dedo del pie cuando
las piernas se doblan hacia la
parte anterior del cuerpo, llego
entre el hombro y la abertura
ótica, raramente más allá de la
apetura ótica) ......................  **35**

B.  Abanico gular en machos café o
alguno color de rojo en vida;
piernas relativamente largas
(punta del cuatro dedo del
pie cuando las piernas se
doblan hacia la parte anterior
del cuerpo, llego entre la
abertura ótica y el anterior
margen del ojo, usualmente
más álla del margen posterior
del ojo) ...............................  **36**

35 A.  Hemipenes bilobulados (Fig.
111); se encuentra en la ver-
tiente del Pacífico en el
sur .. *Norops wellbornae* (p. 205)

B.  Hemipenes unilobulados (Fig.
112); se encuentra en la ver-

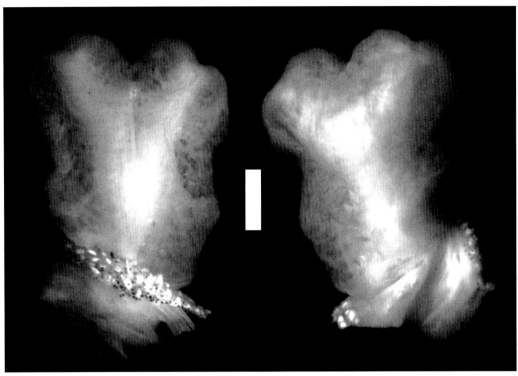

Figure 117. Sulcal branches of hemipenis extending to tip of each lobe in *Norops lemurinus* (SMF 80816) from Caobita, Olancho, Honduras. Scale bar = 1.0 mm. Photograph by Gunther Köhler.

tiente del Caribe, incluyendo varios valles interiores en el sur-central Honduras ............

........ *Norops unilobatus* (p. 189)

36 A. Abanico gular en machos café en vida, usualmente con un punto grande de color café oscuro (Fig. 18); sin unas líneas conspicuas de color café oscuro que radian hacia fuera del ojo; LHC máxima aproximadamente 53 mm; escamas ventrales no mucronadas y variando de subimbricadas a imbricadas (Fig. 113) ............

.............. *Norops cupreus* (p. 55)

B. Abanico gular en machos alguno color de rojo en vida; líneas conspicuas de color café oscuro que radian hacia fuera del ojo presente (Fig. 114); LHC

máxima 62–86 mm; escamas ventrales mucronadas e imbricadas (Fig. 115) ................... **37**

37 A. Abanico gular en machos sin influencia de pigmento negro en la parte central, frecuentemente con escamas gorgetales con un borde negro o café oscuro (Fig. 30); 56 mm LHC promedio en machos; 5–8 (raramente 9) hileras de escamas loreales horizontales; se encuentran en los Cayos Cochinos y el continente...............

.......... *Norops lemurinus* (p. 87)

B. Abanico gular en machos con influencia de pigmento negro en el centro y principalmente con escamas gorgetales blancas (Fig. 54); 8–10 (raramente 6 o 7) hileras de escamas loreales horizontales; se encuentran en la Isla de Roatán o en la Isla de Utila ................................... **38**

38 A. LHC promedio en machos aproximadamente 64 mm, hembras aproximadamente 66 mm; ramas sulcal de hemipenis abiertas a una zona amplia concava distal a punta de bifurcación en cada lóbulo (Fig. 116); baja processus alsulcado presente; endémicos a la Isla de Utila ..... *Norops bicaorum* (p. 27)

B. LHC promedio en machos aproximadamente 56 mm, hembras aproximadamente 58 mm; ramas sulcal de hemipenis continuán para la punta de cada lóbulo (Fig. 117); processus alsulcado ausente; endémicos a la Isla de Roatán ....... *Norops roatanensis* (p. 151)

## DISCUSSION OF HONDURAN ANOLE SPECIES RELATIONSHIPS

The systematic relationships of anoles continue to be a fertile ground for future studies performing phylogenetic analyses of combined morphological and molecular data. Guyer and Savage (1987, 1992) and Savage and Guyer (1989) have been proponents of dividing the large genus *Anolis* (*sensu lato*, with close to 400 recognized species) into eight genera based on cladistic analyses of morphological, karyological, and biochemical data. Recently, Nicholson et al. (2012; also see Pyron et al., 2013) again proposed an eight-genera concept of the anoles (see the section "Justification for the use of the generic name *Norops*" in our Introduction).

Although there are problems with the Nicholson et al. (2012) generic study (i.e., the most obvious is the relatively few number of species utilized in their molecular analyses), we think that those problems will be resolved by incorporating molecular data with more detailed morphological study, including many additional species. In this section, we are only interested in species group relationships based on combined morphological and molecular data.

Of the 39 named species of anoles (Dactyloidae) occurring in Honduras, one is a member of the genus *Anolis* of Nicholson et al. (2012), and the remaining 38 represent the beta group *Norops*. The only nonbeta species in Honduras is *A. allisoni*. *Anolis allisoni* belongs to the *A. carolinensis* subgroup (Glor et al., 2005) or *A. carolinensis* species group (Ruibal, 1964; Williams, 1976a; Nicholson et al., 2012). *Anolis carolinensis* is the type species of the genus *Anolis*.

Nicholson et al. (2012) placed the 175 nominal species of *Norops* they recognized (including one species known only as a fossil) into three species groups, two of which occur in Honduras (the *N. auratus* and *N. sagrei* groups). Nicholson et al. (2012) only had molecular data for 69 of the 150 species (46.0%) included in their *N. auratus* species group. Also, only 18 of 36 (50.0%) species in their *N. auratus* group occurring in Honduras were studied. The numbers of recognized species of the *N. sagrei* group were better represented; 13 of the 18 (72.2%) species in their *N. sagrei* group were subjected to a phylogenetic analysis based on combined data sets.

The *Norops sagrei* group, is represented in Honduras by only two species, *N. nelsoni* and *N. sagrei* (populations of the latter in Honduras are probably introduced). The 36 species of anoles of the *N. auratus* species group occurring in Honduras are placed in 10 species subgroups (31 species), with five species left as incertae sedis within the *N. auratus* group (Table 5). For these results, we used the data in Nicholson (2002—nuclear DNA data), Nicholson et al. (2005—combined nuclear and mtDNA data sets; 2012—combined data sets), Poe (2004—combined data sets), and our morphological data on Honduran species. In total, we recognize one species group for the lone *Anolis* species (*A. carolinensis* group) and two species groups (*N. auratus*

TABLE 5.   PROPOSED SPECIES GROUP AND SUBGROUP RELATIONSHIPS AMONG THE 39 NAMED HONDURAN ANOLE SPECIES.

| Species | Reference |
| --- | --- |
| *Anolis*: 1 | |
| *Anolis carolinensis* species group | Nicholson et al., 2012; also see Ruibal and Williams, 1961; Ruibal, 1964; Williams, 1976a |
| *allisoni* | |
| *Norops*: 38 species | |
| *Norops auratus* species group | Nicholson et al., 2012 |
| *Norops biporcatus* species subgroup | |
| *biporcatus* | |
| *petersii* | |
| *Norops crassulus* species subgroup | |
| *amplisquamosus* | |
| *crassulus* | |
| *heteropholidotus* | |
| *morazani* | |
| *muralla* | |
| *rubribarbaris* | |
| *sminthus* | |
| *Norops cupreus* species subgroup | |
| *cupreus* | |
| *Norops fuscoauratus* species subgroup | |
| *carpenteri* | |
| *limifrons* | |
| *ocelloscapularis* | |
| *rodriguezii* | |
| *yoroensis* | |
| *zeus* | |
| *Norops laeviventris* species subgroup | |
| *cusuco* | |
| *kreutzi* | |
| *laeviventris* | |
| *Norops lemurinus* species subgroup | Williams, 1976b |
| *bicaorum* | |
| *lemurinus* | |
| *roatanensis* | |
| *Norops lionotus* species subgroup | Williams, 1976b |
| *oxylophus* | |
| *Norops pentaprion* species subgroup | Myers, 1971b; Williams, 1976b |
| *beckeri* | |
| *utilensis* | |
| *Norops schiedii* species subgroup | |
| *johnmeyeri* | |
| *loveridgei* | |
| *pijolense* | |
| *purpurgularis* | |
| *Norops sericeus* species subgroup | |
| *unilobatus* | |
| *wellbornae* | |
| Incertae sedis | |
| *capito* | |
| *quaggulus* | |
| *tropidonotus* | |
| *uniformis* | |
| *wampuensis* | |
| *Norops sagrei* species group | Nicholson et al., 2012; also see Williams, 1976a |
| *nelsoni* | |
| *sagrei* | |

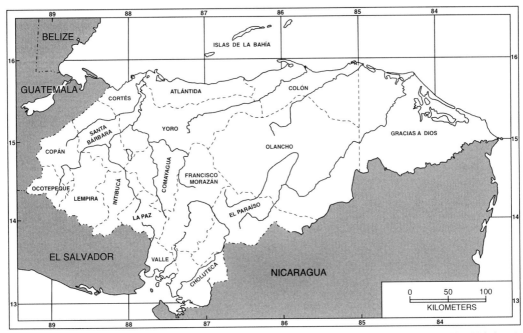

Map 40. Map of Honduras showing the boundaries of the 18 departments. The Islas del Cisne shown on Map 20 belong to Gracias a Dios.

and *N. sagrei*) for the 38 Honduran species of *Norops*. Our *N. sagrei* group includes two species, whereas our *N. auratus* group contains 36 species in 10 subgroups, with five species left as incertae sedis (Table 5). Brief discussions of these placements are also made in the Results for each species account.

## DISTRIBUTION OF ANOLES IN HONDURAS

### Distribution within Departments

The distribution by departments of the 39 anole species of Honduras is shown in Table 6 (Map 40 shows the distributions of the 18 departments). The total number of species by departments ranges from a high of 16 species in Cortés and 15 in Olancho to a low of one species in Valle. Both Cortés and Olancho are relatively large departments that contain terrain that extends from low to intermediate elevations (from near sea level to over 2,200 m). Both Cortés and Olancho also have terrain that includes various ecological regimes (Lowland Moist Forest, Lowland Dry Forest, Premontane Wet Forest, Premontane Moist Forest, and Lower Montane Wet Forest). Other departments with high species numbers include Atlántida and Santa Bárbara (each with 13 species), Copán, Gracias a Dios, and Yoro (each with 12 species), and Colón (with 10 species). Four of these high-species departments (Atlántida, Copán, Santa Bárbara, and Yoro) also contain terrain with extensive elevation ranges and varied ecological regimes. On the other hand, Colón and Gracias a Dios lie largely (Colón) or almost entirely (Gracias a Dios) in the lowlands (below 600 m elevation). All eight high-species departments lie entirely on the Atlantic versant. The three departments with the lowest number of anole species are Valle (one species), Choluteca (two species), and Ocotepeque (two species). Both Valle and Choluteca lie entirely on the depauperate subhumid Pacific versant

TABLE 6. DISTRIBUTION BY DEPARTMENTS OF THE 39 NAMED SPECIES OF ANOLES KNOWN FROM HONDURAS. DEPARTMENT ABBREVIATIONS ARE: ATL = ATLÁNTIDA; CHO = CHOLUTECA; COL = COLÓN; COM = COMAYAGUA; COP = COPAN; COR = CORTÉS; EP = EL PARAISO; FM = FRANCISCO MORAZÁN; GAD = GRACIAS A DIOS; INT = INTIBUCA; IDB = ISLAS DE LA BAHÍA; LAP = LA PAZ; LEM = LEMPIRA; OCO = OCOTEPEQUE; OLA = OLANCHO; SB = SANTA BARBARA; VAL = VALLE; AND YOR = YORO. SYMBOLS IN THE SPECIES LISTS ARE: X = WITH VOUCHER SPECIMENS; I = INTRODUCED POPULATION (WITH VOUCHER SPECIMENS); O = AN ACCEPTED LITERATURE RECORD.

| Species | ATL | CHO | COL | COM | COP | COR | EP | FM | GAD | INT | IDB | LAP | LEM | OCO | OLA | SB | VAL | YOR | Total |
|---|---|---|---|---|---|---|---|---|---|---|---|---|---|---|---|---|---|---|---|
| Anolis allisoni | I | — | — | — | — | — | — | — | — | — | X | — | — | — | — | — | — | — | 2 |
| Norops amplisquamosus | — | — | — | — | — | X | — | — | — | — | — | — | — | — | — | — | — | — | 1 |
| Norops beckeri | X | X | — | — | — | X | — | — | X | — | — | — | — | — | X | X | — | X | 7 |
| Norops bicaorum | — | — | — | — | — | — | — | — | — | — | X | — | — | — | — | — | — | — | 1 |
| Norops biporcatus | X | — | X | — | X | X | X | X | X | — | — | — | — | — | X | — | — | X | 9 |
| Norops capito | — | — | X | — | X | X | — | — | X | — | — | — | — | — | X | X | — | — | 6 |
| Norops carpenteri | — | — | — | — | — | — | — | — | X | — | — | — | — | — | — | — | — | — | 1 |
| Norops crassulus | — | — | X | X | — | — | — | — | — | X | — | X | X | — | — | — | — | — | 5 |
| Norops cupreus | — | X | — | — | X | X | X | X | X | — | — | — | — | — | — | — | — | — | 6 |
| Norops cusuco | — | — | — | — | X | X | — | — | — | — | — | — | — | — | — | X | — | — | 3 |
| Norops heteropholidotus | — | — | — | — | — | — | — | — | — | X | — | — | — | X | — | — | — | — | 2 |
| Norops johnmeyeri | X | — | — | — | — | X | — | — | — | — | — | — | — | — | — | — | — | — | 2 |
| Norops kreutzi | X | — | — | — | — | — | — | — | — | — | — | — | — | — | — | — | — | X | 2 |
| Norops laeviventris | X | — | X | X | X | X | X | X | — | X | — | X | X | — | X | X | — | X | 9 |
| Norops lemurinus | X | X | X | — | X | X | X | — | X | — | X | — | — | — | X | X | — | X | 11 |
| Norops limifrons | X | — | — | — | — | — | — | — | X | — | X | — | — | — | X | X | — | X | 3 |
| Norops loveridgei | X | — | — | — | — | — | — | X | — | — | — | — | — | — | — | — | — | X | 2 |
| Norops morazani | — | — | — | — | — | — | — | X | — | — | — | — | — | — | X | — | — | — | 1 |
| Norops muralla | — | — | — | — | — | — | — | — | — | — | — | — | — | — | X | — | — | — | 1 |
| Norops nelsoni | — | — | — | — | — | — | — | — | X | — | — | — | — | — | — | — | — | — | 1 |
| Norops ocelloscapularis | — | — | — | — | X | X | — | — | — | — | — | — | — | — | — | X | — | — | 3 |
| Norops oxylophus | — | — | — | — | — | X | — | — | X | — | — | — | — | — | X | X | — | — | 2 |
| Norops petersii | — | — | — | — | X | X | — | — | — | — | — | — | — | — | — | X | — | — | 3 |
| Norops pijolense | — | — | — | — | — | — | — | — | — | — | — | — | — | — | — | — | — | X | 1 |
| Norops purpurgularis | X | — | — | — | — | — | — | — | — | — | — | — | — | — | — | — | — | X | 2 |
| Norops quaggulus | — | X | — | — | — | — | — | — | X | — | — | — | — | — | X | — | — | — | 3 |
| Norops roatanensis | — | — | — | — | — | — | — | — | — | — | X | — | — | — | — | — | — | — | 1 |
| Norops rodriguezii | — | — | — | — | X | X | — | — | — | — | — | — | — | — | — | X | — | — | 3 |
| Norops rubribarbaris | I | — | — | — | — | — | — | — | — | — | I | — | — | — | — | X | — | — | 1 |
| Norops sagrei | I | — | — | — | — | I | — | X | — | — | I | — | — | — | — | — | — | — | 3 |
| Norops sminthus | — | — | X | X | — | — | — | X | — | — | — | — | — | — | — | — | — | — | 2 |
| Norops tropidonotus | X | X | X | X | X | X | X | X | X | X | — | X | X | X | X | X | — | X | 15 |
| Norops uniformis | X | — | X | X | X | X | — | — | — | — | — | — | — | — | — | — | — | — | 4 |
| Norops unilobatus | X | X | X | X | X | X | X | X | X | X | X | X | — | — | X | X | — | X | 14 |
| Norops utilensis | — | — | — | — | — | — | — | — | — | — | X | — | — | — | — | — | — | — | 1 |
| Norops wampuensis | — | — | — | — | — | — | — | — | — | — | — | — | — | — | X | — | — | — | 1 |

TABLE 6. CONTINUED.

| Species | ATL | CHO | COL | COM | COP | COR | EP | FM | GAD | INT | IDB | LAP | LEM | OCO | OLA | SB | VAL | YOR | Total |
|---|---|---|---|---|---|---|---|---|---|---|---|---|---|---|---|---|---|---|---|
| *Norops wellbornae* | — | X | — | — | — | — | X | X | — | X | — | — | X | — | — | — | X | — | 6 |
| *Norops yoroensis* | X | — | X | — | X | X | — | X | — | — | — | — | — | — | X | X | — | X | 7 |
| *Norops zeus* | X | — | — | — | — | X | — | — | — | — | — | — | — | — | — | — | — | X | 4 |
| Total | 13 | 2 | 10 | 6 | 12 | 16 | 7 | 9 | 12 | 6 | 7 | 4 | 4 | 2 | 15 | 13 | 1 | 12 | 151 |

and Ocotepeque also contains terrain on the Pacific versant, but additionally contains terrain extending to above 2,200 m elevation on both the Pacific and Atlantic versants. Thus, Ocotepeque appears to be understudied. As demonstrated for the amphibian and snake faunas (McCranie, 2007b, 2011, respectively), Valle also appears to be understudied when compared with the neighboring Choluteca (two species).

## Distribution within Forest Formations and by Elevation

The distribution of the 39 anole species by nine forest formations is shown in Table 7 and Map 41 shows the distributions of these formations. The Montane Rainforest formation is not included because no anole species are known from that formation. The following distributional categories are used: widespread (occurs widely in a particular formation in Honduras, as well as in at least one other formation in Honduras); restricted (restricted to a single formation in Honduras); and peripheral (barely enters a particular forest formation in Honduras). Tables 7 and 8 demonstrate the following numbers of anole species from each formation and their distributional categories (from highest to lowest number of species in each formation; forest formation abbreviations are as defined in Table 7): PWF—23 total species (18 widespread, 5 peripheral); LMF—21 total species (13 widespread, 7 restricted, 1 peripheral); LMWF—17 total species (11 widespread, 3 restricted, 3 peripheral); LDF—9 total species (7 widespread, 2 peripheral); PMF—9 total species (5 widespread, 4 peripheral); LAF—5 total species (5 widespread); LMMF—5 total species (3 widespread, 1 restricted, 1 peripheral); PDF—3 total species (3 widespread); and LDF(WI)—1 total species (1 restricted). The three formations with the highest numbers of species are mesic formations on the Atlantic versant. The mean number of formations inhabited by the 39 anole species is 2.4. The highest total number of

TABLE 7. DISTRIBUTION OF THE HONDURAN ANOLE FAUNA WITHIN NINE ECOLOGICAL FORMATIONS. THE MONTANE RAINFOREST FORMATION IS NOT INCLUDED IN THIS TABLE. ABBREVIATIONS USED IN THIS AND SOME SUBSEQUENT TABLES ARE AS FOLLOWS: LMF = LOWLAND MOIST FOREST; LDF = LOWLAND DRY FOREST; LAF = LOWLAND ARID FOREST; PWF = PREMONTANE WET FOREST; PMF = PREMONTANE MOIST FOREST; PDF = PREMONTANE DRY FOREST; LMWF = LOWER MONTANE WET FOREST; LMMF = LOWER MONTANE MOIST FOREST; LDF (WI) = LOWLAND DRY FOREST, WEST INDIAN SUBREGION; M = METERS; W = WIDESPREAD IN THAT PARTICULAR FORMATION; P = PERIPHERAL IN THAT FORMATION; R = RESTRICTED TO THAT FORMATION.

| Species (39) | LMF | LDF | LAF | PWF | PMF | PDF | LMWF | LMMF | LDF(WI) | Total | Elevational Range (m) |
|---|---|---|---|---|---|---|---|---|---|---|---|
| *Anolis allisoni* | R | — | — | — | — | — | — | — | — | 1 | 0–30 |
| *Norops amplisquamosus* | — | — | — | — | — | — | R | — | — | 1 | 1,530–1,990 |
| *Norops beckeri* | W | P | — | W | — | — | — | — | — | 3 | 0–ca. 1,400 |
| *Norops bicaorum* | R | — | — | — | — | — | — | — | — | 1 | 0–20 |
| *Norops biporcatus* | W | — | — | W | P | — | — | — | — | 3 | 0–1,050 |
| *Norops capito* | W | — | — | W | — | — | — | — | — | 2 | 0–1,300 |
| *Norops carpenteri* | R | — | — | — | — | — | — | — | — | 1 | 30–40 |
| *Norops crassulus* | — | — | — | W | W | — | W | — | — | 3 | 1,200–2,285 |
| *Norops cupreus* | W | W | — | W | W | — | W | — | — | 5 | 0–1,300 |
| *Norops cusuco* | — | — | — | P | — | — | W | — | — | 2 | 1,350–1,990 |
| *Norops heteropholidotus* | — | — | — | — | — | — | — | R | — | 1 | 1,860–2,200 |
| *Norops johnmeyeri* | — | — | — | P | — | — | W | — | — | 2 | 1,300–2,000 |
| *Norops kreutzi* | — | — | — | W | — | — | W | — | — | 2 | 980–1,690 |
| *Norops laeviventris* | — | — | — | W | W | — | W | W | — | 4 | 1,000–2,000 |
| *Norops lemurinus* | W | — | W | W | — | — | — | — | — | 4 | 0–960 |
| *Norops limifrons* | W | — | — | W | — | — | — | — | — | 2 | 0–900 |
| *Norops loveridgei* | P | — | — | W | — | — | W | — | — | 3 | ca. 550–1,600 |
| *Norops morazani* | — | — | — | P | — | — | W | — | — | 2 | 1,275–2,150 |
| *Norops muralla* | — | — | — | P | — | — | W | — | — | 2 | 1,440–1,740 |
| *Norops nelsoni* | — | — | — | — | — | — | — | — | R | 1 | 0–10 |
| *Norops ocelloscapularis* | — | — | — | W | — | — | P | — | — | 2 | 1,040–1,550 |
| *Norops oxylophus* | R | — | — | — | — | — | — | — | — | 1 | 60–225 |
| *Norops petersii* | — | — | — | W | — | — | W | — | — | 2 | 1,300–1,550 |
| *Norops pijolense* | — | — | — | W | — | — | W | — | — | 2 | 1,180–2,050 |
| *Norops purpurgularis* | — | — | — | — | — | — | R | — | — | 1 | 1,550–2,040 |
| *Norops quaggulus* | W | — | — | W | — | — | — | — | — | 2 | 60–840 |
| *Norops roatanensis* | R | — | — | — | — | — | — | — | — | 2 | 0–30 |
| *Norops rodriguezii* | W | W | — | W | P | — | — | — | — | 4 | 0–1,200 |
| *Norops rubribarbaris* | — | — | — | — | — | — | R | — | — | 1 | 1,600–1,800 |
| *Norops sagrei* | W | W | — | — | — | — | — | — | — | 2 | 0–100 |
| *Norops sminthus* | — | — | — | — | — | — | W | W | — | 2 | 1,450–1,900 |
| *Norops tropidonotus* | W | W | W | W | W | W | P | P | — | 8 | 0–1,900 |
| *Norops uniformis* | W | — | — | W | — | W | — | — | — | 3 | 30–1,370 |
| *Norops unilobatus* | W | W | W | P | W | W | — | — | — | 6 | 0–1,320 |
| *Norops utilensis* | R | — | — | — | — | — | — | — | — | 1 | 0–8 |

| Species (39) | LMF | LDF | LAF | PWF | PMF | PDF | LMWF | LMMF | LDF(WI) | Total | Elevational Range (m) |
|---|---|---|---|---|---|---|---|---|---|---|---|
| *Norops wampuensis* | R | — | — | — | — | — | — | — | — | 1 | 95–110 |
| *Norops wellbornae* | — | W | W | — | — | W | W | — | — | 3 | 0–1,000 |
| *Norops yoroensis* | W | P | — | W | P | — | — | — | — | 3 | 650–1,600 |
| *Norops zeus* | W | — | — | W | — | — | — | — | — | 3 | 0–900 |
| Totals | 21 | 9 | 5 | 23 | 9 | 3 | 17 | 5 | 1 | 93 | 0–2,285 |

TABLE 7. CONTINUED.

formations inhabited by an anole species is eight (*Norops tropidonotus*).

Table 8 summarizes the absolute and relative numbers for each of these distributional categories relative to the nine forest formations under study. The following conclusions are based on these data:

1. Widespread species are the most numerous in eight of the nine formations, the exception being the LDF(WI), to which a single anole species is restricted. The percent representations of the widespread species range from 0.0% in the LDF(WI) to 100.0% in the LAF and PDF formations, with an average value of 69.9% for all nine formations.

2. The largest percentage of restricted species is found in the LDF (WI) formation (100.0%, but only one total species). The next largest percentages for restricted species are the LMF (33.3%), LMMF (20.0), and LMWF (17.6). The percentage representations for this group range from 0.0% to 100.0%, with an average value of only 12.9%. No anole species are restricted to a single subhumid formation on the mainland of Honduras, but *N. wellbornae* is restricted in distribution to three subhumid forest formations (LDF, LAF, and PMF formations). Restricted species are the least commonly represented species.

3. Peripheral species are also poorly represented with an average of only 17.2% per formation. They are absent in the LAF, PDF, and LDF(WI) formations and poorly represented in the LMF (4.8%). Only the PMF (44.4%), LDF (22.2%), PWF (21.7%), LMMF (20.0%), and LMWF (17.6%) are relatively well represented by peripheral species.

Anoles in Honduras range from sea level to 2,285 m elevation (Table 7). Placing each anole species into the elevational categories of Stuart (1963) demonstrates the following: (1) low elevations (0–600 m), 21 species; (2) moderate elevations (601–1,500 m), 25 species; (3) intermediate elevations (1,501–

2,700 m), 18 species. Mean elevation span for anoles is 681.3 m.

The number of anole species found at various elevations declines more or less consistently with an increase in elevation (Table 9). However, there are some slight increases between elevations of 600 and 700 m (two species), 1,100 and 1,200 m (one species), 1,200 and 1,300 m (two species), and 1,500 and 1,600 m (two species). Marked decreases in numbers of species occur between 100 and 200 m (seven species), 1,600 and 1,700 m (four species), 1,900 and 2,000 m (two species), and 2,000 and 2,100 m (five species).

## Distribution within Physiographic Regions

The distribution of Honduran anole species in 11 physiographic regions is shown in Table 10 (Map 42 shows the distributions of 10 of these physiographic regions; see Map 20 for the Swan Islands physiographic region). The physiographic regions and number of anole species known for each region are as follows (from highest to lowest number of species; physiographic region abbreviations as in Table 8): NC, 29 species; MC, 11 species; SC, 11 species; NDP, 8 species; UCP, 7 species; ANP, 7 species; BI, 6 species; MP, 4 species; PLR, 2 species; CC, 2 species; and SI, 1 species. The two regions with the highest numbers of species are also those having the greatest areas. Contrariwise, the two with the smallest areas harbor the lowest numbers of species. The Northern Cordillera, in addition to being the largest area, also contains the most complex topography (ranges from near sea level to 2,744 m) and vegetation (containing six of the nine ecological formations), so it is not surprising that it contains the largest number of species.

Each of the 39 anole species occurs in from one to eight physiographic regions, with the average being 2.3 areas.

## Distribution within Ecophysiographic Areas

The distribution of the Honduran anole species by 34 of the 39 Honduran ecophy-siographic areas is indicated in Table 11 (Map 43 shows the distributions of 38 of the 39 ecophysiographic areas; see Map 20 for the Swan Islands ecophysiographic area). The five areas not utilized remain poorly known. The ecophysiographic areas and their known total number of anole species are as follows (from highest to lowest; area names for each number are shown in Table 12): Area 16, 13 species; Area 21, 12 species; Area 15, 11 species; Area 30, 11 species; Area 26, 10 species; Area 22, 9 species; Area 27, 9 species; Area 31, 8 species; Area 9, 7 species; Area 29, 7 species; Area 32, 6 species; Area 13, 5 species; Area 17, 5 species; Area 35, 5 species; Area 8, 4 species; Area 11, 4 species; Area 14, 4 species; Area 10, 3 species; Area 19, 3 species; Area 24, 3 species; Area 28, 3 species; Area 36, 3 species; Area 1, 2 species; Area 3, 2 species; Area 7, 2 species; Area 23, 2 species; Area 33, 2 species; Area 38, 2 species; Area 5, 1 species; Area 12, 1 species; Area 18, 1 species; Area 20, 1 species; Area 37, 1 species; Area 39, 1 species.

The eight areas with the greatest numbers of anole species (8–13 species) are all Atlantic versant mesic areas that lie at either low or moderate elevations. Of two areas with the ninth greatest number (7) of anole species (Areas 9 and 29), one lies in both versants, but four of the seven anole species are known only from the Atlantic versant in that area. The other (Area 29) lies on the Atlantic versant and is a low elevation subhumid forest, but does have several mesic gallery forest corridors. The next four areas with the greatest numbers of anole species (5–6) are all on the Atlantic versant, two of which are mesic (Areas 32 and 17), one of which (Area 13) lies largely in subhumid forest at moderate elevations (with three of its five total species known only from mesic enclaves within that area), and one (Area 35) is an island off the northern coast. The mean number of ecophysiographic areas inhabited by the anole species is 4.2.

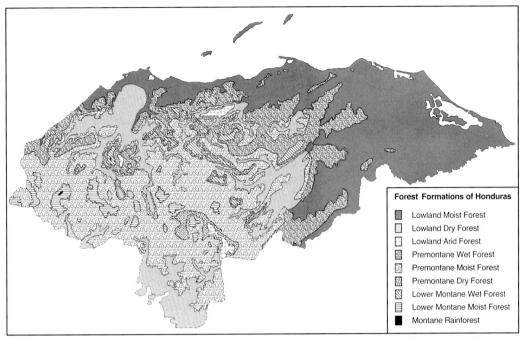

Map 41. Forest formations (modified from Holdridge, 1967) of Honduras. A color version of this map is included in McCranie (2011).

## Broad Patterns of Geographic Distribution

The broad areas of geographic distribution used herein are as follows:

A. northern terminus of the range in Mexico north of the Isthmus of Tehuantepec and southern terminus in South America;

B. northern terminus of the range in Mexico north of the Isthmus of Tehuantepec and southern terminus in Central America south of the Nicaraguan Depression;

C. northern terminus of the range in Mexico north of the Isthmus of Te-

TABLE 8. ANOLE SPECIES NUMBERS FOR THE THREE DISTRIBUTIONAL CATEGORIES IN EACH OF THE NINE ECOLOGICAL FORMATIONS. SEE TABLE 7 FOR DEFINITIONS OF THE ABBREVIATIONS FOR THE ECOLOGICAL FORMATIONS.

| | Distributional Categories | | | | | |
| | Widespread | | Restricted | | Peripheral | |
| Formations | N | % | N | % | N | % |
| --- | --- | --- | --- | --- | --- | --- |
| LMF | 13 | 61.9 | 7 | 33.3 | 1 | 4.8 |
| LDF | 7 | 77.8 | 0 | 0.0 | 2 | 22.2 |
| LAF | 5 | 100.0 | 0 | 0.0 | 0 | 0.0 |
| PWF | 18 | 78.3 | 0 | 0.0 | 5 | 21.7 |
| PMF | 5 | 55.6 | 0 | 0.0 | 4 | 44.4 |
| PDF | 3 | 100.0 | 0 | 0.0 | 0 | 0.0 |
| LMWF | 11 | 64.7 | 3 | 17.6 | 3 | 17.6 |
| LMMF | 3 | 60.0 | 1 | 20.0 | 1 | 20.0 |
| LDF(WI) | 0 | 0.0 | 1 | 100.0 | 0 | 0.0 |
| Totals | 65 | 69.9 | 12 | 12.9 | 16 | 17.2 |

TABLE 9.    NUMBERS OF ANOLE SPECIES FOUND AT VARIOUS ELEVATIONS IN HONDURAS.

| Elevational Segments | Totals |
|---|---|
| 0–100 | 22 |
| 101–200 | 15 |
| 201–300 | 14 |
| 301–400 | 14 |
| 401–500 | 13 |
| 501–600 | 13 |
| 601–700 | 15 |
| 701–800 | 15 |
| 801–900 | 15 |
| 901–1,000 | 13 |
| 1,001–1,100 | 13 |
| 1,101–1,200 | 14 |
| 1,201–1,300 | 16 |
| 1,301–1,400 | 15 |
| 1,401–1,500 | 14 |
| 1,501–1,600 | 16 |
| 1,601–1,700 | 12 |
| 1,701–1,800 | 12 |
| 1,801–1,900 | 11 |
| 1,901–2,000 | 9 |
| 2,001–2,100 | 4 |
| 2,101–2,200 | 3 |
| 2,201–2,300 | 1 |

huantepec and southern terminus in Nuclear Middle America;

D. northern terminus of range in Nuclear Middle America and southern terminus in Central America south of the Nicaraguan Depression;

E. restricted to Nuclear Middle America (exclusive of Honduran endemics);

F. endemic to Honduras (including insular endemics);

G. insular species (not including endemics);

H. introduced species.

The allocation of anole species to these categories is as follows:

A. *Norops biporcatus*;

B. *Norops lemurinus*;

C. *Norops laeviventris, N. petersii, N. tropidonotus, N. uniformis*;

D. *Norops capito, N. carpenteri, N. cupreus, N. limifrons, N. oxylophus, N. quaggulus, N. unilobatus*;

E. *Norops beckeri, N. crassulus, N. heteropholidotus, N. ocelloscapularis, N. rodriguezii, N. wellbornae*;

F. *Norops amplisquamosus, N. bicaorum, N. cusuco, N. johnmeyeri, N. kreutzi, N. loveridgei, N. morazani, N. muralla, N. nelsoni, N. pijolense, N. purpurgularis, N. roatanensis, N. rubribarbaris, N. sminthus, N. utilensis, N. wampuensis, N. yoroensis, N. zeus*;

G. *Anolis allisoni*;

H. *Norops sagrei.*

The number of anole species in each category is as follows (highest to lowest): F, 18 species (46.2% of total anole fauna); D, 7 species (17.9%); E, 6 species (15.4%); C, 4 species (10.3%); A, 1 species (2.6%); B, 1 species (2.6%); G, 1 species (2.6%); H, 1 species (2.6%).

This summary indicates that the largest category (F) contains those species endemic to Honduras (18 species). Category E (restricted to Nuclear Middle America), with six species, contains the third highest total. Anole species, as a whole, are more restricted in distribution than are the snakes (McCranie, 2011). The Honduran endemic anoles (46.2% of the total) and those species otherwise restricted to Nuclear Middle America (17.9%) make up well over half (61.5%) of the total Honduran anole fauna (versus 14.7% and 12.5% for a total of only 27.2% of the entire snake fauna).

## HISTORICAL UNITS OF HONDURAN ANOLE GENERA AND SPECIES

Savage (1966, 1982), in pioneering works discussing historical units of amphibian and reptilian genera, placed the beta anoles (*Norops*) in his Middle American Historical Unit, thus postulating that their ancestral beginnings were on the mainland of what is now Central America. Savage (1982: 511) defined that historical unit as "derivative groups of a generalized tropical American biota isolated in tropical North and Central America during most of Cenozoic; developed *in situ* north of the Panamanian Portal and restricted by mountain building and climatic change in late Cenozoic to Middle America." Savage (1966, 1982) also placed

TABLE 10. DISTRIBUTION OF THE HONDURAN ANOLE FAUNA BY 11 PHYSIOGRAPHIC REGIONS. ABBREVIATIONS USED IN THIS AND SOME SUBSEQUENT TABLES ARE AS FOLLOWS: PLR = PACIFIC LOWLAND REGION; MP = MOTAGUA PLAIN; UCP = ULÚA–CHAMELECÓN PLAIN; NDP = NOMBRE DE DIOS PIEDMONT; ANP = AGUÁN–NEGRO PLAIN; MC = MOSQUITO COAST; NC = NORTHERN CORDILLERA; SC = SOUTHERN CORDILLERA; BI = BAY ISLANDS; CC = CAYOS COCHINOS; SI = SWAN ISLANDS.

| Species (39) | PLR | MP | UCP | NDP | ANP | MC | NC | SC | BI | CC | SI | Total |
|---|---|---|---|---|---|---|---|---|---|---|---|---|
| *Anolis allisoni* | — | — | — | X | — | — | — | — | X | X | — | 3 |
| *Norops amplisquamosus* | — | — | — | — | — | — | X | — | — | — | — | 1 |
| *Norops beckeri* | — | — | X | X | X | X | X | — | — | — | — | 5 |
| *Norops bicaorum* | — | — | — | — | — | — | — | — | X | — | — | 1 |
| *Norops biporcatus* | — | — | — | X | X | X | X | X | — | — | — | 5 |
| *Norops capito* | — | — | — | — | X | X | X | X | — | — | — | 4 |
| *Norops carpenteri* | — | — | — | — | — | X | — | — | — | — | — | 1 |
| *Norops crassulus* | — | — | — | — | — | — | X | X | — | — | — | 2 |
| *Norops cupreus* | X | — | — | — | — | X | X | X | — | — | — | 4 |
| *Norops cusuco* | — | — | — | — | — | — | X | — | — | — | — | 1 |
| *Norops heteropholidotus* | — | — | — | — | — | — | — | X | — | — | — | 1 |
| *Norops johnmeyeri* | — | — | — | — | — | — | X | — | — | — | — | 1 |
| *Norops kreutzi* | — | — | — | — | — | — | X | — | — | — | — | 1 |
| *Norops laeviventris* | — | — | — | — | — | — | X | X | — | — | — | 2 |
| *Norops lemurinus* | — | X | X | X | X | X | X | X | — | X | — | 8 |
| *Norops limifrons* | — | — | — | — | X | X | X | — | — | — | — | 3 |
| *Norops loveridgei* | — | — | — | — | — | — | X | — | — | — | — | 1 |
| *Norops morazani* | — | — | — | — | — | — | X | — | — | — | — | 1 |
| *Norops muralla* | — | — | — | — | — | — | X | — | — | — | — | 1 |
| *Norops nelsoni* | — | — | — | — | — | — | — | — | — | — | X | 1 |
| *Norops ocelloscapularis* | — | — | — | — | — | — | X | — | — | — | — | 1 |
| *Norops oxylophus* | — | — | — | — | — | X | X | — | — | — | — | 2 |
| *Norops petersii* | — | — | — | — | — | — | X | — | — | — | — | 1 |
| *Norops pijolense* | — | — | — | — | — | — | X | — | — | — | — | 1 |
| *Norops purpurgularis* | — | — | — | — | — | — | X | — | — | — | — | 1 |
| *Norops quaggulus* | — | — | — | — | — | X | X | — | — | — | — | 2 |
| *Norops roatanensis* | — | — | — | — | — | — | — | — | X | — | — | 1 |
| *Norops rodriguezii* | — | X | X | — | — | — | X | — | — | — | — | 3 |
| *Norops rubribarbaris* | — | — | — | — | — | — | X | — | — | — | — | 1 |
| *Norops sagrei* | — | — | X | X | — | — | — | — | X | — | — | 3 |
| *Norops sminthus* | — | — | — | — | — | — | — | X | — | — | — | 1 |
| *Norops tropidonotus* | — | X | X | X | — | X | X | X | — | — | — | 6 |
| *Norops uniformis* | — | — | X | — | — | — | X | — | — | — | — | 2 |
| *Norops unilobatus* | — | X | X | X | X | X | X | X | X | — | — | 8 |
| *Norops utilensis* | — | — | — | — | — | — | — | — | X | — | — | 1 |
| *Norops wampuensis* | X | — | — | — | — | — | — | — | — | — | — | 1 |
| *Norops wellbornae* | — | — | — | — | — | — | X | X | — | — | — | 2 |
| *Norops yoroensis* | — | — | — | — | — | — | X | — | — | — | — | 1 |
| *Norops zeus* | — | — | — | X | X | — | X | — | — | — | — | 3 |
| Totals | 2 | 4 | 7 | 8 | 7 | 11 | 29 | 11 | 6 | 2 | 1 | 88 |

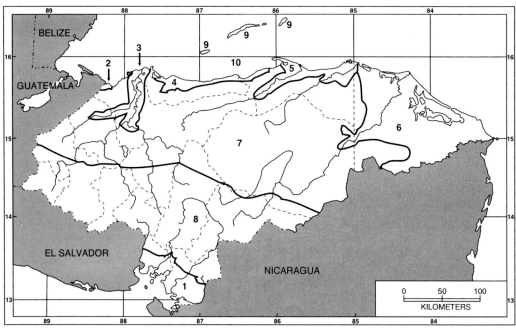

Map 42.   Physiographic regions of Honduras. (1) Pacific lowland region; (2) Motagua Plain of Caribbean lowland region; (3) Ulúa-Chamelecón Plain of Caribbean lowland region; (4) Nombre de Dios Piedmont of Caribbean lowland region; (5) Aguán-Negro Plain of Caribbean lowland region; (6) Mosquitia Coast of Caribbean lowland region; (7) Northern Cordillera of Serranía region; (8) Southern Cordillera of Serranía region; (9) Bay Islands; (10) Cayos Cochinos. See Map 20 for the location of region 11 (Swan Islands or Islas del Cisne).

the alpha anoles in his South American Historical Unit, postulating that their ancestors were "derivatives of a generalized tropical American biota that evolved *in situ* in isolation in South America during most of Cenozoic" (Savage, 1982: 511). The affinities of the South American Historical Unit were thought to be Gondwanian. Studies of fossil records of pleurodont iguanians by Estes (1983a, 1983b) led him to postulate that the earliest ancestors of anoles had a South American origin with later dispersal northward, thus agreeing in general with Savage's conclusions.

Nicholson et al. (2005) recovered evidence of a West Indian origin for the ancestors of the mainland and Bay Island beta anoles (*Norops*). Nicholson et al. (2005) postulated that the *Norops* clade ancestors colonized Central and South America from the West Indies and the *A. carolinensis* ancestor colonized the southeastern United States from the West Indies.

Nicholson et al. (2012) studied a larger suite of anoles using both morphological and molecular data, and in the process returned to the Savage (1966, 1982) conclusion that the anoles (Dactyloidae) originated in South America about 130 Ma. Nicholson et al. (2012: 4) further postulated, "The complicated divergence and accretion events that generated the current conformation of the Antillean islands, and eventually closed the Panamanian Portal, transported six island genera to their current center of diversity" on the Caribbean Islands and leaving "two genera on the mainland (*Dactyloa* and *Norops*)." Nicholson et al. (2012: 4) went on to say, "Our historical reconstruction makes *Norops* a much older radiation than previous reconstructions, allowing basal diversification of this species-rich lineage to occur on mainland terrains that eventually separated from the mainland to become parts of Cuba and Jamaica. This early diversification extended

TABLE 11. Anole distributional records for 35 ecophysiographic areas in Honduras. Areas 2 and 3 are combined for this analysis. See Table 12 for the corresponding area name for each area number.

| Species (39) | 1 | 3 | 5 | 7 | 8 | 9 | 10 | 11 | 12 | 13 | 14 | 15 | 16 | 17 | 18 | 19 | 20 | 21 | 22 | 23 | 24 | 26 | 27 | 28 | 29 | 30 | 31 | 32 | 33 | 35 | 36 | 37 | 38 | 39 | Total |
|---|---|---|---|---|---|---|---|---|---|---|---|---|---|---|---|---|---|---|---|---|---|---|---|---|---|---|---|---|---|---|---|---|---|---|---|
| *Anolis allisoni* | | | | | | | | | | | | | | | | | | | | | | X | | | | | | | | X | X | X | X | | 5 |
| *Norops amplisquamosus* | | | | | | | | | | | | | | | | | | | | | | | | | | | | X | | | | | | | 1 |
| *Norops beckeri* | | | | | | | | | | | | X | | | | | | | X | | | X | | | X | X | X | X | | | | | | | 8 |
| *Norops bicaorum* | | | | | | | | | | | | | | | | | | | | | | | | | | | | | | X | | | | | 1 |
| *Norops biporcatus* | | | | | | X | | | | X | X | X | X | | | | | X | X | X | X | X | X | | X | X | X | | | | | | | | 11 |
| *Norops capito* | | | | | | | | | | X | X | X | | | | | X | X | X | | | X | X | | | X | X | | | | | | | | 7 |
| *Norops carpenteri* | | | | | | | | | | | | | | | | | X | | | | | | | | | | | | | | | | | | 1 |
| *Norops crassulus* | X | X | | | X | X | | X | | X | | X | | | | | | X | | | | X | | | | | | | | | | | | | 3 |
| *Norops cupreus* | | | | | | | | | | | | X | | | | X | | X | X | | | X | | | | X | | X | | | X | | | | 8 |
| *Norops cusuco* | | | | | | | | | | | | | | | | | | | | | | | | | | X | | X | | | | | | | 2 |
| *Norops heteropholidotus* | | | | | | | | X | | | | | | | | | | | | | | | | | | | | | | | | | | | 1 |
| *Norops johnmeyeri* | | | | | | | | | | | | | | | | | | | | | | | | | | X | | X | | | | | | | 2 |
| *Norops kreutzi* | | | | | | | | | | | | | X | X | | | | | | | | | | | | | | | | | | | | | 2 |
| *Norops laeviventris* | | | | | X | X | X | X | | | | X | X | | | | | X | X | | X | X | X | X | X | X | X | | | | | | | | 9 |
| *Norops lemurinus* | | | | | X | X | X | X | | | | X | X | | | X | | X | X | | X | X | X | X | X | X | X | | X | | | | X | | 12 |
| *Norops limifrons* | | | | | | | | | | | | X | X | X | | | | | | | | X | | | | | | | | | | | | | 3 |
| *Norops loveridgei* | | | | | | | | | | | | | X | X | | | | | | | | X | | | | | | | | | | | | | 3 |
| *Norops morazani* | | | | | | | | | | | | | X | | X | | | | | | | | | | | | | | | | | | | | 2 |
| *Norops muralla* | | | | | | | | | | | | X | X | | | X | | | | | | | | | | | | | | | | | | | 2 |
| *Norops nelsoni* | | | | | | | | | | | | | | | | | | | | | | | | | | | | | | | | | | X | 1 |
| *Norops ocelloscapularis* | | | | | | | | | | | | | | | | | | X | | | | | X | | | X | X | X | | | | | | | 2 |
| *Norops oxylophus* | | | | | | | | | | | | | | | | | | | | | | | | | | X | X | X | | | | | | | 1 |
| *Norops petersii* | | | | | | | | | | | | | X | | | | | | | | | | X | | | X | X | X | | | | | | | 2 |
| *Norops pijolense* | | | | | | | | | | | | | X | X | | | | | | | | | | | | | | | | | | | | | 2 |
| *Norops purpurgularis* | | | | | | | | | | | | | X | X | | | | | | | | | | | | | | | | | | | | | 1 |
| *Norops quaggulus* | | | | | | | | | | | | X | | | | | | X | X | | | | | | | | | | | | | | | | 3 |
| *Norops roatanensis* | | | | | | | | | | | | | | | | | | | | | | | | | | | | | | | X | | | | 1 |
| *Norops rodriguezii* | | | | | | | | | | | X | | | | | | | | | | | | X | X | X | X | | | | | | | | | 5 |
| *Norops rubribarbaris* | | | | | | | | | | | | | | | | | | | | | | | | | | | | | X | | | | | | 1 |
| *Norops sagrei* | | | | | | | | | | | | | | | | | | | | | | X | X | X | X | | | | | X | X | | | | 5 |
| *Norops sminthus* | | | | | | X | X | | | | | | | | | | | | | | | | | | | | | | | | | | | | 2 |
| *Norops tropidonotus* | | | X | X | X | X | X | X | X | X | | X | X | | | X | | X | X | X | X | X | X | X | X | X | X | X | | | | | | | 19 |
| *Norops uniformis* | | | X | X | | | | | | | | X | X | | | | | | | | | | | | | X | X | X | | | | | | | 5 |
| *Norops unilobatus* | | | | | | X | | | | X | X | X | X | | | | | X | X | X | X | | | | X | X | X | X | | X | X | | | | 16 |
| *Norops utilensis* | | | | | | | | | | | | | | | | | | | | | | | | | | | | | | X | | | | | 1 |
| *Norops wampuensis* | | | | | | | | | | | | | | | | | | X | | | | | | | | | | | | | | | | | 1 |
| *Norops wellbornae* | X | X | | X | X | X | | | | | | | | | | | | | | | | | | | | | | | | | | | | | 4 |
| *Norops yoroensis* | | | | | | | | | | | | | X | X | | | | | | | | X | | | X | X | X | | | | | | | | 3 |
| *Norops zeus* | | | | | | | | | | | | X | X | X | | X | | | | | | X | X | | | | X | | | | | | | | 5 |
| Totals | 2 | 2 | 1 | 2 | 4 | 7 | 3 | 4 | 1 | 5 | 4 | 11 | 13 | 5 | 1 | 3 | 1 | 12 | 9 | 2 | 3 | 10 | 9 | 3 | 7 | 11 | 8 | 6 | 2 | 5 | 3 | 1 | 2 | 1 | 163 |

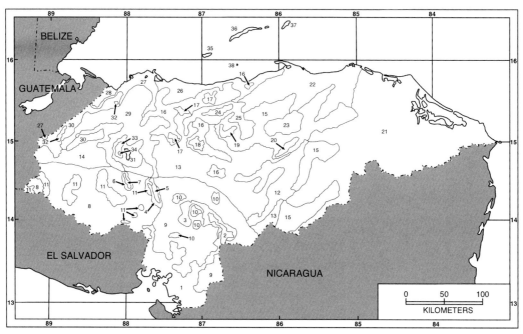

Map 43.    Ecophysiographic areas of Honduras: 1 = Pacific Lowlands; 2 = Middle Choluteca Valley; 3 = Upper Choluteca Valley; 4 = Comayagua Valley Rim; 5 = Comayagua Valley; 6 = Otoro Valley Rim; 7 = Otoro Valley; 8 = Southwestern Uplands; 9 = Southeastern Uplands; 10 = Southeastern Highlands; 11 = Southwestern Highlands; 12 = Guayape-Guayambre Valley; 13 = Northeastern Uplands; 14 = Northwestern Uplands; 15 = Eastern Caribbean Slope; 16 = Central Caribbean Slope; 17 = North-central Highlands; 18 = Yoro Highlands; 19 = Ocote Highlands; 20 = Agalta Highlands; 21 = Eastern Caribbean Lowlands; 22 = East-central Caribbean Lowlands; 23 = San Esteban Valley; 24 = Aguán Valley; 25 = Aguán Valley Rim; 26 = West-central Caribbean Lowlands; 27 = Western Caribbean Lowlands; 28 = Lower Motagua Valley; 29 = Sula Valley; 30 = Western Caribbean Slope; 31 = Yojoa Uplands; 32 = Northwestern Highlands; 33 = Santa Bárbara Highlands; 34 = Santa Bárbara Peak; 35 = Utila Island; 36 = Roatán Island; 37 = Guanaja Island; 38 = Cayos Cochinos. See Map 20 for the location of area 39 (Swan Islands or Islas del Cisne).

into northern South America, where a basal lineage of *Norops* coevolved with *Dactyloa* prior to the mainland-island separation." Subsequently, Castañeda et al. (2014) provided evidence that the 130 Ma origin for the anoles (Savage, 1966, 1982; Nicholson et al., 2012) was in error and returned to "somewhere in the range of 40–70" Ma as previously estimated (see references in Castañeda et al., 2014). Thus, the vicariant explanation for the current distribution of anoles on Caribbean islands is invalidated in favor of "overwater dispersal accounting for their [the anoles] occurrence on these [Caribbean] islands today remains [remaining] as the most robust hypothesis" (Castañeda et al., 2014).

In regard to the more recent history of Honduran anoles, Glor et al. (2005; also see Schoener, 1988) concluded that *A. allisoni*

originated on Cuba. *Anolis allisoni* currently occurs on Cuba, but the possibility exists that the Central American and Cuban populations might not be conspecific. However, if the one species concept is correct, then that species apparently reached the Islas de la Bahía and Belizean and Mexican islands by overwater dispersal, probably aided by early human boat traffic. The recent discoveries of *A. allisoni* on Isla de Utila and the Honduran mainland at La Ceiba are certainly of recent human transport. A genetic study of the Central American island and mainland populations of *A. allisoni* should solve the questions of its origin in the region.

Regarding the more recent history of Honduran components of the *N. sagrei* species group, most workers (i.e., Nicholson et al., 2005, 2012) consider the Central

TABLE 12.    CHARACTERISTICS OF 39 ECOPHYSIOGRAPHIC AREAS OF HONDURAS WITH DOMINANT FOREST FORMATIONS AND VERSANTS INDICATED. AREAS INDICATED WITH AN ASTERISK (*) ARE NOT INCLUDED IN THE ANALYSIS HEREIN. SEE TABLE 7 FOR EXPLANATION OF FOREST FORMATION ABBREVIATIONS, EXCEPT FOR MR (MONTANE RAINFOREST) FOR WHICH NO ANOLES ARE KNOWN (SEE TEXT).

| Area No. | Area Name | Forest Formation | Versant |
|---|---|---|---|
| 1 | Pacific Lowlands | LDF | Pacific |
| 2 | Middle Choluteca Valley | PDF | Pacific |
| 3 | Upper Choluteca Valley | PDF | Pacific |
| 4* | Comayagua Valley Rim | PDF | Atlantic |
| 5 | Comayagua Valley | LDF | Atlantic |
| 6* | Otoro Valley Rim | PDF | Atlantic |
| 7 | Otoro Valley | LDF | Atlantic |
| 8 | Southwestern Uplands | PMF | Pacific[1] |
| 9 | Southeastern Uplands | PMF | Pacific[1] |
| 10 | Southeastern Highlands | LMMF | Pacific[1] |
| 11 | Southwestern Highlands | LMMF | Pacific[1] |
| 12 | Guayape-Guayambre Valley | LDF | Atlantic |
| 13 | Northeastern Uplands | PMF | Atlantic |
| 14 | Northwestern Uplands | PMF | Atlantic |
| 15 | Eastern Caribbean Slope | PWF | Atlantic |
| 16 | Central Caribbean Slope | PWF | Atlantic |
| 17 | North-central Highlands | LMWF | Atlantic |
| 18 | Yoro Highlands | LMWF | Atlantic |
| 19 | Ocote Highlands | LMWF | Atlantic |
| 20 | Agalta Highlands | LMWF | Atlantic |
| 21 | Eastern Caribbean Lowlands | LMF | Atlantic |
| 22 | East-central Caribbean Lowlands | LMF | Atlantic |
| 23 | San Esteban Valley | LDF | Atlantic |
| 24 | Aguán Valley | LAF | Atlantic |
| 25* | Aguán Valley Rim | LDF | Atlantic |
| 26 | West-central Caribbean Lowlands | LMF | Atlantic |
| 27 | Western Caribbean Lowlands | LMF | Atlantic |
| 28 | Lower Motagua Valley | LMF | Atlantic |
| 29 | Sula Valley | LDF | Atlantic |
| 30 | Western Caribbean Slope | PWF | Atlantic |
| 31 | Yojoa Uplands | PWF | Atlantic |
| 32 | Northwestern Highlands | LMWF | Atlantic |
| 33 | Santa Bárbara Highlands | LMWF | Atlantic |
| 34* | Santa Bárbara Peak | MR | Atlantic |
| 35 | Utila Island | LMF | Atlantic |
| 36 | Roatán Island | LMF | Atlantic |
| 37 | Guanaja Island | LMF | Atlantic |
| 38 | Cayos Cochinos | LMF | Atlantic |
| 39 | Swan Islands | LDF (WI) | Atlantic |

[1] The northernmost portions of these areas are on the Atlantic versant.

American mainland and associated island populations (i.e., Cozumel, Mexico; Islas de la Bahía, Honduras) of N. sagrei to be human-caused introductions from the West Indies. As 15 of the 18 species (the exceptions being N. luteosignifer [Garman, 1888], N. nelsoni, and N. ordinatus [Cope, 1864]) of the N. sagrei group (Nicholson et al., 2012) currently occur on Cuba (although all at some point have been considered subspecies of N. sagrei), that island was likely the source of some of those introduc-

tions to the Central American mainland and associated islands. Similar genetic studies for the N. sagrei species group to those just discussed for A. allisoni would test that scenario.

## HONDURAS AS A DISTRIBUTIONAL ENDPOINT

In addition to the 18 Honduran anole species endemic to Honduras, analysis of the overall geographic distribution of the 20 remaining anole species (the probably

TABLE 13.   ANOLE SPECIES HAVING THEIR KNOWN GEOGRAPHICAL DISTRIBUTION ENDING ON THE MAINLAND OF HONDURAS OR ON HONDURAN ISLANDS (EXCEPT THE 18 HONDURAN ENDEMICS AND THE POSSIBLY INTRODUCED *NOROPS SAGREI*). SEE TEXT.

| |
|---|
| From west and north |
| *Norops crassulus* |
| *Norops heteropholidotus* |
| *Norops laeviventris* |
| *Norops ocelloscapularis* |
| *Norops petersii* |
| *Norops rodriguezii* |
| *Norops uniformis* |
| Total 7 |
| From east and south |
| *Norops carpenteri* |
| *Norops cupreus* |
| *Norops limifrons* |
| *Norops oxylophus* |
| *Norops quaggulus* |
| Total 5 |

introduced species *N. sagrei* is not included) occurring in the country shows that an additional 12 species have their known distributional ranges terminating somewhere in Honduras (Table 13). A total of 30 (including the Honduran endemics) of the 38 nonintroduced anole species (78.9%) occurring in Honduras have their ranges terminating in the country. Further analysis of the 12 nonendemic species with their ranges terminating in Honduras provide the following patterns:

Category 1. From west and/or north. Four of these seven species have their distributional ranges terminating in northwestern Honduras (*Norops ocelloscapularis*, *N. petersii*, *N. rodriguezii*, and *N. uniformis*). Two species (*N. crassulus* and *N. laeviventris*) occur in isolated montane habitats extending to central Honduras. One species (*N. heteropholidotus*) occurs in isolated montane habitats only in the southwestern portion of the country.

Category 2. From east and south. Two of these five species (*N. carpenteri* and *N. oxylophus*) are restricted to the Mosquitia region of northeastern Honduras, whereas two others (*N. limifrons* and *N. quaggulus*) barely occur outside the Mosquitia in

north-central Honduras. The fifth species in this category (*N. cupreus*) extends through the Honduran Mosquitia to north-central and southern Honduras.

## CONSERVATION STATUS OF ANOLES OF HONDURAS

### Vulnerability Gauges

We use a modified environmental vulnerability gauge for Honduran anole populations, based on those developed for amphibian and other reptilian populations (see Wilson et al., 2010; McCranie, 2011). Those gauges were designed for establishment of a set of conservation priorities for the Honduran reptiles and amphibians. Each gauge has three components, with those for the anole gauge discussed as follows:

The first component deals with the extent of the geographic distribution of each species as follows: 1 = widespread in and outside of Honduras; 2 = distribution peripheral to Honduras, but widespread elsewhere; 3 = distribution restricted to Nuclear Middle America (exclusive of Honduran endemics); 4 = distribution restricted to Honduras; 5 = known only from the vicinity of the type locality.

The second component deals with the extent of the ecological distribution of each species, based on a slightly modified version of the forest formations of Holdridge (1967), using the following scale (omitting the Montane Rainforest, from which no reptile species are known): 1 = occurs in eight formations; 2 = occurs in seven formations; 3 = occurs in six formations; 4 = occurs in five formations; 5 = occurs in four formations; 6 = occurs in three formations; 7 = occurs in two formations; 8 = occurs in one formation.

The third component considers the degree of human persecution. Since humans in Honduras generally ignore all anole species, we deviate from the environmental gauges used previously and place all anole species in a single category with a score of 1.

TABLE 14.  ENVIRONMENTAL VULNERABILITY SCORES (EVS) FOR THE 39 NAMED SPECIES OF ANOLES KNOWN FROM HONDURAS. NUMBERS FOR EACH ENVIRONMENTAL GAUGE ARE EXPLAINED IN THE TEXT. THE TABLE IS BROKEN INTO THREE PARTS: LOW-VULNERABILITY SPECIES (EVS OF 3–9), MEDIUM-VULNERABILITY SPECIES (EVS OF 10–12), AND HIGH-VULNERABILITY SPECIES (EVS OF 13–14).

| Species | Geographic Distribution | Ecological Distribution | Human Persecution | Total Score |
|---|---|---|---|---|
| Low (13 species): | | | | |
| *Norops beckeri* | 1 | 6 | 1 | 8 |
| *Norops biporcatus* | 1 | 6 | 1 | 8 |
| *Norops capito* | 1 | 7 | 1 | 9 |
| *Norops cupreus* | 1 | 4 | 1 | 6 |
| *Norops laeviventris* | 1 | 5 | 1 | 7 |
| *Norops lemurinus* | 1 | 5 | 1 | 7 |
| *Norops limifrons* | 1 | 7 | 1 | 9 |
| *Norops rodriguezii* | 2 | 5 | 1 | 8 |
| *Norops sagrei* | 1 | 7 | 1 | 9 |
| *Norops tropidonotus* | 1 | 1 | 1 | 3 |
| *Norops uniformis* | 2 | 6 | 1 | 9 |
| *Norops unilobatus* | 1 | 3 | 1 | 5 |
| *Norops wellbornae* | 1 | 6 | 1 | 8 |
| Medium (16 species): | | | | |
| *Anolis allisoni* | 1 | 8 | 1 | 10 |
| *Norops carpenteri* | 2 | 8 | 1 | 11 |
| *Norops crassulus* | 3 | 6 | 1 | 10 |
| *Norops cusuco* | 4 | 7 | 1 | 12 |
| *Norops heteropholidotus* | 3 | 8 | 1 | 12 |
| *Norops johnmeyeri* | 4 | 7 | 1 | 12 |
| *Norops kreutzi* | 4 | 7 | 1 | 12 |
| *Norops loveridgei* | 4 | 6 | 1 | 11 |
| *Norops morazani* | 4 | 7 | 1 | 12 |
| *Norops ocelloscapularis* | 3 | 7 | 1 | 11 |
| *Norops oxylophus* | 2 | 8 | 1 | 11 |
| *Norops petersii* | 2 | 7 | 1 | 10 |
| *Norops quaggulus* | 2 | 7 | 1 | 10 |
| *Norops sminthus* | 4 | 7 | 1 | 12 |
| *Norops yoroensis* | 4 | 6 | 1 | 11 |
| *Norops zeus* | 4 | 6 | 1 | 11 |
| High (10 species): | | | | |
| *Norops amplisquamosus* | 5 | 8 | 1 | 14 |
| *Norops bicaorum* | 5 | 8 | 1 | 14 |
| *Norops muralla* | 5 | 7 | 1 | 13 |
| *Norops nelsoni* | 5 | 8 | 1 | 14 |
| *Norops pijolense* | 5 | 7 | 1 | 13 |
| *Norops purpurgularis* | 4 | 8 | 1 | 13 |
| *Norops roatanensis* | 5 | 8 | 1 | 14 |
| *Norops rubribarbaris* | 5 | 8 | 1 | 14 |
| *Norops utilensis* | 5 | 8 | 1 | 14 |
| *Norops wampuensis* | 5 | 8 | 1 | 14 |

All Honduran anole species are placed in one of three categories, as indicated in Table 14, with 13 species of low vulnerability (33.3% of total anole species), 16 species of medium vulnerability (41.0%), and 10 species of high vulnerability (25.6%).

Of the 18 Honduran endemics, eight (*Norops cusuco, N. johnmeyeri, N. kreutzi, N. loveridgei, N. morazani, N. sminthus, N. yoroensis,* and *N. zeus*) are classified as having medium vulnerability. Seven of these eight species (only lacking *N. kreutzi;* see

TABLE 15.   CURRENT STATUS OF POPULATIONS OF HONDURAN ANOLE ENDEMICS AND SPECIES OTHERWISE RESTRICTED TO NUCLEAR MIDDLE AMERICA. STABLE = AT LEAST SOME POPULATIONS STABLE; DECLINING = ALL POPULATIONS BELIEVED TO BE DECLINING.

| Species | Stable | Declining |
|---|---|---|
| Honduran endemics (18 species) | | |
| *Norops amplisquamosus* | | X |
| *Norops bicaorum* | X | |
| *Norops cusuco* | X | |
| *Norops johnmeyeri* | X | |
| *Norops kreutzi* | | X |
| *Norops loveridgei* | X | |
| *Norops morazani* | X | |
| *Norops muralla* | X | |
| *Norops nelsoni* | X | |
| *Norops pijolense* | X | |
| *Norops purpurgularis* | X | |
| *Norops roatanensis* | X | |
| *Norops rubribarbaris* | X | |
| *Norops sminthus* | X | |
| *Norops utilensis* | X | |
| *Norops wampuensis* | | X |
| *Norops yoroensis* | X | |
| *Norops zeus* | X | |
| Honduran species otherwise restricted to Nuclear Middle America (6 species) | | |
| *Norops beckeri* | X | |
| *Norops crassulus* | X | |
| *Norops heteropholidotus* | X | |
| *Norops ocelloscapularis* | X | |
| *Norops rodriguezii* | X | |
| *Norops wellbornae* | X | |

Table 15) are thought to have stable populations somewhere within their known ranges. *Norops kreutzi* is thought to have all populations declining despite having some pristine forest remaining within its range. The remaining 10 Honduran endemics (*N. amplisquamosus, N. bicaorum, N. muralla, N. nelsoni, N. pijolense, N. purpurgularis, N. roatanensis, N. rubribarbaris, N. utilensis,* and *N. wampuensis*; see Table 14) fall into the high-vulnerability category. However, eight of these 10 species (only lacking *N. amplisquamosus* and *N. wampuensis*) are thought to have some remaining stable populations (see Table 15) because of either having some forest remaining where they occur or having adapted to alternative habitats. The remaining two Honduran endemics in the high-vulnerabil-

ity category (*N. amplisquamosus* and *N. wampuensis*) appear to have all populations declining. Pristine forest or seemingly suitable habitat remains for *N. amplisquamosus,* whereas the few known localities of the *N. wampuensis* have had their forest habitat completely destroyed. The reason(s) for the decline(s) of *N. amplisquamosus* is(are) not understood, but climate change might have had an effect. Forest in the vicinity of the road where individuals of *N. amplisquamosus* were extremely common during the 1980s are no longer as moss covered as in the past (JRM, personal observation); a sign of the drying out of those forests.

## IUCN Red List Categories

Each of the 39 anole species known to occur in Hondurans was placed into one of five categories (Table 16) using the criteria developed by the IUCN. Examination of Table 16 shows that three species (*Norops amplisquamosus, N. utilensis,* and *N. wampuensis*) are classified as Critically Endangered, two species (*N. cusuco* and *N. kreutzi*) as Endangered, and five species (*N. morazani, N. muralla, N. pijolense, N. purpurgularis,* and *N. rubribarbaris*) as vulnerable. All 10 of those species are Honduran endemics. Six other Honduran endemics (*N. bicaorum, N. johnmeyeri, N. loveridgei, N. nelsoni, N. roatanensis,* and *N. yoroensis*) are classified as Near Threatened, whereas two Honduran endemics (*N. sminthus* and *N. zeus*) are classified as species of Least Concern.

## Anole Species Occurring in Protected Areas

As stated by McCranie (2011: 572, 576), "At first glance, Honduras appears to have in place a robust system of protected areas, especially when compared to nearby countries. However, most of those areas exist on paper only. Usually only when a nongovernmental organization (NGO), financed largely by donations from people living outside of Honduras, is founded with the purpose of conservation of a given area, does some sort of an infrastructure exists. Also, the distributional limits of many of

TABLE 16. IUCN RED LIST CATEGORIES FOR THE 39 NAMED HONDURAN ANOLE SPECIES.

| Species | Critically Endangered | Endangered | Vulnerable | Near Threatened | Least Concern |
|---|---|---|---|---|---|
| *Anolis allisoni* | — | — | — | — | X |
| *Norops amplisquamosus* | B1ab(v) | — | — | — | — |
| *Norops beckeri* | — | — | — | — | X |
| *Norops bicaorum* | — | — | — | X | — |
| *Norops biporcatus* | — | — | — | — | X |
| *Norops capito* | — | — | — | — | X |
| *Norops carpenteri* | — | — | — | — | X |
| *Norops crassulus* | — | — | — | — | X |
| *Norops cupreus* | — | — | — | — | X |
| *Norops cusuco* | — | B2ab(iii) | — | — | — |
| *Norops heteropholidotus* | — | — | — | — | X |
| *Norops johnmeyeri* | — | — | — | X | — |
| *Norops kreutzi* | — | B2ab(iii) | — | — | — |
| *Norops laeviventris* | — | — | — | — | X |
| *Norops lemurinus* | — | — | — | — | X |
| *Norops limifrons* | — | — | — | — | X |
| *Norops loveridgei* | — | — | — | X | — |
| *Norops morazani* | — | — | B2ab(iv) | — | — |
| *Norops muralla* | — | — | B2ab(iv) | — | — |
| *Norops nelsoni* | — | — | — | X | — |
| *Norops ocelloscapularis* | — | — | — | X | — |
| *Norops oxylophus* | — | — | — | — | X |
| *Norops petersii* | — | — | — | — | X |
| *Norops pijolense* | — | — | B2ab(iii) | — | — |
| *Norops purpurgularis* | — | — | B2ab(iii) | — | — |
| *Norops quaggulus* | — | — | — | — | X |
| *Norops roatanensis* | — | — | — | X | — |
| *Norops rodriguezii* | — | — | — | — | X |
| *Norops rubribarbaris* | — | — | B2ab(iii) | — | — |
| *Norops sagrei* | — | — | — | — | X |
| *Norops sminthus* | — | — | — | — | X |
| *Norops tropidonotus* | — | — | — | — | X |
| *Norops uniformis* | — | — | — | — | X |
| *Norops unilobatus* | — | — | — | — | X |
| *Norops utilensis* | B2ab(iii) | — | — | — | — |
| *Norops wampuensis* | B2ab(iii) | — | — | — | — |
| *Norops wellbornae* | — | — | — | — | X |
| *Norops yoroensis* | — | — | — | X | — |
| *Norops zeus* | — | — | — | — | X |
| Totals | 3 | 2 | 5 | 7 | 22 |

those areas are poorly defined. Despite those and other problems, there is still some good forest remaining in many protected areas in Honduras." Some of the problems alluded to by McCranie (2011) was the continued deforestation occurring in many, if not all, of those areas. Unfortunately, that deforestation seems to have accelerated in the last five years. McCranie (2011) compiled a list of 30 areas in which some forest remains and herpetological collections have been made. One of those areas, Rus Rus (area 23 in McCranie, 2011), is not included herein because it is no longer being considered for proposal as a protected area; in addition 11 other areas not discussed in McCranie (2011) are added. Table 17 includes a list of main forest type(s), sizes, and comments on 40 areas and Table 18 shows the distribution of each anole species in those areas. Map 44 shows the general location of 39 of those

TABLE 17.   HONDURAN PROTECTED AREAS WITH SOME PRISTINE FOREST OR SUITABLE HABITAT REMAINING. AREA ABBREVIATIONS: PN = PARQUE NACIONAL (NATIONAL PARK); R = RESERVA (RESERVE); RA = RESERVA ANTOPOLÓGICA (ANTHROPOLOGICAL RESERVE); RB = RESERVA BIOLÓGICA (BIOLOGICAL RESERVE); RVS = RESERVA DE VIDA SILVESTRE (WILDLIFE REFUGE); PNM = PARQUE NACIONAL MARINO (NATIONAL MARINE PARK); AUM = ÁREA DE USO MÚLTIPLE (MULTIPLE USE AREA); JB = JARDÍN BOTÁNICA (BOTANICAL GARDEN). FOREST TYPE ABBREVIATIONS: LDF(WI) = LOWLAND DRY FOREST, WEST INDIAN SUBREGION; LMF = LOWLAND MOIST FOREST; LMMF = LOWER MONTANE MOIST FOREST; LMWF = LOWER MONTANE WET FOREST; MR = MONTANE RAINFOREST; PMF = PREMONTANE MOIST FOREST; AND PWF = PREMONTANE WET FOREST. AN ASTERISK (*) PRECEDING A NUMBER INDICATES THAT WE HAVE NOT COLLECTED IN THAT AREA. AREA SIZES IN HECTARES (HA) ARE FROM SÁNCHEZ ET AL. (2002).

| Area (ha) | Main Forest Type(s) | Comments |
|---|---|---|
| 1-PN Capiro y Calentura (5,566) | LMF, PWF | Locked gate across only road accessing this park offers protection to existing forest. |
| 2-PN Celaque (26,639) | PMF, LMMF | Large tract of forest S of visitor's center. Western section in Ocotepeque badly deforested. |
| 3-PN Cerro Azul (15,574) | PWF, LMMF | Best tract of forest remaining is along western side above San Isidro. Much of eastern portion badly deforested, except highest peak above about 1,600 m elevation near Quebrada Grande. |
| 4-PN Cerro Azul Meámbar (20,789) | PWF, LMWF | Large tracts of both forest types present. |
| 5-PN Cusuco (17,908) | PWF, PMF, LMWF | Core zone around El Cusuco well protected until recently, much of lower elevations around flanks of core zone under heavy human pressure. |
| 6-PN El Merendón (35,182) | LMF, PWF | Some primary or old secondary forest remaining in more interior areas. |
| 7-PN La Muralla (14,941) | PWF, LMWF | LMWF above visitor's center well protected; some of harder-to-reach PWF still preserved. |
| 8-PN La Tigra (8,768) | PMF, LMMF | LMMF well protected, but much of PMF damaged and/or burned annually. |
| 9-PN Montaña Botaderos (38,214) | PWF, LMWF | Largely deforested with primary or old secondary forest remaining only in highest reaches. |
| 10-PN Montaña de Comayagua (18,273) | PWF, PMF, LMMF | Much of park highly disturbed; some pine forest remaining. |
| 11-PN Montaña de Yoro (15,468) | LMWF | Large tracts of forest remaining in harder-to-reach areas. |
| 12-PN Montecristo-Trifinio (1,534) | LMMF | Some of higher reaches still forested. |
| 13-PN Patuca (37,6448) | LMF | Areas along rivers and trails heavily deforested. Forest remaining in more interior areas. |
| 14-PN Pico Bonito (56,473) | LMF, PWF, LMWF | Northern and western slopes and upper reaches retain much forest, but interior along Río Viejo and tributaries heavily impacted by humans and Hurricane Mitch. |
| 15-PN Pico Pijol (11,453) | PWF, LMWF | Forest still remaining in some upper reaches and in a few pockets in lower elevations. |
| *16-PN Punta Izopo (6,405) | LMF | Little forest remaining. Lagoons and swamps make up much of park. |
| *17-PN Punta Sal, also called PN Jeannette Kawas (37,996) | LMF | Lagoons and swamps make up much of park. Apparently some pockets of forest remain. |
| 18-PN Santa Bárbara (13,236) | LMWF, MR | Tracts of forest remain only above about 2,100 m elevation. |
| 19-PN Sierra de Agalta (51,837) | PWF, LMWF | Much forest remains in core zone; area around flanks of core zone much disturbed. |
| 20-R Biósfera Río Plátano (829,779) | LMF | Large core zone of broadleaf forest remains, but becoming more impacted every year; all of buffer zone badly impacted, including broadleaf forest, pine savanna, and freshwater lagoons and swamps. |
| 21-RA Tawahka (252,079) | LMF | Forest remains in places, but forests along all rivers and trails badly impacted. |
| 22-RB Cerro El Uyuca (817) | PMF, LMMF | Pine and broadleaf forest remains at highest elevations. |
| 23-RB Cordillera de Montecillos (13,191) | PMF, LMMF | A little forest remaining in highest reaches. |
| *24-RB El Chile (6,280) | PMF, LMMF | Apparently much of reserve still forested. |

TABLE 17. CONTINUED.

| Area (ha) | Main Forest Type(s) | Comments |
|---|---|---|
| 25-RB El Pital (1,799) | LMMF | Almost entirely deforested for agriculture. Unusually heavy pesticide use. |
| 26-RB Guajiquiro (7,368) | LMMF | Limited patches of forest remaining. |
| 27-RB Guisayote (hectares not given in Sánchez et al.) | LMMF | Limited patches of forest remaining. |
| 28-RB Hierba Buena (3,522) | PMF, LMMF | Heavily impacted for agriculture; primary or secondary forest remaining on higher elevations of Cerro Cantagallo. |
| 29-RB Monserrat (2,241) | PMF, LMMF | Limited patches of forest remaining, except on top at communications tower. |
| 30-RB Opalaca (14,953) | LMMF | Almost entirely deforested for agriculture. |
| 31-RB Río Kruta (115,107) | LMF | Area along Río Kruta and Río Coco denuded, but more inland broadleaf swamp forest and freshwater marshes of little use to humans. Rising sea levels a threat. |
| 32-RVS Cuero y Salado (7,948) | LMF | Lagoons and swamps make up much of park. Little original forest remaining. |
| 33-RVS Erapuca (7,317) | LMMF | Tracts of little disturbed forest remain above about 2,000 m elevation. |
| 34-RVS Laguna de Caratasca (133,749) | LMF | Reserve made up largely of pine savanna and cocotales, although largely disturbed, still offers habitat for several species otherwise occurring only in denuded subhumid forest in south and interior valleys of country. |
| 35-RVS Mixcure (7,766) | LMMF | Higher portions show little human impact. |
| 36-RVS Texíguat (15,810) | LMWF | Little primary forest remaining. Much of reserve now crop fields. Heavy logging also present. |
| *37-PNM Cayos Cochinos (hectares not given in Sánchez et al.) | LWF | Marine park also protects two largest islands. Primary or old second growth forest remaining on parts of both main islands. |
| 38-PNM Islas del Cisne (hectares not given in Sánchez et al.) | LDF(WI) | Marine park also includes these difficult-to-reach islands. Isla Pequeña lacks good landing beaches and mostly covered by karsted limestone of little use to humans. Much shrub forest remaining. |
| 39-AUM Isla del Tigre (588) | LDF | Much of slopes covered with second growth forest; lagoon on naval base is restricted access. |
| 40-JB Lancetilla (1,010) | LMF | Botanical gardens with many species of introduced trees and some old secondary forest preserved. |

areas (see Map 20 for the Islas del Cisne). Average protected area occurrence for the 39 anole species in Honduras is 3.8.

Examination of Table 18 demonstrates that from 0 (Area 16 [not visited by us]; see Table 17 for area names) to 11 (Area 21) anole species are known from a given area. Since most anole species are easily collected where they occur, we consider many of these areas to be well studied regarding their anole fauna. However, there are 15 areas (Areas 11, 15–17, 22–25, 27–30, 32, 33, and 35) that could use additional survey work to get a more accurate assessment of

the actual number of anole species in those areas. It is expected that one or two additional anole species will be shown to occur in each of those 15 areas.

Five of the 39 anole species known from Honduras have not been collected from any of the 40 protected areas (Table 18). One of these five species is the probably introduced *N. sagrei*; thus, that species is of little concern for conservation issues (in addition *N. sagrei* thrives in edificarian situations, so would be of no concern even if native). Three of the four remaining species not known from any of the 40 protected areas

TABLE 18.   OCCURRENCE OF THE 39 SPECIES OF *NOROPS* IN THE PROTECTED AREAS INCLUDED HEREIN. REFER TO TABLE 17 FOR THE CORRESPONDING AREA FOR EACH NUMBER.

| Species (39) | 1 | 2 | 3 | 4 | 5 | 6 | 7 | 8 | 9 | 10 | 11 | 12 | 13 | 14 | 15 | 16 | 17 |
|---|---|---|---|---|---|---|---|---|---|---|---|---|---|---|---|---|---|
| *Anolis allisoni* | — | — | — | — | — | — | — | — | — | — | — | — | — | — | — | — | — |
| *Norops amplisquamosus* | — | — | — | — | X | — | — | — | — | — | — | — | — | — | — | — | — |
| *Norops beckeri* | X | — | — | X | — | — | X | — | — | — | — | — | — | — | — | — | X |
| *Norops bicaorum* | — | — | — | — | — | — | — | — | — | — | — | — | — | — | — | — | — |
| *Norops biporcatus* | — | — | X | X | — | X | X | — | X | — | — | — | — | X | — | — | — |
| *Norops capito* | — | — | X | X | X | X | X | — | X | — | — | — | X | — | — | — | — |
| *Norops carpenteri* | — | — | — | — | — | — | — | — | — | — | — | — | — | — | — | — | — |
| *Norops crassulus* | — | X | — | — | — | — | — | — | — | — | — | — | — | — | — | — | — |
| *Norops cupreus* | X | — | — | — | — | — | X | — | X | — | — | — | X | — | — | — | — |
| *Norops cusuco* | — | — | X | — | X | — | — | — | — | — | — | — | — | — | — | — | — |
| *Norops heteropholidotus* | — | — | — | — | — | — | — | — | — | — | — | X | — | — | — | — | — |
| *Norops johnmeyeri* | — | — | X | — | X | — | — | — | — | — | — | — | — | — | — | — | — |
| *Norops kreutzi* | — | — | — | — | — | — | — | — | — | — | — | — | — | — | — | — | — |
| *Norops laeviventris* | — | X | — | X | — | — | X | X | X | X | — | — | — | — | — | — | — |
| *Norops lemurinus* | X | — | — | X | — | X | — | — | — | — | — | — | X | X | — | — | — |
| *Norops limifrons* | — | — | — | — | — | — | — | — | X | — | — | — | X | — | — | — | — |
| *Norops loveridgei* | — | — | — | — | — | — | — | — | — | — | — | — | — | X | — | — | — |
| *Norops morazani* | — | — | — | — | — | — | — | — | — | — | X | — | — | — | — | — | — |
| *Norops muralla* | — | — | — | — | — | — | X | — | — | — | — | — | — | — | — | — | — |
| *Norops nelsoni* | — | — | — | — | — | — | — | — | — | — | — | — | — | — | — | — | — |
| *Norops ocelloscapularis* | — | — | X | — | X | — | — | — | — | — | — | — | — | — | — | — | — |
| *Norops oxylophus* | — | — | — | — | — | — | — | — | — | — | — | — | X | — | — | — | — |
| *Norops petersii* | — | — | X | — | X | — | — | — | — | — | — | — | — | — | — | — | — |
| *Norops pijolense* | — | — | — | — | — | — | — | — | — | — | — | — | — | — | X | — | — |
| *Norops purpurgularis* | — | — | — | — | — | — | — | — | — | — | — | — | X | — | — | — | — |
| *Norops quaggulus* | — | — | — | — | — | — | — | — | — | — | — | — | X | — | — | — | — |
| *Norops roatanensis* | — | — | — | — | — | — | — | — | — | — | — | — | — | — | — | — | — |
| *Norops rodriguezii* | — | — | X | — | — | X | — | — | — | — | — | — | — | — | — | — | — |
| *Norops rubribarbaris* | — | — | — | — | — | — | — | — | — | — | — | — | — | — | — | — | — |
| *Norops sagrei* | — | — | — | — | — | — | — | — | — | — | — | — | — | — | — | — | — |
| *Norops sminthus* | — | — | — | — | — | — | — | X | — | X | — | — | — | — | — | — | — |
| *Norops tropidonotus* | X | — | — | X | — | X | X | X | X | X | — | X | — | X | — | — | — |
| *Norops uniformis* | — | — | X | — | — | X | — | — | — | — | — | — | — | X | — | — | — |
| *Norops unilobatus* | — | — | — | — | — | X | — | — | — | — | — | — | — | X | — | — | X |
| *Norops utilensis* | — | — | — | — | — | — | — | — | — | — | — | — | — | — | — | — | — |
| *Norops wampuensis* | — | — | — | — | — | — | — | — | — | — | — | — | — | — | — | — | — |
| *Norops wellbornae* | — | — | — | — | — | — | — | — | — | — | — | — | — | — | — | — | — |
| *Norops yoroensis* | — | — | X | — | — | X | — | — | — | — | — | — | — | X | X | — | — |
| *Norops zeus* | X | — | — | X | — | — | — | — | — | — | — | — | — | X | — | — | X |
| Totals | 5 | 2 | 9 | 7 | 6 | 8 | 7 | 3 | 6 | 3 | 1 | 2 | 6 | 9 | 2 | 0 | 3 |

(*N. bicaorum, N. roatanensis*, and *N. utilensis*) are Honduran endemics that are restricted to either Isla de Roatán or Isla de Utila in the Islas de la Bahía. There are no land areas on either Roatán or Utila officially designated as protected, and the NGO "Bay Island Conservation Association" has been unable to slow down the increased development of these islands that exploded during the 1990s and continues to this day (Stonich, 2000; McCranie et al., 2005; JRM, personal observation). However, one of those Bay Island endemic species (*N. roatanensis*) has adapted well to altered habitats and will probably retain healthy populations into at least the near future. *Norops bicaorum*, a close relative of *N. roatanensis*, on the other hand, appears to be declining in its similarly altered habitat on Isla de Utila (but still remains common, but seemingly down from its former exceedingly abundant category). *Norops uti-*

TABLE 18. EXTENDED.

| 18 | 19 | 20 | 21 | 22 | 23 | 24 | 25 | 26 | 27 | 28 | 29 | 30 | 31 | 32 | 33 | 34 | 35 | 36 | 37 | 38 | 39 | 40 | Totals |
|---|---|---|---|---|---|---|---|---|---|---|---|---|---|---|---|---|---|---|---|---|---|---|---|
| — | — | — | — | — | — | — | — | — | — | — | — | — | — | — | — | X | — | — | X | — | — | — | 2 |
| — | — | — | — | — | — | — | — | — | — | — | — | — | — | — | — | — | — | — | — | — | — | — | 1 |
| — | — | X | X | — | — | — | — | — | — | — | — | — | — | — | — | — | — | — | — | — | — | X | 7 |
| — | — | — | — | — | — | — | — | — | — | — | — | — | — | — | — | — | — | — | — | — | — | — | 0 |
| — | X | X | X | — | — | X | — | — | — | — | — | — | X | — | — | — | — | — | — | — | — | X | 12 |
| — | X | X | X | — | — | — | — | — | — | — | X | — | — | — | — | — | — | — | — | — | — | — | 11 |
| — | — | — | — | — | — | — | — | — | — | — | — | — | — | — | — | — | — | — | — | — | — | — | 0 |
| — | X | — | — | X | — | — | X | — | — | — | — | — | — | — | — | — | X | — | — | — | — | — | 5 |
| — | X | — | X | — | — | X | — | — | — | — | — | — | X | — | — | — | — | — | — | — | — | — | 8 |
| — | — | — | — | — | — | — | — | — | — | — | — | — | — | — | — | — | — | — | — | — | — | — | 2 |
| — | — | — | — | — | — | X | — | X | — | — | — | — | — | X | — | — | — | — | — | — | — | — | 4 |
| — | — | — | — | — | — | — | — | — | — | — | — | — | — | — | — | — | — | — | — | — | — | — | 2 |
| — | — | — | — | — | — | — | — | — | — | — | — | — | — | — | — | — | X | — | — | — | — | — | 1 |
| X | X | — | — | X | — | — | — | X | — | — | — | — | — | — | — | — | — | — | — | — | — | — | 10 |
| — | — | X | X | — | — | — | — | — | — | — | — | — | X | — | — | — | — | X | — | — | — | X | 10 |
| — | — | X | X | — | — | — | — | — | — | — | — | — | X | — | — | — | — | — | — | — | — | — | 5 |
| — | — | — | — | — | — | — | — | — | — | — | — | — | — | — | — | — | X | — | — | — | — | — | 2 |
| — | — | — | — | — | — | — | — | — | — | — | — | — | — | — | — | — | — | — | — | — | — | — | 1 |
| — | — | — | — | — | — | — | — | — | — | — | — | — | — | — | — | — | — | — | — | — | — | — | 1 |
| — | — | — | — | — | — | — | — | — | — | — | — | — | — | — | — | — | — | — | — | X | — | — | 1 |
| — | — | — | — | — | — | — | — | — | — | — | — | — | — | — | — | — | — | — | — | — | — | — | 2 |
| — | — | — | X | — | — | — | — | — | — | — | — | — | — | — | — | — | — | — | — | — | — | — | 2 |
| — | — | — | — | — | — | — | — | — | — | — | — | — | — | — | — | — | — | — | — | — | — | — | 2 |
| — | — | — | — | — | — | — | — | — | — | — | — | — | — | — | — | — | — | — | — | — | — | — | 1 |
| — | — | — | — | — | — | — | — | — | — | — | — | — | — | — | — | — | X | — | — | — | — | — | 2 |
| — | — | X | X | — | — | — | — | — | — | — | — | — | — | — | — | — | — | — | — | — | — | — | 3 |
| — | — | — | — | — | — | — | — | — | — | — | — | — | — | — | — | — | — | — | — | — | — | — | 0 |
| — | — | — | — | — | — | — | — | — | — | — | — | — | — | — | — | — | — | — | — | — | — | — | 2 |
| X | — | — | — | — | — | — | — | — | — | — | — | — | — | — | — | — | — | — | — | — | — | — | 1 |
| — | — | — | — | — | — | — | — | — | — | — | — | — | — | — | — | — | — | — | — | — | — | — | 0 |
| — | — | — | — | — | — | — | — | — | — | X | — | — | — | — | — | — | — | — | — | — | — | — | 3 |
| — | X | X | X | X | X | X | — | — | — | — | X | — | — | — | — | — | X | X | — | — | — | X | 20 |
| — | — | — | — | — | — | — | — | — | — | — | — | — | — | — | — | — | — | — | — | — | — | — | 3 |
| — | X | X | X | — | — | X | — | — | — | — | — | — | X | — | — | X | — | — | — | — | — | X | 10 |
| — | — | — | — | — | — | — | — | — | — | — | — | — | — | — | — | — | — | — | — | — | — | — | 0 |
| — | — | — | X | — | — | — | — | — | — | — | — | — | — | — | — | — | — | — | — | — | — | — | 1 |
| — | — | — | — | — | — | — | — | — | — | — | — | — | — | — | — | — | — | — | — | — | X | — | 1 |
| — | — | — | — | — | — | — | — | — | — | — | — | — | — | — | — | — | — | X | — | — | — | — | 5 |
| — | — | — | — | — | — | — | — | — | — | — | — | — | — | X | — | — | — | — | — | — | — | X | 6 |
| 2 | 7 | 8 | 11 | 2 | 2 | 4 | 1 | 2 | 1 | 1 | 1 | 1 | 4 | 2 | 1 | 2 | 2 | 5 | 2 | 1 | 1 | 6 | 148 |

*lensis* needs to have its remaining mangrove habitat on Utila protected from human destruction, or it might not survive for the long term (*N. utilensis* has recently been found several times outside of mangrove habitat; hopefully that signifies an adaptation to other habitat types as its primary habitat is altered). The remaining species not known from any of the 40 protected areas (*N. carpenteri*) is known from two nearby localities in the Honduran Mosqui-

tia, where much primary forest remains, and it likely occurs in at least one of the protected areas in that region.

McCranie (2011) and Townsend and Wilson (2010a) provided additional discussions of conservation issues affecting Honduras. The interested reader is referred to those publications for more Honduran conservation information. However, and unfortunately, deforestation throughout Honduras has accelerated in recent years,

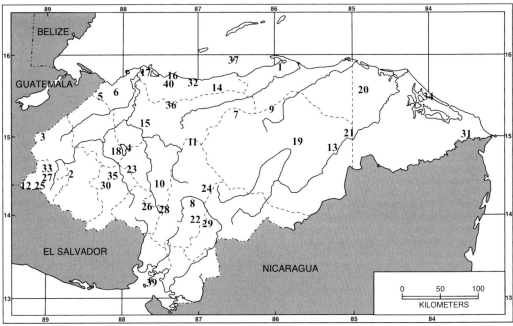

Map 44.   The general locations of some protected areas of Honduras with forest remaining: 1 = Capiro and Calentura National Park; 2= Celaque National Park; 3 = Cerro Azul National Park; 4 = Cerro Azul Meámbar National Park; 5 = Cusuco National Park; 6 = El Merendón National Park; 7 = La Muralla National Park; 8 = La Tigra National Park; 9 = Montaña Botaderos National Park; 10 = Montaña de Comayagua National Park; 11 = Montaña de Yoro National Park; 12 = Montecristo-Trifinio National Park; 13 = Patuca National Park; 14 = Pico Bonito National Park; 15 = Pico Pijol National Park; 16 = Punta Izopo National Park; 17 = Punta Sal National Park; 18 = Santa Bárbara National Park; 19 = Sierra de Agalta National Park; 20 = Río Plátano Biosphere Reserve; 21 = Tawahka Anthropological Reserve; 22 = Cerro El Uyuca Biological Reserve; 23 = Cordillera de Montecillos Biological Reserve; 24 = El Chile Biological Reserve; 25 = El Pital Biological Reserve; 26 = Guajiquiro Biological Reserve; 27 = Guisayote Biological Reserve; 28 = Hierba Buena Biological Reserve; 29 = Monserrat Biological Reserve; 30 = Opalaca Biological Reserve; 31 = Río Kruta Biological Reserve; 32 = Cuero and Salado Wildlife Refuge; 33 = Erapuca Wildlife Refuge; 34 = Laguna de Caratasca Wildlife Refuge; 35 = Mixcure Wildlife Refuge; 36 = Texíguat Wildlife Refuge; 37 = Cayos Cochinos Marine National Park; 39 = Islas del Tigre Multiple Use Area; 40 = Lancetilla Botanical Gardens. See Map 20 for the location of area 38 (Swan Islands or Islas del Cisne).

including inside the 40 "protected" areas. Despite the illegality of cutting down forests inside "protected" areas and those along water courses throughout the entire country, those laws are not enforced and are completely ignored (from the *campesinos* to those at the top of ICF). Thus, there is no deterrent to stop those illegal activities and with the increasing human population growth, matters will only get worse.

## ACKNOWLEDGMENTS

Collecting and exportation permits have been provided over the years by personnel of the Dirección General de Recursos Naturales Renovables, the Departamentos de Áreas Protegidas y Vida Silvestre de la Corporación Hondureña de Desarrollo Forestal (DAPVS/AFE-COHDEFOR), and the Instituto Nacional de Conservación y Desarrollo Forestal, Áreas Protegidas y Vida Silvestre (ICF). Lic. Iris Acosta, Lic. Carla Cárcamo de Martínez, Lic. Franklin E. Castañeda, Lic. Saíd Laínez, and especially Leonardo Valdés Orellana have been helpful during the last 10 years.

Field assistance in our pursuit of Honduran anoles has been provided over the years by the following individuals: Damian Almendarez, Breck Bartholomew, Franklin E. Castañeda, Juan R. Collart, Gustavo A. Cruz, Gary Dodge, Mario R. Espinal,

Gerardo A. Flores, Steve W. Gotte, Dalmacia Green, the late Emiliano Green, the late Mario Guiffaro, Alexander Gutsche, Alexis Harrison, Eric Hedl, John Himes, Elke Köhler, Jonathan Losos, Tomás Manzanares, Emiliano Meráz, Kirsten E. Nicholson, Louis Porras, John Rindfleish, Javier Rodriguez, José M. Solis, Josiah H. Townsend, Leonardo Valdés Orellana, Rony Valle Ocho, Kenneth L. Williams, and Larry D. Wilson.

The following curators and other museum personnel facilitated loans or provided laboratory space for our examinations of specimens: Margaret Arnold, David Dickey, Darrel Frost, David Kizirian (AMNH); Ted Daeschler, Ned S. Gilmore (ANSP); Jack W. Sites (BYU); Jens Vindum (CAS); the late Clarence J. McCoy, Stephen P. Rogers (CM); Kathleen M. Kelly, Alan Resetar (FMNH); Chris Mayer, John E. Petzing, Chris A. Phillips (INHS); Andrew Campbell, William E. Duellman, Jamie Oaks, John Simmons, Linda Trueb (KU); Rick Feeney, Jeff Seigel (LACM); Christopher Austin (LSUMZ); James Hanken, José Rosado (MCZ); Carol Spencer, David B. Wake (MVZ); Laura Abraczinskas (MSUM); Ross MacCullogh, Robert W. Murphy (ROM); Brad Hollingsworth, Robert E. Lovich (SDSNH); Toby Hibbits (TCWC); David C. Cannatella, Travis La Duc (TNHC); Kenneth L. Krysko (UF); Chris Phillips, Steven D. Sroka (UIMNH); Ronald Nussbaum, Gregory Schneider (UMMZ); Gustavo A. Cruz, Julio Mérida (UNAH); Steve W. Gotte, W. Ron Heyer, James A. Poindexter (USNM); and Jonathan A. Campbell, Carl J. Franklin, Eric N. Smith (UTA). James A. Poindexter also helped with some difficult-to-obtain literature, as did Alexander Gutsche and Steven Poe.

The Marshall Field Fund (FMNH), through the efforts of Alan Resetar, paid part of McCranie's expenses for fieldwork during 2011–2012, and Jonathan Losos and Melissa Woodley Aja were also helpful in acquiring funds from the Barbour Fund at the Museum of Comparative Zoology (MCZ) for a trip to the Swan Islands during December 2012. McCranie is extremely grateful to those persons for their financial assistance.

Jay Savage reviewed an earlier draft of the manuscript and Robert Lovich reexamined the SDSNH series of *Norops cupreus* for us to verify the presence of calyculate hemipenes in that series. Oscar Flores-Villela translated the English version of the identification keys into Spanish, and Leonardo Valdés Orellana made comments on the Spanish version of those keys.

We are especial grateful to Kraig Ader, Tim Perry, and the Society for the Study of Amphibians and Reptiles (SSAR) for permission to use several maps, including the base map to plot the localities for the species maps. Jonathan B. Losos and Steven Poe thoroughly reviewed an earlier draft of this manuscript and made many corrections and helpful comments. Steven Poe, Josiah H. Townsend, and Leonardo Valdés Orellana also provided one or more photographs used herein.

## LITERATURE CITED

ADALSTEINSSON, S. A., W. R. BRANCH, S. TRAPE, L. J. VITT, AND S. B. HEDGES. 2009. Molecular phylogeny, classification, and biogeography of snakes of the family Leptotyphlopidae (Reptilia, Squamata). *Zootaxa* **2244:** 1–50.

AHL, E. 1939. Ueber eine Sammlung von Reptilien aus El Salvador. *Sitzungsberichte der Gesellschaft Naturforschender Freunde zu Berlin* **1939:** 245–249.

ALFÖLDI, J., F. DI PALMA, M. GRABHERR, C. WILLIAMS, L. KONG, E. MAUCELI, P. RUSSELL, C. B. LOWE, R. E. GLOR, J. D. JAFFE, AND OTHERS. 2011. The genome of the green anole lizard and a comparative analysis with birds and mammals. *Nature* **477:** 587–591.

ÁLVAREZ DEL TORO, M. 1983 (dated 1982). *Los Reptiles de Chiapas. Tercera Edición, corregida y aumentada.* Tuxtla Gutiérrez, Chiapas, Mexico: Publicación del Instituto de Historia Natural.

ANONYMOUS. 1986. Opinion 1385. *Anolis carolinensis* Voigt, 1832 designated as type species of *Anolis* Daudin, 1802 (Reptilia, Sauria). *The Bulletin of Zoological Nomenclature* **43:** 125–127.

ANONYMOUS. 1994. *Evaluación Ecológica Rápida (EER) Parque Nacional "El Cusuco" y Cordillera del Merendón.* San Pedro Sula, Honduras: Fundación Ecologista "Hector Rodrigo Pastor Fasquelle."

ANONYMOUS. 2002. Opinion 2015 (Case 3145). *Dactyloa biporcata* Wiegmann, 1834 (currently *Anolis biporcatus*) and *Anolis petersii* Bocourt, 1873 (Reptilia, Sauria): specific names conserved by the designation of a neotype for *A. biporcatus*. *The Bulletin of Zoological Nomenclature* **59:** 230–231.

ANONYMOUS. 2007. *Análisis del Potencial Desarrollo en Islas del Cisne*. Tegucigalpa, Honduras: Reporte para la Secretaria de Turismo e Instituto Hondureño de Turismo y Secretaria de Recursos Naturales y Ambiente.

BARBOUR, T. 1914. A contribution to the zoögeography of the West Indies, with especial reference to amphibians and reptiles. *Memoirs of the Museum of Comparative Zoölogy* **44:** 209–359, 1 pl.

BARBOUR, T. 1920. A note on *Xiphocercus*. *Proceedings of the New England Zoölogical Club* **7:** 61–63.

BARBOUR, T. 1928. Reptiles from the Bay Islands. *Proceedings of the New England Zoölogical Club* **10:** 55–61.

BARBOUR, T. 1930. The anoles I. The forms known to occur on the neotropical islands. *Bulletin of the Museum of Comparative Zoölogy* **70(3):** 105–144.

BARBOUR, T. 1934. The anoles II. The mainland species from Mexico southward. *Bulletin of the Museum of Comparative Zoölogy* **77(4):** 121–155.

BARBOUR, T., AND A. LOVERIDGE. 1929a. Vertebrates from the Corn Islands. Reptiles and amphibians. *Bulletin of the Museum of Comparative Zoölogy* **69(7):** 138–146.

BARBOUR, T., AND A. LOVERIDGE. 1929b. Typical reptiles and amphibians in the Museum of Comparative Zoölogy. *Bulletin of the Museum of Comparative Zoölogy* **69(10):** 205–360.

BARBOUR, T., AND A. LOVERIDGE. 1946. First supplement to typical reptiles and amphibians. *Bulletin of the Museum of Comparative Zoölogy* **96(2):** 59–214.

BARTLETT, R. D., AND P. P. BARTLETT. 1999. *A Field Guide to Florida Reptiles and Amphibians*. Houston, Texas: Gulf Publishing Company.

BAUER, A. M., R. GÜNTHER, AND M. KLIPFEL (EDS.). 1995. Synopsis of the herpetological taxa described by Wilhelm Peters, pp. 39–81. *In The Herpetological Contributions of Wilhelm C. H. Peters (1815–1883)*. Ithaca, New York: Society for the Study of Amphibians and Reptiles.

BECKERS, H. 2009. *Anolis (Norops) quaggulus* (Cope, 1885). *Iguana Rundschreiben* **22(2):** 18–22.

BEEST, P. VAN, AND M. HARTMAN. 2003. De eerste kweek van *Norops bicaorum*—Köhler 1996. *Anolis lemurinus*—Wilson & Hahn 1973. *Lacerta* **61(1):** 3–9.

BERMINGHAM, E., A. COATES, G. CRUZ DÍAZ, L. EMMONS, R. B. FOSTER, R. LESCHEN, G. SEUTIN, S. THORN, W. WCISLO, AND B. WERFEL. 1998. Geology and terrestrial flora and fauna of Cayos Cochinos, Honduras. *Revista de Biología Tropical* **46(Suppl. 4):** 15–37.

BONNATERRE, P. J. 1789. *Tableau Encyclopédique et Méthodique des Trois Règnes de la Nature. Dédie et Présenté a M. Necker, Ministre d État, & Directeur Géneral des Finances. Erpétologie*. Paris: PANCKOUCKE.

BOULENGER, G. A. 1882 (dated 1881). Description of a new species of *Anolis* from Yucatan. *Proceedings of the Zoological Society of London* **1881:** 921–922 (also reprinted by the Society for the Study of Amphibians and Reptiles, 1971).

BOULENGER, G. A. 1885. *Catalogue of the Lizards in the British Museum (Natural History)*. 2nd Ed. Vol. 2. *Iguanidae, Xenosauridae, Zonuridae, Anguidae, Anniellidae, Helodermatidae, Varanidae, Xantusiidae, Teiidae, Amphisbaenidae*. London: Printed by Order of Trustees of British Museum (Natural History).

BRATTSTROM, B. H., AND T. R. HOWELL. 1954. Notes on some collections of reptiles and amphibians from Nicaragua. *Herpetologica* **10:** 114–123.

BRYGOO, E. R. 1989. Les types d'Iguanidés (Reptiles, Sauriens) du Muséum national d'Histoire naturelle. Catalogue critique. *Bulletin du Muséum National d'Histoire Naturelle, Paris*. Section A, Zoologie, Biologie et Ecologie Animales, Ser. 4, **11(Suppl. 3):** 1–112.

CABRERA-GUZMÁN, E., AND V. H. REYNOSO. 2010. Use of sleeping perches by the lizard *Anolis uniformis* (Squamata: Polychrotidae) in the fragmented tropical rainforest at Los Tuxtlas, Mexico. *Revista Mexicana de Biodiversidad* **81:** 921–924.

CACERES, D. A. 1993. *Representantes más Communes de la Herpetofauna del "Parque Nacional La Tigre"* [thesis]. Tegucigalpa, Honduras: Universidad Pedagogica Nacional.

CALDERÓN-MANDUJANO, R. R., H. BAHENA BASAVE, AND S. CALMÉ. 2008. *Guía de Los Anfibios y Reptiles de la Reserva de la Biósfera de Sian Ka'an y zonas aledañas/Amphibians and Reptiles of Sian Ka'an Biosphere Reserve and surrounding areas. Segunda edición*. Quintana Roo, Mexico: COMPACT, ECOSUR, CONABIO, y SHM A. C.

CAMPBELL, H. W., AND T. R. HOWELL. 1965. Herpetological records from Nicaragua. *Herpetologica* **21:** 130–140.

CAMPBELL, J. A. 1998. *Amphibians and Reptiles of Northern Guatemala, the Yucatán, and Belize*. Norman: University of Oklahoma Press.

CAMPBELL, J. A., D. M. HILLIS, AND W. W. LAMAR. 1989. A new lizard of the genus *Norops* (Sauria: Iguanidae) from the cloud forest of Hidalgo, México. *Herpetologica* **45:** 232–242.

CANNATELLA, D. C., AND K. DE QUEIROZ. 1989. Phylogenetic systematics of the anoles: is a new taxonomy warranted? *Systematic Zoology* **38:** 57–69.

CASTAÑEDA, F. E. 2002. *Anfibios y Reptiles del Área Protegida Propuesta Rus-Rus, La Mosquitia*. Located at: Tegucigalpa, Honduras: Report Submitted to Corporación Hondureña Desarrolo Forestal (AFE COHDEFOR).

CASTAÑEDA, F. E. 2006. *Herpetofauna del Parque Nacional Sierra de Agalta, Honduras*. Located at: Washington, D.C.: Report to International Resource Group.

CASTAÑEDA, F. E., AND L. MARINEROS. 2006. La herpetofauna de la zona Río Amarillo, Copán, Honduras, pp. 3-1–3-13. *In* O. Komar, J. P. Arce, C. Begley, F. E. Castañeda, K. Eisermann, R. J.

Gallardo, and L. Marineros (eds.), *Evaluación de la biodiversidad del Parque Arqueológico y Reserva Forestal Río Amarillo (Copán, Honduras)*. San Salvador, El Salvador: Salva NATURA.

CASTAÑEDA, M. D. R., AND K. DE QUEIROZ. 2011. Phylogenetic relationships of the *Dactyloa* clade of *Anolis* lizards based on nuclear and mitochondrial DNA sequence data. *Molecular Phylogenetics and Evolution* **61**: 784–800.

CASTAÑEDA, M. D. R., AND K. DE QUEIROZ. 2013. Phylogeneny of the *Dactyloa* clade of *Anolis* lizards: new insights from combining morphological and molecular data. *Bulletin of the Museum of Comparative Zoology* **160(7)**: 345–398.

CASTAÑEDA, M. D. R., E. SHERRATT, AND J. B. LOSOS. 2014. The Mexican amber anole, *Anolis electrum*, within a phylogenetic context: implications for the origins of Caribbean anoles. *Zoological Journal of the Linnean Society*. In Press.

COCHRAN, D. M. 1934. Herpetological collections made in Hispaniola by the Utowana Expedition, 1934. *Occasional Papers of the Boston Society of Natural History* **8**: 163–188.

COCHRAN, D. M. 1961. Type specimens of reptiles and amphibians in the U.S. National Museum. *United States National Museum Bulletin* **220**: i–xv, 1–291.

COPE, E. D. 1861. Notes and descriptions of anoles. *Proceedings of the Academy of Natural Sciences of Philadelphia* **13**: 208–215.

COPE, E. D. 1862. Contributions to neotropical saurology. *Proceedings of the Academy of Natural Sciences of Philadelphia* **14**: 176–188.

COPE, E. D. 1864. Contributions to the herpetology of tropical America. *Proceedings of the Academy of Natural Sciences of Philadelphia* **16**: 166–181.

COPE, E. D. 1875. *On the Batrachia and Reptilia of Costa Rica. With notes on the Herpetology and Ichthyology of Nicaragua and Peru*. Philadelphia: Published by the author (also published in 1877 [but dated 1876] in *Journal of the Academy of Natural Sciences of Philadelphia* [2]**8**: 93–154, pls. 23–28).

COPE, E. D. 1885. A contribution to the herpetology of Mexico. *Proceedings of the American Philosophical Society* **22**: 379–404.

CROTHER, B. I. 1999. Evolutionary relationships, pp. 269–334. *In* B. I. Crother (ed.), *Caribbean Amphibians and Reptiles*. San Diego: Academic Press.

CRUZ, G. A., L. GIRÓN, S. FLORES, AND V. HENRÍQUEZ. 2006. Evaluación de la herpetofauna en las partes que formarán el Área Protegida Trinacional Montecristo en territorio guatemalteco y hondureño, pp. 3-1–3-15. *In Consultoría para Ejecutar una Evaluación Ecológica Rápida (EER) en las Partes que Formarán el Área Protegida Trinacional de Montecristo en Territorio Guatemalteco y Hondureño*. San Salvador: SalvaNATURA.

CRUZ, G. A., V. J. LOPEZ, AND S. RODRIGUEZ. 1993. Primer inventario de mamíferos, reptiles y anfibios del Parque Nacional de Celaque. *Serie Miscellana de Conservación Forestales de Honduras (CONSEFORH)* **34-16/93**: i–ix, 1–30.

CRUZ DÍAZ, G. A. 1978. *Herpetofauna del Río Plátano* [dissertation]. Tegucigalpa, Honduras: Universidad Nacional Autónoma de Honduras.

CUVIER, G. 1816 (dated 1817). *Le Règne Animal. Distribué d'Après son Orgnisation, pour Servir de Base a l'Histoire Naturelle des Animaux et d'Introduction a l'Anatomie Comparée. Tome II, contenant les Reptiles, les Poissons, contenant les Mollusques et les Annélides*. Paris: Deterville.

D'ANGIOLELLA, A. B., T. GAMBLE, T. C. S. AVILA-PIRES, G. R. COLLI, B. P. NOONAN, AND L. J. VITT. 2011. *Anolis chrysolepis* Duméril & Bibron, 1837 (Squamata: Iguanidae), revisited: molecular phylogeny and taxonomy of the *Anolis chrysolepis* species group. *Bulletin of the Museum of Comparative* **160(2)**: 35–63.

DAUDIN, F. M. 1802. *Histoire Naturelle, Générale et Particulière des Reptiles; Ouvrage faisant suite à l'Histoire Naturelle générale et particulière, composée par Leclerc de Buffon, et rédigée par C. S. Sonnini, membre de plusieurs Sociétés savantes. Tome Quatrième*. Paris: F. Dufart.

D'CRUZE, N. 2005. Natural history observations of sympatric *Norops* (beta *Anolis*) in a subtropical mainland community. *Herpetological Bulletin* **91**: 10–18.

DE QUEIROZ, K., AND P. D. CANTINO. 2001. Phylogenetic nomenclature and the PhyloCode. *Bulletin of Zoological Nomenclature* **58**: 254–271.

DIENER, E. 2008. Freilandbeobachtungen an einigen Saumfingerarten (*Anolis* sensu lato) im Honduras. *Iguana Rundschreiben* **21(2)**: 10–19.

DUELLMAN, W. E., AND B. BERG. 1962. Type specimens of amphibians and reptiles in the Museum of Natural History, The University of Kansas. *University of Kansas Publications, Museum of Natural History* **15**: 183–204.

DUMÉRIL, A. H. A., M. F. BOCOURT, AND F. MOCQUARD. 1870–1909. *Études sur les Reptiles. Recherches Zoologiques pour servir à l'Histoire de la Faune de l'Amérique Centrale et du Mexique. Mission Scientifique au Mexique et dans l'Amérique Centrale. Troisième Partie*. Paris: Imprimerie Nationale.

DUMÉRIL, A. M. C., AND G. BIBRON. 1837. *Erpétologie Générale ou Histoire Naturelle Complète des Reptiles. Tome Quatrième. Contenant l'Histoire de Quarante-Six Genres et de Cent Quarante-Six Espèces de la Famille des Iguaniens, de l'Ordre des Sauriens*. Paris: Librairie Encyclopédique de Roret (also reprinted by Society for the Study of Amphibians and Reptiles, 2012).

DUNN, E. R., AND J. T. EMLEN, JR. 1932. Reptiles and amphibians from Honduras. *Proceedings of the Academy of Natural Sciences of Philadelphia* **84**: 21–32.

ECHELLE, A. A., A. F. ECHELLE, AND H. S. FITCH. 1971. A new anole from Costa Rica. *Herpetologica* **27**: 354–362.

EIFLER, D. A. 1995. Natural history notes. *Anolis uniformis* (NCN). Feeding behavior. *Herpetological Review* **26**: 204.

ESPINAL, M. R. 1993. *Anfibios y reptiles del Parque Nacional "La Muralla."* Located at: Tegucigalpa, Honduras: Report Submitted to Corporación Hondureña Desarrolo Forestal (AFE COHDE-FOR).

ESPINAL, M. R., J. R. McCRANIE, AND L. D. WILSON. 2001. The herpetofauna of Parque Nacional La Muralla, Honduras, pp. 100–108. *In* J. D. Johnson, R. G. Webb, and O. A. Flores-Villela (eds.). *Mesoamerican Herpetology: Systematics, Zoogeography, and Conservation. Special Publication, Centennial Museum, University of Texas at El Paso* **1**: i–iv, 1–200.

ESTES, R. 1983a. The fossil record and early distribution of lizards, pp. 365–398. *In* A. G. J. Rhodin and K. Miyata (eds.), *Advances in Herpetology and Evolutionary Biology. Essays in Honor of Ernest E. Williams*. Cambridge, Massachusetts: Museum of Comparative Zoology, Harvard University.

ESTES, R. 1983b. *Sauria terrestria, Amphisbaenia* (Encyclopedia of Paleoherpetology Series, Part 10a). Stuttgart: Gustav Fischer Verlag.

ETHERIDGE, R. E. 1959. The relationships of the anoles (Reptilia: Sauria: Iguanidae). An interpretation based on skeletal morphology [Ph.D. dissertation]. Ann Arbor, Michigan: University of Michigan.

ETHERIDGE, R. 1967. Lizard caudal vertebrae. *Copeia* **1967**: 699–721.

ETHERIDGE, R., AND E. E. WILLIAMS. 1985. Notes on *Pristidactylus* (Squamata: Iguanidae). *Breviora* **483**: 1–18.

FITCH, H. S. 1973. A field study of Costa Rican lizards. *University of Kansas Science Bulletin* **50(2)**: 39–126.

FITCH, H. S. 1975. Sympatry and interrelationships in Costa Rican anoles. *Occasional Papers of the Museum of Natural History, The University of Kansas* **40**: 1–60.

FITCH, H. S., A. A. ECHELLE, AND A. F. ECHELLE. 1972. Variation in the Central American iguanid lizard, *Anolis cupreus*, with the description of a new subspecies. *Occasional Papers of the Museum of Natural History, The University of Kansas* **8**: 1–20.

FITCH, H. S., AND D. M. HILLIS. 1984. The *Anolis* dewlap: interspecific variability and morphological associations with habitat. *Copeia* **1984**: 315–323.

FITCH, H. S., AND R. A. SEIGEL. 1984. Ecological and taxonomic notes on Nicaraguan anoles. *Milwaukee Public Museum, Contributions in Biology and Geology* **57**: 1–13.

FITZINGER, L. 1826. *Neue Classification der Reptilien nach ihren Natürlichen Verwandtschaften. Nebst einer Verwandtschafts—Tafel und einem Verzeichnisse der Reptilien—Sammlung des K. K. Zoologischen Museum's zu Wien*. Vienna: Verlage von J. G. Heubner (also reprinted by the Society for the Study of Amphibians and Reptiles, 1997).

FITZINGER, L. 1843. *Systema Reptilium. Fasciculus Primus. Amblyglossae*. Vindobonae, Austria: Braumüller et Seidel Bibliopolas (also reprinted by the Society for the Study of Amphibians and Reptiles, 1973).

FLÄSCHENDRÄGER, A., AND L. C. M. WIJFFELS. 1996. *Anolis: Im Biotop und Terrarium*. Münster, Germany: Natur und Tier—Verlag.

FLORES-VILLELA, O., AND L. CANSECO-MÁRQUEZ. 2004. Nuevas especies y cambios taxonómicos para la herpetofauna de México. *Acta Zoologica Mexicana (Nueva Serie)* **20**: 115–144.

FLOWER, S. S. 1928. Reptilia and Amphibia. *Zoological Record* **65(14)**: 1–72.

FRANKLIN, C. J., AND J. FRANKLIN. 1999a. Geographic distribution. *Anolis biporcatus* (giant green anole). *Herpetological Review* **30**: 109.

FRANKLIN, C. J., AND J. FRANKLIN. 1999b. Geographic distribution. *Anolis capito* (big-headed anole). *Herpetological Review* **30**: 109.

FROST, D. R., AND R. ETHERIDGE. 1989. A phylogenetic analysis and taxonomy of iguanian lizards (Reptilia: Squamata). *The University of Kansas Museum of Natural History, Miscellaneous Publication* **81**: 1–65.

FROST, D. R., R. ETHERIDGE, D. JANIES, AND T. A. TITUS. 2001. Total evidence, sequence alignment, evolution of polychrotid lizards, and a reclassification of the Iguania (Squamata: Iguania). *American Museum Novitates* **3343**: 1–38.

FUGLER, C. M. 1968. The distributional status of *Anolis sagrei* in Central America and northern South America. *Journal of Herpetology* **1**: 96–98.

GAMBLE, T., A. J. GENEVA, R. E. GLOR, AND D. ZARKOWER. 2014. *Anolis* sex chromosomes are derived from a single ancestral pair. *Evolution* **68**: 1027–1041.

GARMAN, S. 1888. Reptiles and batrachians from the Caymans and from the Bahamas. Collected by Prof. C. J. Maynard for the Museum of Comparative Zoology at Cambridge, Mass. *Bulletin of the Essex Institute* **20**: 1–13.

GARRIDO, O. H., AND M. L. JAUME. 1984. Catálogo descriptivo de los anfibios y reptiles de Cuba. *Doñana, Acta Vertebrata* **11(2)**: 5–128.

GLOR, R. E., J. B. LOSOS, AND A. LARSON. 2005. Out of Cuba: overwater dispersal and speciation among lizards in the *Anolis carolinensis* subgroup. *Molecular Ecology* **14**: 2419–2432.

GLOR, R. E., L. J. VITT, AND A. LARSON. 2001. A molecular phylogenetic analysis of diversification in Amazonian *Anolis* lizards. *Molecular Ecology* **10**: 2661–2668.

GRAY, J. E. 1827. A synopsis of the genera of saurian reptiles, in which some new genera are indicated, and the others reviewed by actual examination. *The Philosophical Magazine* **(2)2**: 54–58.

GRISMER, L. L., L. L. GRISMER, K. M. MARSON, A. B. MATTESON, E. J. R. SIHOTANG, K. M. CRANE, J. DAYOV, T. A. MAYER, A.-L. SIMPSON, AND H. KAISER. 2001. New herpetological records for the Islas de la

Bahía, Honduras. *Herpetological Review* **32:** 134–135.

GUNDY, G. C., AND G. Z. WURST. 1976. The occurrence of parietal eyes in Recent Lacertilia (Reptilia). *Journal of Herpetology* **10:** 113–121.

GUTSCHE, A. 2005. The world's most endangered anole? *Iguana* **12:** 240–243.

GUTSCHE, A. 2012. Bewohner des Elfenwaldes: herpetofauna am Pico La Picucha im Nationalpark Sierra de Agalta, Honduras. *Elaphe* **2012(5):** 68–71.

GUTSCHE, A., J. R. MCCRANIE, AND K. E. NICHOLSON. 2004. Field observations on a nesting site of *Norops utilensis* Köhler, 1996 (Reptilia, Squamata) with comments about its conservation status. *Salamandra* **40:** 297–302.

GUYER, C., AND M. A. DONNELLY. 2005. *Amphibians and Reptiles of La Selva, Costa Rica, and the Caribbean Slope. A Comprehensive Guide.* Berkley: University of California Press.

GUYER, C., AND J. M. SAVAGE. 1987 (dated 1986). Cladistic relationships among anoles (Sauria: Iguanidae). *Systematic Zoology* **35:** 509–531.

GUYER, C., AND J. M. SAVAGE. 1992. Anole systematics revisited. *Systematic Biology* **41:** 89–110.

GUZMÁN, E. C., AND V. H. REYNOSO. 2008. Natural history notes. *Anolis uniformis* (Lesser Scaly Anole). Diet. *Herpetological Review* **39:** 348.

HAHN, D. E. 1971. Noteworthy herpetological records from Honduras. *Herpetological Review* **3:** 111–112.

HALLMEN, M., AND A. HUY. 2012. Natural history notes. *Anolis utilensis* (Utila Anole). Habitat. *Herpetological Review* **43:** 642–643.

HALLMEN, S. 2011. Neues aus Utila—reloaded. *Iguana Rundschreiben* **24(2):** 5–16.

HALLOWELL, E. 1857 (dated 1856). Notes on the reptiles in the collection of the Academy of Natural Sciences of Philad'a. *Proceedings of the Academy of Natural Sciences of Philadelphia* **8:** 221–238.

HALLOWELL, E. 1861 (dated 1860). Report upon the Reptilia of the North Pacific Exploring Expedition, under command of Capt. John Rogers [sic], U.S.N. *Proceedings of the Academy of Natural Sciences of Philadelphia* **12:** 480–510.

HARVEY, M. B., G. N. UGUETO, AND R. L. GUTBERLET, JR. 2012. Review of teiid morphology with a revised taxonomy and phylogeny of the Teiidae (Lepidosauria: Squamata). *Zootaxa* **3459:** 1–156.

HEDGES, S. B. 2013. Revision shock in taxonomy. *Zootaxa* **3681:** 297–298.

HEDGES, S. B., AND C. E. CONN. 2012. A new skink fauna from Caribbean islands (Squamata, Mabuyidae, Mabuyinae). *Zootaxa* **3288:** 1–244.

HEDGES, S. B., A. COULOUX, AND N. VIDAL. 2009. Molecular phylogeny, classification, and biogeography of West Indian racer snakes of the tribe Alsophiini (Squamata, Dipsadidae, Xenodontinae). *Zootaxa* **2067:** 1–28.

HEDGES, S. B., A. B. MARION, K. M. LIPP, J. MARIN, AND N. VIDAL. 2014. A taxonomic framework for typhlopid snakes from the Caribbean and other regions (Reptilia, Squamata). *Caribbean Herpetology* **49:** 1–61.

HENDERSON, R. W. 1972. Notes on the reproduction of a giant anole, *Anolis biporcatus* (Sauria, Iguanidae). *Journal of Herpetology* **6:** 239–240.

HENDERSON, R. W., AND R. POWELL. 2004. Thomas Barbour and the Utowana voyages (1929–1934) in the West Indies. *Bonner zoologische Beiträge* **52:** 297–309.

HENDERSON, R. W., AND R. POWELL. 2009. *Natural History of West Indian Reptiles and Amphibians.* Gainesville: University Press of Florida.

HERREL, A., B. VANHOOYDONCK, R. JOACHIM, AND D. J. IRSCHICK. 2004. Frugivory in polychrotid lizards: effects of body size. *Oecology* **140:** 160–168.

HOLBROOK, J. D. 2012. Natural history notes. *Anolis sagrei* (Brown Anole). Prey. *Herpetological Review* **43:** 641.

HOLDRIDGE, L. R. 1967. *Life Zone Ecology.* Revised Ed. San José, Costa Rica: Tropical Science Center.

HOLM, P. A., AND G. A. CRUZ D. 1994: A new species of *Rhadinaea* (Colubridae) from a cloud forest in northern Honduras. *Herpetologica* **50:** 15–23.

HUDSON, D. M. 1981. Blood parasitism incidence among reptiles of Isla de Roatan, Honduras. *Journal of Herpetology* **15:** 377–379.

IRSCHICK, D. J., L. T. VITT, P. A. ZANI, AND J. B. LOSOS. 1997. A comparison of evolutionary radiations in mainland and Caribbean *Anolis* lizards. *Ecology* **78:** 2191–2203.

JACKSON, J. F. 1973. Notes on the population biology of *Anolis tropidonotus* in a Honduran highland pine forest. *Journal of Herpetology* **7:** 309–311.

KLUGE, A. G. 1984. Type-specimens of reptiles in the University of Michigan Museum of Zoology. *Miscellaneous Publications Museum of Zoology, University of Michigan* **167:** i–ii, 1–85.

KLÜTSCH, C. F. C., B. MISOF, W.-R. GROSSE, AND R. F. A. MORITZ. 2007. Genetic and morphometric differentiation among island populations of two *Norops* lizards (Reptilia: Sauria: Polychrotidae) on independently colonized islands of the Islas de Bahia (sic) (Honduras). *Journal of Biogeography* **34:** 1124–1135.

KÖHLER, G. 1991. Das Portrait. *Norops capito* (Peters). *Sauria* **13:** 1–2.

KÖHLER, G. 1994. *Sobre la Sistematica y Ecologia de* Ctenosaura bakeri *y* C. oedirhina *(Sauria: Iguanidae). Estudios de Campo Realizados en las Islas de la Bahía, Honduras.* Located at: Tegucigalpa, Honduras: Report Submitted to Corporación Hondureña Desarrolo Forestal (AFE COHDEFOR).

KÖHLER, G. 1995. Freilanduntersuchungen zur Morphologie und Ökologie von *Ctenosaura bakeri* und *C. oedirhina* auf den Islas de la Bahia, Honduras, mit Bemerkungen zur Schutzproblematik. *Salamandra* **31:** 93–106.

KÖHLER, G. 1996a. A new species of anole of the *Norops pentaprion* group from Isla de Utila, Honduras

(Reptilia: Sauria: Iguanidae). *Senckenbergiana biologica* **75**: 23–31.

KÖHLER, G. 1996b. Additions to the known herpetofauna of Isla de Utila (Islas de la Bahia [sic], Honduras) with the description of a new species of the genus *Norops* (Reptilia: Sauria: Iguanidae). *Senckenbergiana biologica* **76**: 19–28.

KÖHLER, G. 1996c. Notes on a collection of reptiles from El Salvador collected between 1951 and 1956. *Senckenbergiana biologica* **76**: 29–38.

KÖHLER, G. 1996d. [Cover photographs]. *Iguana Rundschreiben* **9(1)**: front and back covers.

KÖHLER, G. 1998a. Das Schutz- und Forschungsprojekt Utila-Schwarzleguan. *Natur und Museum* **128**: 44–49.

KÖHLER, G. 1998b. Herpetologische Beobachtungen in Honduras I. Die Islas de la Bahía. *Natur und Museum* **128**: 372–383.

KÖHLER, G. 1999a. Eine neue Saumfingerart der Gattung *Norops* von der Pazifikseite des nördlichen Mittelamerika. *Salamandra* **35**: 37–52.

KÖHLER, G. 1999b. Herpetologische Beobachtungen in Honduras II. Das Comayagua–Becken. *Natur und Museum* **129**: 212–217.

KÖHLER, G. 1999c. The amphibians and reptiles of Nicaragua. A distributional checklist with keys. *Courier Forschungsinstitut Senckenberg* **213**: 1–121.

KÖHLER, G. 1999d. Amphibien und Reptilien im Hochland von Nicaragua. *Die Aquarien und Terrarien Zeitschrift* **52(4)**: 48–54.

KÖHLER, G. 2000. *Reptilien und Amphibien Mittelamerikas. Band 1: Krokodile, Schildkröten, Echsen.* Offenbach, Germany: Herpeton, Verlag Elke Köhler.

KÖHLER, G. 2001a. Type material and use of the name *Anolis bourgeaei* Bocourt (Sauria: Polychrotidae). *Copeia* **2001**: 274–275.

KÖHLER, G. 2001b. *Anfibios y Reptiles de Nicaragua.* Offenbach, Germany: Herpeton, Verlag Elke Köhler.

KÖHLER, G. 2003. *Reptiles of Central America.* Offenbach: Herpeton, Germany: Verlag Elke Köhler.

KÖHLER, G. 2008. *Reptiles of Central America.* 2nd ed. Offenbach, Germany: Herpeton, Verlag Elke Köhler.

KÖHLER, G. 2010. A revision of the Central American species related to *Anolis pentaprion* with the resurrection of *A. beckeri* and the description of a new species (Squamata: Polychrotidae). *Zootaxa* **2354**: 1–18.

KÖHLER, G. 2012. *Color Catalogue for Field Biologists.* Offenbach, Germany: Herpeton, Verlag Elke Köhler.

KÖHLER, G. 2014. Characters of external morphology used in *Anolis* taxonomy—definition of terms, advice on usage, and illustrated examples. *Zootaxa* **3774**: 201–257.

KÖHLER, G., AND M. ACEVEDO. 2004. The anoles (genus *Norops*) of Guatemala. I. The species of the Pacific versant below 1500 m elevation. *Salamandra* **40**: 113–140.

KÖHLER, G., S. ALT, C. GRÜNFELDER, M. DEHLING, AND J. SUNYER. 2006. Morphological variation in Central American leaf-litter anoles: *Norops humilis*, *N. quaggulus* and *N. uniformis*. *Salamandra* **42**: 239–254.

KÖHLER, G., AND A. M. BAUER. 2001. *Dactyloa biporcata* Wiegmann, 1834 (currently *Anolis biporcatus*) and *Anolis petersii* Bocourt, 1873 (Reptilia, Sauria): proposed conservation of the specific names and designation of a neotype for *A. biporcatus. The Bulletin of Zoological Nomenclature* **58**: 122–125.

KÖHLER, G., AND J. KREUTZ. 1999. *Norops macrophallus* (Werner, 1917), a valid species of anole from Guatemala and El Salvador (Squamata: Sauria: Iguanidae). *Herpetozoa* **12**: 57–65.

KÖHLER, G., AND J. R. MCCRANIE. 1998. Zur Kenntnis von *Norops heteropholidotus* (Mertens, 1952). *Herpetofauna* **20(113)**: 12–13.

KÖHLER, G., AND J. R. MCCRANIE. 2001. Two new species of anoles from northern Honduras (Reptilia, Squamata, Polychrotidae). *Senckenbergiana biologica* **81**: 235–245.

KÖHLER, G., J. R. MCCRANIE, AND K. E. NICHOLSON. 2000. Eine herpetologische Expedition in den Patuca-Nationalpark, Honduras. *Natur und Museum* **130**: 421–425.

KÖHLER, G., J. R. MCCRANIE, K. E. NICHOLSON, AND J. KREUTZ. 2003. Geographic variation in hemipenial morphology in *Norops humilis* (Peters 1863), and the systematic status of *Norops quaggulus* (Cope 1885) (Reptilia, Squamata, Polychrotidae). *Senckenbergiana biologica* **82**: 213–222.

KÖHLER, G., J. R. MCCRANIE, AND L. D. WILSON. 1999. Two new species of anoles of the *Norops crassulus* group from Honduras (Reptilia: Sauria: Polychrotidae). *Amphibia-Reptilia* **20**: 279–298.

KÖHLER, G., J. R. MCCRANIE, AND L. D. WILSON. 2001. A new species of anole from western Honduras (Squamata: Polychrotidae). *Herpetologica* **57**: 247–255.

KÖHLER, G., AND M. OBERMEIER. 1998. A new species of anole of the *Norops crassulus* group from central Nicaragua (Reptilia: Sauria: Iguanidae). *Senckenbergiana biologica* **77**: 127–137.

KÖHLER, G., M. PONCE, J. SUNYER, AND A. BATISTA. 2007. Four new species of anoles (Genus *Anolis*) from the Serranía de Tabasará, west-central Panama (Squamata: Polychrotidae). *Herpetologica* **63**: 375–391.

KÖHLER, G., A. SCHULZE, AND M. VESELY. 2005a. Morphological variation in *Norops capito* (Peters, 1863), a wide-spread species in southeastern Mexico and Central America. *Salamandra* **41**: 129–136.

KÖHLER, G., AND E. N. SMITH. 2008. A new species of anole of the *Norops schiedei* group from western Guatemala (Squamata: Polychrotidae). *Herpetologica* **64**: 216–223.

KÖHLER, G., AND J. SUNYER. 2008. Two new species of anoles formerly referred to as *Anolis limifrons*

(Squamata: Polychrotidae). *Herpetologica* **64**: 92–108.

KÖHLER, G., AND M. VESELY. 2003. A comparison of *Norops petersii* (Bocourt) and Central American *N. biporcatus* (Wiegmann), with notes on the holotype of *D[actyloa] biporcata* Wiegmann (Reptilia, Squamata, Polychrotidae). *Senckenbergiana biologica* **82**: 223–233.

KÖHLER, G., AND M. VESELY. 2010. A revision of the *Anolis sericeus* complex with the resurrection of *A. wellbornae* and the description of a new species (Squamata: Polychrotidae). *Herpetologica* **66**: 207–228.

KÖHLER, G., M. VESELY, AND E. GREENBAUM. 2005b (dated 2006). *The Amphibians and Reptiles of El Salvador*. Malabar, Florida: Krieger Publishing Company.

KRAMER, J. 2010. Utilas *Anolis*. *Iguana Rundschreiben* **23(1)**: 5–9.

KRAUS, F. 2009. *Alien Reptiles and Amphibians. A Scientific Compendium and Analysis*. New York: Springer Science.

LEE, J. C. 1980. Variation and systematics of the *Anolis sericeus* complex (Sauria: Iguanidae). *Copeia* **1980**: 310–320.

LEE, J. C. 1983. *Anolis sericeus. Catalogue of American Amphibians and Reptiles* **340**: 1–2.

LEE, J. C. 1992. *Anolis sagrei* in Florida: phenetics of a colonizing species III. West Indian and Middle American comparisons. *Copeia* **1992**: 942–954.

LEE, J. C. 1996. *The Amphibians and Reptiles of the Yucatán Peninsula*. Ithaca, New York: Cornell University Press, Comstock Publishing Associates.

LEE, J. C. 2000. *A Field Guide to the Amphibians and Reptiles of the Maya World. The Lowlands of Mexico, Northern Guatemala, and Belize*. Ithaca, New York: Cornell University Press, Comstock Publishing Associates.

LEENDERS, T. A. A. M., AND G. J. WATKINS-COLWELL. 2003. Natural history notes. *Norops heteropholidotus* (NCN). Dewlap coloration. *Herpetological Review* **34**: 369–370.

LEENDERS, T. A. A. M., AND G. J. WATKINS-COLWELL. 2004. Notes on a collection of amphibians and reptiles from El Salvador. *Postilla* **231**: 1–31.

LEMOS-ESPINAL, J. A., AND J. R. DIXON. 2013. *Amphibians and Reptiles of San Luis Potosí*. Eagle Mountain, Utah: Eagle Mountain Publishing, LC.

LEVITON, A. E., AND R. H. GIBBS, JR. 1988. Standards in herpetology and ichthyology. Standard symbolic codes for institution resource collections in herpetology and ichthyology. Supplement No. 1: additions and corrections. *Copeia* **1988**: 280–282.

LEVITON, A. E., R. H. GIBBS, JR., E. HEAL, AND C. E. DAWSON. 1985. Standards in herpetology and ichthyology: part I. Standard symbolic codes for institutional resource collections in herpetology and ichthyology. *Copeia* **1985**: 802–832.

LIEB, C. S. 1981. *Biochemical and Karyological Systematics of the Mexican Lizards of the* Anolis gadovi *and* A. nebulosus *Species Groups (Reptilia: Iguanidae)* [Ph.D. dissertation]. Los Angeles: University of California.

LISTER, B. C. 1976a. The nature of niche expansion in West Indian *Anolis* lizards I: ecological consequences of reduced competition. *Evolution* **30**: 659–676.

LISTER, B. C. 1976b. The nature of niche expansion in West Indian *Anolis* lizards II: evolutionary components. *Evolution* **30**: 677–692.

LOGAN, M. L., C. E. MONTGOMERY, S. M. BOBACK, R. N. REED, AND J. A. CAMPBELL. 2012. Divergense in morphology, but not habitat use, despite low genetic differentiation among insular populations of the lizard *Anolis lemurinus* in Honduras. *Journal of Tropical Ecology* **28**: 215–222.

LOSOS, J. B. 2009. *Lizards in an Evolutionary Tree. Ecology and Adaptive Radiation of Anoles*. Berkeley: University of California Press.

LOSOS, J. B. 2012. Of ecomodes and ecomorphs: I. Are the data available to categorize the habitat use of all anoles? *Anole Annals* [Internet] [cited 2014 January 19]. Available from: http://www.anoleannals.org/2012/10/06/of-ecomodes-and-ecomorphs-i-are-the-data-available-to-categorize-the-habitat-use-of-all-anoles/

LOSOS, J. B. 2013. Lizard superfamily contains 4,000+ species. *Anole Annals* [Internet] [cited 2014 January 19]. Available from: http://www.anoleannals.org/2013/06/05/lizard-super-phylogeny-contains-4000-species/

LOSOS, J. B. 2014. Available now: a new, large phylogeny of anoles. *Anole Annals* [Internet] [cited 2014 April 2]. Available from: http://www.anoleannals.org/2014/03/14/available-now-a-new-large-phylogeny-of-anoles/

LOSOS, J. B., AND K. DE QUEIROZ. 1997. Evolutionary consequences of ecological release in Caribbean *Anolis* lizards. *Biological Journal of the Linnean Society* **61**: 459–483.

LOVICH, R., T. AKRE, M. RYAN, N. SCOTT, AND R. FORD. 2006. *Herpetofaunal surveys of Cerro Guanacaure, Montaña La Botija, and Isla del Tigre protected areas in southern Honduras*. Located at: Washington, D.C.: Report to International Resources Group.

LOVICH, R. E., T. AKRE, M. RYAN, S. NUÑEZ, G. CRUZ, G. BORJAS, N. J. SCOTT, S. FLORES, W. DEL CID, A. FLORES, C. RODRIGUEZ, I. R. LUQUE-MONTES, AND R. FORD. 2010. New herpetofaunal records from southern Honduras. *Herpetological Review* **41**: 112–115.

LUNDBERG, M. 2000. Herpetofaunan på Isla de Utila, Honduras. *Snoken* **30(4)**: 2–8.

LUNDBERG, M. 2001. Herpetofaunan på Roatán, Honduras. *Snoken* **31(2)**: 20–29.

LUNDBERG, M. 2002a. Herpetofaunan på Hog Islands, Honduras. *Snoken* **32(1)**: 4–13.

LUNDBERG, M. 2002b. Herpetofaunan på Guanaja, Honduras. *Snoken* **32(2)**: 4–12.

LUNDBERG, M. 2003. Besöki nationalparken La Tigra, Honduras. *Snoken* **33(2)**: 25–29.

LYNN, W. G. 1944. Notes on some reptiles and amphibians from Ceiba, Honduras. *Copeia* **1944:** 189–190.

MACLEAN, W. P., R. KELLNER, AND H. DENNIS. 1977. Island lists of West Indian amphibians and reptiles. *Smithsonian Herpetological Information Service* **40:** 1–47.

MALNATE, E. V. 1971. A catalog of primary types in the herpetological collections of the Academy of Natural Sciences, Philadelphia (ANSP). *Proceedings of the Academy of Natural Sciences of Philadelphia* **123:** 345–375.

MARTINEZ, A. M., AND S. M. CLAYSON. 2013. Geographic distribution. *Anolis allisoni* (Allison's Anole). *Herpetological Review* **44:** 624.

MARX, H. 1958. Catalogue of type specimens of reptiles and amphibians in Chicago Natural History Museum. *Fieldiana: Zoology* **36:** 409–496.

MCCOY, C. J. 1975. Reproduction in Guatemalan *Anolis biporcatus* (Sauria: Iguanidae). *Herpetologica* **31:** 65–66.

MCCRANIE, J. R. 1993. Additions to the herpetofauna of Honduras. *Caribbean Journal of Science* **29:** 254–255.

MCCRANIE, J. R. 1996. Geographic distribution. *Norops purpurgularis* (Middle American Anole). *Herpetological Review* **27:** 32.

MCCRANIE, J. R. 2005 (dated 2004). The herpetofauna of Parque Nacional Cerro Azul, Honduras (Amphibia, Reptilia). *Herpetological Bulletin* **90:** 10–21.

MCCRANIE, J. R. 2007a. Herpetological fieldwork in the lowland rainforests of northeastern Honduras: pleasure or how quickly we forget? *Iguana* **14:** 172–183.

MCCRANIE, J. R. 2007b. Distribution of the amphibians of Honduras by departments. *Herpetological Review* **38:** 35–39.

MCCRANIE, J. R. 2009. *Amphibians and Reptiles of Honduras. Listas Zoológicas Actualizadas UCR.* San Pedro, Costa Rica: Museum de Zoológica Universidad de Costa Rica. [Internet] [cited 2014 January 17]. Available from: http://museo.biologia.ucr.ac.cr/Listas/LZAPublicaciones.htm

MCCRANIE, J. R. 2011. The snakes of Honduras: systematics, distribution, and conservation. *Society for the Study of Amphibians and Reptiles, Contributions to Herpetology* **26:** i–x, 1–714, pls. 1–20.

MCCRANIE, J. R. (in preparation). *The Lizards (Exclusive of the Anoles), Crocodyles, and Turtles of Honduras. Systematics, Distribution, and Conservation.* Expected publication date 2015.

MCCRANIE, J. R., AND F. E. CASTAÑEDA. 2005. The herpetofauna of Parque Nacional Pico Bonito, Honduras. *Phyllomedusa* **4:** 3–16.

MCCRANIE, J. R., F. E. CASTAÑEDA, AND K. E. NICHOLSON. 2002a. Preliminary results of herpetofaunal survey work in the Rus Rus region, Honduras: a proposed biological reserve. *Herpetological Bulletin* **81:** 22–29.

MCCRANIE, J. R., AND G. A. CRUZ. 1992. Rediscovery of the Honduran giant anole—*Norops loveridgei* (Sauria: Polychridae). *Caribbean Journal of Science* **28:** 233–234.

MCCRANIE, J. R., G. A. CRUZ, AND P. A. HOLM. 1993a. A new species of cloud forest lizard of the *Norops schiedei* group (Sauria: Polychrotidae) from northern Honduras. *Journal of Herpetology* **27:** 386–392.

MCCRANIE, J. R., AND A. GUTSCHE. 2009. Geographic distribution. *Anolis allisoni* (Green Anole). *Herpetological Review* **40:** 112.

MCCRANIE, J. R., AND G. KÖHLER. 2001. A new species of anole from eastern Honduras related to *Norops tropidonotus* (Reptilia, Squamata, Polychrotidae). *Senckenbergiana biologica* **81:** 227–233.

MCCRANIE, J. R., AND G. KÖHLER. 2012. Geographic distribution. *Norops carpenteri. Herpetological Review* **43:** 103.

MCCRANIE, J. R., G. KÖHLER, AND L. D. WILSON. 2000. Two new species of anoles from northwestern Honduras related to *Norops laeviventris* (Wiegmann 1834) (Reptilia, Squamata, Polychrotidae). *Senckenbergiana biologica* **80:** 213–223.

MCCRANIE, J. R., K. E. NICHOLSON, AND G. KÖHLER. 2002b (dated 2001). A new species of *Norops* (Squamata: Polychrotidae) from northwestern Honduras. *Amphibia-Reptilia* **22:** 465–473.

MCCRANIE, J. R., AND S. NUÑEZ. 2014. Geographic distribution. *Anolis sagrei* (Brown Anole; Abaniquillo Costero). *Herpetological Review* **45:** 91.

MCCRANIE, J. R., AND J. M. SOLÍS. 2013. Additions to the amphibians and reptiles of Parque Nacional Pico Bonito, Honduras, with an updated nomenclatural list. *Herpetology Notes* **6:** 239–243.

MCCRANIE, J. R., J. H. TOWNSEND, AND L. D. WILSON. 2006. *The Amphibians and Reptiles of the Honduran Mosquitia.* Malabar, Florida: Krieger Publishing Company.

MCCRANIE, J. R., AND L. VALDÉS ORELLANA. 2012. Geographic distribution. *Anolis yoroensis. Herpetological Review* **43:** 304–305.

MCCRANIE, J. R., AND L. VALDÉS ORELLANA. 2014. New island records and updated nomenclature of amphibians and reptiles from the Islas de la Bahía, Honduras. *Herpetology Notes* **7:** 41–49.

MCCRANIE, J. R., AND L. D. WILSON. 1985. *Plectrohyla matudai* Hartweg and *Norops petersi* (Bocourt): additions to the herpetofauna of Honduras. *Herpetological Review* **16:** 107–108.

MCCRANIE, J. R., L. D. WILSON, AND G. KÖHLER. 2005. *The Amphibians & Reptiles of the Bay Islands and Cayos Cochinos, Honduras.* Salt Lake City, Utah: Bibliomania!

MCCRANIE, J. R., L. D. WILSON, AND K. L. WILLIAMS. 1984. *Anolis johnmeyeri* Wilson and McCranie: additional specimens and a new locality. *Journal of Herpetology* **18:** 337–338.

MCCRANIE, J. R., L. D. WILSON, AND K. L. WILLIAMS. 1992. A new species of anole of the *Norops crassulus* group (Sauria: Polychridae) from northwestern Honduras. *Caribbean Journal of Science* **28:** 208–215.

McCRANIE, J. R., L. D. WILSON, AND K. L. WILLIAMS. 1993b. Another new species of lizard of the *Norops schiedei* group (Sauria: Polychrotidae) from northern Honduras. *Journal of Herpetology* **27:** 393–399.

MERTENS, R. 1952a. Neues über die Reptilienfauna von El Salvador. *Zoologischer Anzeiger* **148(3/4):** 87–93.

MERTENS, R. 1952b. Die Amphibien und Reptilien von El Salvador, auf Grund der Reisen von R. Mertens und A. Zilch. *Abhandlungen der Senckenbergischen Naturforschenden Gesellschaft* **487:** 1–120, pls. 1–20.

MEYER, J. R. 1966. Records and observations on some amphibians and reptiles from Honduras. *Herpetologica* **22:** 172–181.

MEYER, J. R. 1969. *A Biogeographic Study of the Amphibians and Reptiles of Honduras* [Ph.D. dissertation]. Los Angeles: The University of Southern California.

MEYER, J. R., AND L. D. WILSON. 1972 (dated 1971). Taxonomic studies and notes on some Honduran amphibians and reptiles. *Bulletin of the Southern California Academy of Sciences* **70:** 106–114.

MEYER, J. R., AND L. D. WILSON. 1973. A distributional checklist of the turtles, crocodilians, and lizards of Honduras. *Contributions in Science, Natural History Museum, Los Angeles County* **244:** 1–39.

MONTGOMERY, C. E., E. J. G. RODRIGUEZ, H. L. ROSS, AND K. R. LIPS. 2011. Communal nesting in the anoline lizard *Norops lionotus* (Polychrotidae) in central Panama. *Southwestern Naturalist* **56:** 83–88.

MONZEL, M. 1998. *Zoogeographische Untersuchungen zur Herpetofauna der Islas de la Bahía (Honduras)* [Diplomarbeit]. Saarbrüken, Germany: Universität des Saarlandes.

MONZEL, M. 2001. Die *Anolis* der Bay Islands. Zoogeographie und Evolutionsökologie. *Reptilia (Münster)* **6:** 25–34.

MORA, J. M., E. TORAL, AND J. C. CALDERON. 2012. Natural history notes. *Anolis capito* (Pug-nosed Anole). Diet. *Herpetological Review* **43:** 129–130.

MORGAN, G. S. 1985. Taxonomic status and relationships of the Swan Island Hutia, *Geocapromys thoracatus* (Mammalia: Rodentia: Capromyidae), and the zoogeography of the Swan Islands vertebrate fauna. *Proceedings of the Biological Society of Washington* **98:** 29–46.

MYERS, C. W. 1971a. A new species of green anole (Reptilia, Sauria) from the north coast of Veraguas, Panama. *American Museum Novitates* **2470:** 1–14.

MYERS, C. W. 1971b. Central American lizards related to *Anolis pentaprion*: two new species from the Cordillera de Talamanca. *American Museum Novitates* **2471:** 1–40.

MYERS, C. W., E. E. WILLIAMS, AND R. W. McDIARMID. 1993. A new anoline lizard (*Phenacosaurus*) from the highland of Cerro de la Neblina, southern Venezuela. *American Museum Novitates* **3070:** 1–15.

NICHOLSON, K. E. 2001. *Phylogenetic Analysis of the Nominal Genus* Norops *(Reptilia: Sauria): Classification, Evolution, and Biogeography* [Ph.D. dissertation]. Coral Gables, Florida: University of Miami.

NICHOLSON, K. E. 2002. Phylogenetic analysis and a test of the current infrageneric classification of *Norops* (beta *Anolis*). *Herpetological Monographs* **16:** 93–120.

NICHOLSON, K. E., B. I. CROTHER, C. GUYER, AND J. M. SAVAGE. 2012. It is time for a new classification of anoles (Squamata: Dactyloidae). *Zootaxa* **3477:** 1–108.

NICHOLSON, K. E., B. I. CROTHER, C. GUYER, AND J. M. SAVAGE. 2014. Anole classification: a response to Poe. *Zootaxa* **3814:** 109–120.

NICHOLSON, K. E., R. E. GLOR, J. J. KOLBE, A. LARSON, S. B. HEDGES, AND J. B. LOSOS. 2005. Mainland colonization by island lizards. *Journal of Biogeography* **32:** 929–938.

NICHOLSON, K. E., L. J. HARMON, AND J. B. LOSOS. 2007. Evolution of *Anolis* lizard dewlap diversity. *PLoS One* **2:** e274 (doi:10.1371/journal.pone.0000274).

NICHOLSON, K. E., J. R. McCRANIE, AND G. KÖHLER. 2000. Herpetofaunal expedition to Parque Nacional Patuca: a newly established park in Honduras. *Herpetological Bulletin* **72:** 26–31.

NICHOLSON, K. E., A. MIJARES-URRUTIA, AND A. LARSON. 2006. Molecular phylogenetics of the *Anolis onca* series: a case history in retrograde evolution revisited. *Journal of Experimental Zoology* **360B:** 450–459.

NIETO-MONTES DE OCA, A. 1994a. Rediscovery and redescription of *Anolis schiedii* (Wiegmann) (Squamata: Polychridae) from central Veracruz, Mexico. *Herpetologica* **50:** 325–335, 377.

NIETO-MONTES DE OCA, A. 1994b. *A Taxonomic Review of the* Anolis schiedii *Group (Squamata: Polychrotidae)* [Ph.D. dissertation]. Lawrence: University of Kansas.

NIETO-MONTES DE OCA, A. 1996. A new species of *Anolis* (Squamata: Polychrotidae) from Chiapas, México. *Journal of Herpetology* **30:** 19–27.

NIETO-MONTES DE OCA, A. 2001. The systematics of *Anolis hobartsmithi* (Squamata: Polychrotidae), another species of the *Anolis schiedii* group from Chiapas, Mexico, pp. 44–52. *In* J. D. Johnson, R. G. Webb, and O. A. Flores-Villela (eds.), *Mesoamerican Herpetology: Systematics, Zoogeography, and Conservation. Centennial Museum, University of Texas at El Paso, Special Publication* **1:** i–iv, 1–200.

O'SHAUGHNESSY, A. W. E. 1875. List and revision of the species of Anolidae in the British-Museum collection, with descriptions of new species. *Annals and Magazine of Natural History* **15:** 270–281.

O'SHEA, M. 1986. *Operation Raleigh. Herpetological Survey of the Rio Paulaya and Laguna Bacalar Regions Northeastern Honduras, Central America.*

*April–June 1985.* Located at: West Midlands, England: Deposited in British Museum (Natural History), London.

O'SHEA, M. T. 1989. New departmental records for northeastern Honduran herpetofauna. *Herpetological Review* **20:** 16.

PAEMELAERE, E. A. D., C. GUYER, AND F. S. DOBSON. 2011. Survival of alternative dorsal-pattern morphs in females of the anole *Norops humilis*. *Herpetologica* **67:** 420–427.

PAEMELAERE, E. A. D., C. GUYER, AND F. S. DOBSON. 2013. The role of microhabitat in predation on females with alternative dorsal patterns in a small Costa Rican anole (Squamata: Dactyloidae). *Revista de Biología Tropical* **61:** 887–895.

PEREZ-HIGAREDA, G., H. M. SMITH, AND D. CHISZAR. 1997. Natural history notes. *Anolis pentaprion* (Lichen Anole). Frugivory and Cannibalism. *Herpetological Review* **28:** 201–202.

PETERS, J. A. 1952. Catalogue of type specimens in the herpetological collections of the University of Michigan Museum of Zoology. *Occasional Papers of the Museum of Zoology University of Michigan* **539:** 1–55.

PETERS, J. A., AND R. DONOSO-BARROS. 1970. Catalogue of the neotropical Squamata: part II. Lizards and amphisbaenians. *United States National Museum Bulletin* **297:** i–viii, 1–293 (also reprinted by Smithsonian Institution Press, 1986).

PETERS, W. 1863. Über einige neue Arten der Sauria-Gattung *Anolis. Monatsberichte der königlich [preussischen] Akademie der Wissenschaften zu Berlin* **1863:** 135–149 (also reprinted by the Society for the Study of Amphibians and Reptiles, 1995).

PIANKA, E. R., AND L. J. VITT. 2003. *Lizards. Windows to the Evolution of Diversity*. Berkeley: University of California Press.

POE, S. 2004. Phylogeny of anoles. *Herpetological Monographs* **18:** 37–89.

POE, S. 2013. 1986 Redux: new genera of anoles (Sauria: Dactyloidae) are unwarranted. Zootaxa **3626:** 295–299 (the publication Poe is referring to in his title was published during 1987).

POE, S., J. VELASCO, K. MIYATA, AND E. E. WILLIAMS. 2009. Descriptions of two nomen nudum species of *Anolis* lizards from northwestern South America. *Breviora* **516:** 1–16.

POWELL, R. 2003. Species profile: Utila's reptiles. *Iguana* **10(2):** 36–38.

POWELL, R., AND R. W. HENDERSON. 2012. Swan Islands, pp. 91–92. *In* R. Powell and R. W. Henderson (eds.), Island Lists of West Indian Amphibians and Reptiles. *Bulletin of the Florida Museum of Natural History* **51:** 85–166.

PYRON, R. A., F. T. BURBRINK, AND J. J. WIENS. 2013. A phylogeny and revised classification of Squamata, including 4161 species of lizards and snakes. *BMC Evolutionary Biology* **13:** e93 (doi:10.1186/1471-2148-13-93).

REGAN, C. T. 1916. Reptilia and Batrachia. *Zoological Record* **51(16):** 1–20.

RODRÍGUEZ SCHETTINO, L. 1999. Systematic accounts of the species, pp. 104–380. *In* L. Rodríguez Schettino (ed.), *The Iguanid Lizards of Cuba*. Gainesville: University Press of Florida.

RODRÍGUEZ SCHETTINO, L. (ED.). 2003. *Anfibios y Reptiles de Cuba*. La Habana, Cuba: Instituto de Ecología y Systemática.

ROSSMAN, D. A., AND D. A. GOOD. 1993. Herpetological type specimens in the Museum of Natural Science, Louisiana State University. *Occasional Papers of the Museum of Natural Science, Louisiana State University* **66:** 1–18.

RUIBAL, R. 1964. An annotated checklist and key to the anoline lizards of Cuba. *Bulletin of the Museum of Comparative Zoology* **130(8):** 473–520.

RUIBAL, R., AND E. E. WILLIAMS. 1961. Two sympatric Cuban anoles of the *carolinensis* group. *Bulletin of the Museum of Comparative Zoology* **125(7):** 183–208.

SAGRA, R. DE LA. 1838. *Histoire Physique, Politique et Naturelle de L'ile de Cuba. Tome VIII. Atlas. Reptiles*. Paris: Arthus Bertrand.

SÁNCHEZ, A., I. OVIEDO, P. R. HOUSE, AND D. VREUGDENHIL. 2002. *Racionalización del Systema Nacional de Areas Protegidas de Honduras. Volumen V: Estado Legal de las Areas Protegidas de Honduras, Actualizacion 2002*. Tegucigalpa, Honduras: World Institute for Conservation and Environment.

SAVAGE, J. M. 1966. The origins and history of the Central American herpetofauna. *Copeia* **1966:** 719–766.

SAVAGE, J. M. 1973. *A Preliminary Handlist of the Herpetofauna of Costa Rica*. Los Angeles: Dept. of Biological Sciences and Allan Hancock Foundation, University of Southern California.

SAVAGE, J. M. 1982. The enigma of the Central American herpetofauna: dispersals or vicariance. *Annals of the Missouri Botanical Gardens* **69:** 464–547.

SAVAGE, J. M. 1997. On terminology for the description of the hemipenes of squamate reptiles. *The Herpetological Journal* **7:** 23–25.

SAVAGE, J. M. 2002. *The Amphibians and Reptiles of Costa Rica. A Herpetofauna between Two Continents, between Two Seas*. Chicago: University of Chicago Press.

SAVAGE, J. M., AND C. GUYER. 1989. Infrageneric classification and species composition of the anole genera, *Anolis, Ctenonotus, Dactyloa, Norops* and *Semiurus* (Sauria: Iguanidae). *Amphibia-Reptilia* **10:** 105–116.

SAVAGE, J. M., AND C. GUYER. 1991. Nomenclatural notes on anoles (Sauria: Polychridae): stability over priority. *Journal of Herpetology* **25:** 365–366.

SAVAGE, J. M., AND J. T. TALBOT. 1978. The giant anoline lizards of Costa Rica and western Panama. *Copeia* **1978:** 480–492.

SAVAGE, J. M., AND J. VILLA R. 1986. Introduction to the/ Introducción a la Herpetofauna of/de Costa Rica.

*Society for the Study of Amphibians and Reptiles, Contributions to Herpetology* **3**: i–viii, 1–207.

SCHAAD, E. W., AND S. POE. 2010. Patterns of ecomorphological convergence among mainland and island *Anolis* lizards. *Biological Journal of the Linnean Society* **101**: 852–859.

SCHMIDT, K. P. 1919. Descriptions of new amphibians and reptiles from Santo Domingo and Navassa. *Bulletin of the American Museum of Natural History* **41(Art. 12)**: 519–525.

SCHMIDT, K. P. 1936. New amphibians and reptiles from Honduras in the Museum of Comparative Zoology. *Proceedings of the Biological Society of Washington* **49**: 43–50.

SCHMIDT, K. P. 1941. The amphibians and reptiles of British Honduras. *Zoological Series, Field Museum of Natural History* **22**: 475–510.

SCHMIDT, W., AND F.-W. HENKEL. 1995. *Leguane. Biologie, Haltung und Zucht.* Stuttgart, Germany: Verlag Eugen Ulmer.

SCHOENER, T. W. 1988. Testing for non-randomness in sizes and habitats of West Indian lizards: choice of species pool affects conclusions from null models. *Evolutionary Ecology* **2**: 1–26.

SCHWARTZ, A., AND R. W. HENDERSON. 1988. West Indian amphibians and reptiles: a check-list. *Milwaukee Public Museum, Contributions in Biology and Geology* **74**: 1–264.

SCHWARTZ, A., AND R. W. HENDERSON. 1991. *Amphibians and Reptiles of the West Indies: Descriptions, Distributions, and Natural History.* Gainesville: University of Florida Press.

SCHWARTZ, A., AND R. THOMAS. 1975. A check-list of West Indian amphibians and reptiles. *Carnegie Museum of Natural History Special Publication* **1**: 1–216.

SMITH, H. M. 1967. *Handbook of Lizards. Lizards of the United States and of Canada.* Ithaca, New York: Cornell University Press, Comstock Publishing Associates.

SMITH, H. M., AND W. L. BURGER. 1949. A new subspecies of *Anolis sagrei* from the Atlantic coast of tropical America. *Anales del Instituto de Biología* **20**: 407–410.

SMITH, H. M., F. W. BURLEY, AND T. H. FRITTS. 1968. A new anisolepid *Anolis* (Reptilia: Lacertilia) from Mexico. *Journal of Herpetology* **2**: 147–151.

SMITH, H. M., AND W. I. FOLLETT. 1960. *Anolis nannodes* Cope, 1864: request for a ruling on lectotype selection (Class Reptilia). *The Bulletin of Zoological Nomenclature* **18**: 75–78.

SMITH, H. M., AND H. W. KERSTER. 1955. New and noteworthy Mexican lizards of the genus *Anolis*. *Herpetologica* **11**: 193–201.

SMITH, H. M., D. A. LANGEBARTEL, AND K. L. WILLIAMS. 1964. Herpetological type-specimens in the University of Illinois Museum of Natural History. *Illinois Biological Monographs* **32**: 1–80.

SMITH, M. A. 1933. Amphibia and Reptilia. *Zoological Record* **69(14)**: 1–39.

SMITH, M. A. 1937. Amphibia and Reptilia. *Zoological Record* **73(14)**: 1–47.

SMITH, P. W. 1950. *Thecadactylus rapicaudus* in Honduras. *Herpetologica* **6**: 55.

SMITHE, F. B. 1975–1981. *Naturalist's Color Guide. Part I. Color Guide.* New York: American Museum of Natural History.

SNYDER, A. 2011. Cusuco National Park Honduras. *Herp Nation* **2**: 6–15.

STAFFORD, P. J. 1991. Amphibians and reptiles of the Joint Services Scientific Expedition to the Upper Raspaculo, Belize, 1991. *British Herpetological Society Bulletin* **38**: 10–17.

STAFFORD, P. J. 1994. Amphibians and reptiles of the Upper Raspaculo River Basin, Maya Mountains, Belize. *British Herpetological Society Bulletin* **47**: 23–29.

STAFFORD, P. J., AND J. R. MEYER. 1999 (dated 2000). *A Guide to the Reptiles of Belize.* San Diego: Academic Press.

STONICH, S. C. 2000. *The Other Side of Paradise: Tourism, Conservation, and Development in the Bay Islands.* Elmsford, New York: Cognizant Communication Corporation.

STUART, L. C. 1942. Comments on several species of *Anolis* from Guatemala, with descriptions of three new forms. *Occasional Papers of the Museum of Zoology, University of Michigan* **464**: 1–10.

STUART, L. C. 1948. The amphibians and reptiles of Alta Verapaz Guatemala. *Miscellaneous Publications Museum of Zoology, University of Michigan* **69**: 1–109.

STUART, L. C. 1955. A brief review of the Guatemalan lizards of the genus *Anolis*. *Miscellaneous Publications Museum of Zoology, University of Michigan* **91**: 1–31.

STUART, L. C. 1963. A checklist of the herpetofauna of Guatemala. *Miscellaneous Publications Museum of Zoology, University of Michigan* **122**: 1–150.

TELFORD, S. R., JR. 1977. The distribution, incidence and general ecology of saurian malaria in Middle America. *International Journal for Parasitology* **7**: 299–314.

TIEDEMANN, F., AND H. GRILLITSCH. 1999. Ergänzungen zu den Katalogen der Typusexemplare der Herpetologischen Sammlung des Naturhistorischen Museums in Wien (Amphibia, Reptilia). *Herpetozoa* **12**: 147–156.

TOWNSEND, J. H. 2006. *Inventory and Conservation Assessment of the Herpetofauna of the Sierra de Omoa, Honduras, with a Review of the* Geophis *(Squamata: Colubridae) of Eastern Nuclear Central America* [master's thesis]. Gainesville: University of Florida.

TOWNSEND, J. H. 2009. Morphological variation in *Geophis nephodrymus* (Squamata: Colubridae), with comments on conservation of *Geophis* in eastern nuclear Central America. *Herpetologica* **65**: 292–302.

TOWNSEND, J. H., J. M. BUTLER, L. D. WILSON, AND J. D. AUSTIN. 2010. A distinctive new species of moss salamander (Caudata: Plethodontidae: *Nototriton*) from an imperiled Honduran endemic hotspot. *Zootaxa* **2434:** 1–16.

TOWNSEND, J. H., S. M. HUGHES, AND T. L. PLENDERLEITH. 2005. Geographic distribution. *Anolis (Norops) ocelloscapularis* (NCN). *Herpetological Review* **36:** 466.

TOWNSEND, J. H., AND T. L. PLENDERLEITH. 2005. Geographic distribution. *Anolis (Norops) petersii* (Peters' Anole). *Herpetological Review* **36:** 466–467.

TOWNSEND, J. H., AND L. D. WILSON. 2006. Denizens of the dwarf forest: the herpetofauna of the elfin forests of Cusuco National Park, Honduras. *Iguana* **13:** 242–251.

TOWNSEND, J. H., AND L. D. WILSON. 2008. *Guide to the Amphibians & Reptiles of Cusuco National Park, Honduras. Guía de los Anfibios y Reptiles del Parque Nacional Cusuco, Honduras.* (Kulstad, P. M., translator). Salt Lake City, Utah: Bibliomania!

TOWNSEND, J. H., AND L. D. WILSON. 2009. New species of cloud forest *Anolis* (Squamata: Polychrotidae) in the *crassulus* group from Parque Nacional Montaña de Yoro, Honduras. *Copeia* **2009:** 62–70.

TOWNSEND, J. H., AND L. D. WILSON. 2010a. Conservation of the Honduran herpetofauna: issues and imperatives, pp. 460–487. *In* L. D. Wilson, J. H. Townsend, and J. D. Johnson (eds.). *Conservation of Mesoamerican Amphibians and Reptiles.* Eagle Mountain, Utah: Eagle Mountain Publishing LC.

TOWNSEND, J. H., AND L. D. WILSON. 2010b. Biogeography and conservation of the Honduran subhumid forest herpetofauna, pp. 686–705. *In* L. D. Wilson, J. H. Townsend, and J. D. Johnson (eds.), *Conservation of Mesoamerican Amphibians and Reptiles.* Eagle Mountain, Utah: Eagle Mountain Publishing LC.

TOWNSEND, J. H., L. D. WILSON, I. R. LUQUE-MONTES, AND L. P. KETZLER. 2008. Redescription of *Anolis rubribarbaris* (Köhler, McCranie, & Wilson, 1999), a poorly known Mesoamerican cloud forest anole (Squamata: Polychrotidae). *Zootaxa* **1918:** 39–44.

TOWNSEND, J. H., L. D. WILSON, M. MEDINA-FLORES, E. AGUILAR-URBINA, B. K. ATKINSON, C. A. CERRATO-MENDOZA, A. CONTRERAS-CASTRO, L. N. GRAY, L. A. HERRERA-B., I. R. LUQUE-MONTES, M. MCKEWY-MEJÍA, A. PORTILLO-AVILEZ, A. L. STUBBS, AND J. D. AUSTIN. 2012. A premontane hotspot for herpetological endemism on the windward side of Refugio de Vida Silvestre Texíguat, Honduras. *Salamandra* **48:** 92–114.

TOWNSEND, J. H., L. D. WILSON, M. MEDINA-FLORES, AND L. A. HERRERA-B. 2013. A new species of centipede snake in the *Tantilla taeniata* group (Squamata: Colubridae) from premontane rainforest in Refugio de Vida Silvestre Texíguat, Honduras. *Journal of Herpetology* **47:** 191–200.

TOWNSEND, J. H., L. D. WILSON, AND J. I. RESTREPO. 2007. *Investigaciones Sobre la Herpetofauna en el Parque Nacional Montaña de Yoro y la Reserva Biológica Cerro Uyuca, Honduras.* Located at: Gainesville: University of Florida.

TOWNSEND, J. H., L. D. WILSON, B. L. TALLEY, D. C. FRASER, T. L. PLENDERLEITH, AND S. M. HUGHES. 2006. Additions to the herpetofauna of Parque Nacional El Cusuco, Honduras. *Herpetological Bulletin* **96:** 29–39.

TOWNSEND, T. M., D. G. MULCAHY, B. P. NOONAN, J. W. SITES, JR., C. A. KUCZYNSKI, J. J. WIENS, AND T. W. REEDER. 2011. Phylogeny of iguanian lizards inferred from 29 nuclear loci, and a comparison of concatenated and species-tree approaches for an ancient, rapid radiation. *Molecular Phylogenetics and Evolution* **61:** 363–380.

VANHOOYDONCK, B., A. HERREL, J. J. MEYERS, AND D. J. IRSCHICK. 2008. What determines dewlap diversity in *Anolis* lizards? An among-island comparison. *Journal of Evolutionary Biology* **22:** 293–305.

VANZOLINI, P. E. 1986. Addenda and corrigenda to part II lizards and amphisbaenians. Preface for reprint of J. A. PETERS and R. DONOSO-BARROS. 1970. *Catalogue of the Neotropical Squamata: Part II. Lizards and Amphisbaenians.* No. 297. Washington, D.C.: Smithsonian Institution Press (also published in *Smithsonian Herpetological Information Service* **70:** 1–25).

VILLA, J., L. D. WILSON, AND J. D. JOHNSON. 1988. *Middle American Herpetology. A Bibliographic Checklist.* Columbia: University of Missouri Press.

VILLARREAL BENÍTEZ, J.-L. 1997. Historia natural del género *Anolis*, pp. 495–500. *In* E. González Soriano, R. Dirzo, and R. C. Vogt (eds.), *Historia Natural de Los Tuxtlas.* Ciudad de México: Universidad Nacional Autónoma de México.

VILLARREAL BENÍTEZ, J.-L., AND L. HERAS LARA. 1997. *Anolis uniformis* (largatija de monte, largatija del abanico, perritos), pp. 476–478. *In* E. González Soriano, R. Dirzo, and R. C. Vogt (eds.), *Historia Natural de Los Tuxtlas.* Ciudad de México: Universidad Nacional Autónoma de México.

VITT, L. J., AND P. A. ZANI. 2005. Ecology and reproduction of *Anolis capito* in rain forest of southeastern Nicaragua. *Journal of Herpetology* **39:** 36–42.

VITT, L. J., P. A. ZANI, AND R. D. DURTSCHE. 1995. Ecology of the lizard *Norops oxylophus* (Polychrotidae) in lowland forest of southeastern Nicaragua. *Canadian Journal of Zoology* **73:** 1918–1927.

VOIGT, F. S. 1832. *Das Thierreich geordnet nach seiner Organisation, aus Grundlage der Naturgebschichte der Thiere und Ginleitung in die vergleischende Anatomie. Vom Baron von Cuvier. Nach der zweiten, bermehrten Ausgabe übersezt und durch Zusätze erweitert. Zweiter Band, die Reptilien und Fische enthalten.* Leipzig, Germany: F. A. Brodhaus.

VOSJOLI, P. DE. 1992. *The General Care and Maintenance of Green Anoles. Including Notes on Other Anoles and Vivarium Design.* Lakeside, California: The Herpetocultural Library.

WAGLER, J. 1830. *Natürliches System der Amphibien, mit vorangehender Classification der Säugthiere und Vögel. Ein Beitrag zur vergleichenden Zoologie.* München, Germany: J. G. Cotta'schen Buchhandlund.

WERNER, F. 1896. Beiträge zur Kenntniss der Reptilien und Batrachier von Centralamerika und Chile, sowie einiger seltenerer Schlangenarten. *Verhandlungen der k. k. zoologisch–botanischen Gesellschaft in Wien* **46:** 344–365, pl. VI.

WIEGMANN, A. F. A. 1834. *Herpetologia Mexicana, seu Descriptio Amphibiorum Novae Hispaniae, quae Itineribus comitis de Sack, Ferdinandi Deppe et Chr. Guil. Schiede in Museum Zoologicum Berolinense Pervenerunt. Pars Prima, Saurorum Species Amplectens. Adiecto Systematis Saurorum Prodromo, Additisque multis in hunc Amphibiorum Ordinem Observationibus.* Berolini: Sumptibus C. G. Lüderitz (also reprinted by the Society for the Study of Amphibians and Reptiles, 1969).

WILLIAMS, E. E. 1960. Notes on Hispaniolan herpetology 1. *Anolis christophei,* new species, from the Citadel of King Christophe, Haiti. *Breviora* **117:** 1–7.

WILLIAMS, E. E. 1963. Studies on South American anoles. Description of *Anolis mirus,* new species, from Rio San Juan, Colombia, with comment on digital dilation and dewlap as generic and specific characters in the anoles. *Bulletin of the Museum of Comparative Zoology* **129(9):** 463–480.

WILLIAMS, E. E. 1966. South American anoles: *Anolis biporcatus* and *Anolis fraseri* (Sauria, Iguanidae) compared. *Breviora* **239:** 1–14.

WILLIAMS, E. E. 1969. The ecology of colonization as seen in the zoogeography of anoline lizards on small islands. *Quarterly Review of Biology* **44:** 345–389.

WILLIAMS, E. E. 1970. South American anoles: *Anolis apollinaris* Boulenger 1919, a relative of *A. biporcatus* Wiegmann (Sauria, Iguanidae). *Breviora* **358:** 1–11.

WILLIAMS, E. E. 1976a. West Indian anoles: a taxonomic and evolutionary summary 1. Introduction and a species list. *Breviora* **440:** 1–21.

WILLIAMS, E. E. 1976b. South American anoles: the species groups. *Papéis Avulsos de Zoologia, São Paulo* **29:** 259–268.

WILLIAMS, E. E. 1984. New or problematic *Anolis* from Colombia. III. Two new semiaquatic anoles from Antioquia and Chocó, Colombia. *Breviora* **478:** 1–22.

WILLIAMS, E. E. 1989. A critique of Guyer and Savage (1986): cladistic relationships among anoles (Sauria: Iguanidae): are the data available to reclassify the anoles?, pp. 433–477. *In* C. A. Woods (ed.), *Biogeography of the West Indies. Past, Present, and Future.* Gainesville, Florida: Sandhill Crane Press, Incorporation.

WILLIAMS, E. E., H. RAND, A. S. RAND, AND R. J. O'HARA. 1995. A computer approach to the comparison and identification of species in difficult taxonomic groups. *Breviora* **502:** 1–47.

WILSON, L. D., AND G. A. CRUZ DÍAZ. 1993. The herpetofauna of the Cayos Cochinos, Honduras. *Herpetological Natural History* **1:** 13–23.

WILSON, L. D., AND D. E. HAHN. 1973. The herpetofauna of the Islas de la Bahía, Honduras. *Bulletin of the Florida State Museum Biological Sciences* **17:** 93–150.

WILSON, L. D, I. R. LUQUE-MONTES, A. B. ALEGRÍA, AND J. H. TOWNSEND. 2013. El componente endémico de la herpetofauna Hondureña en peligro crítico: priorización y estrategias de conservación. *Revista Latinoamericana de Conservación* **2(2)–(3)1:** 47–67.

WILSON, L. D., AND J. R. MCCRANIE. 1982. A new cloud forest *Anolis* (Sauria: Iguanidae) of the *schiedei* group from Honduras. *Transactions of the Kansas Academy of Science* **85:** 133–141.

WILSON, L. D., AND J. R. MCCRANIE. 1994. Comments on the occurrence of a salamander and three lizard species in Honduras. *Amphibia-Reptilia* **15:** 416–421.

WILSON, L. D., AND J. R. MCCRANIE. 1998. The biogeography of the herpetofauna of the subhumid forests of Middle America (Isthmus of Tehuantepec to northwestern Costa Rica). *Royal Ontario Museum Life Sciences Contribution* **163:** 1–50.

WILSON, L. D., AND J. R. MCCRANIE. 2003. Herpetofaunal indicator species as measures of environmental stability in Honduras. *Caribbean Journal of Science* **39:** 50–67.

WILSON, L. D., AND J. R. MCCRANIE. 2004a. The conservation status of the herpetofauna of Honduras. *Amphibian and Reptile Conservation* **3:** 6–33.

WILSON, L. D., AND J. R. MCCRANIE. 2004b. The herpetofauna of the cloud forests of Honduras. *Amphibian and Reptile Conservation* **3:** 34–48.

WILSON, L. D., AND J. R. MCCRANIE. 2004c. The herpetofauna of Parque Nacional El Cusuco, Honduras (Reptilia, Amphibia). *Herpetological Bulletin* **87:** 13–24.

WILSON, L. D., J. R. MCCRANIE, AND M. R. ESPINAL. 2001. The ecogeography of the Honduran herpetofauna and the design of biotic reserves, pp. 109–158. *In* J. D. Johnson, R. G. Webb, and O. A. Flores-Villela (eds.), *Mesoamerican Herpetology: Systematics, Zoogeography, and Conservation.* Centennial Museum, University of Texas at El Paso, Special Publication **1**.

WILSON, L. D., J. R. MCCRANIE, AND L. PORRAS. 1979a. New departmental records for reptiles and amphibians from Honduras. *Herpetological Review* **10:** 25.

WILSON, L. D., J. R. MCCRANIE, AND L. PORRAS. 1979b. *Rhadinaea montecristi* Mertens: an addition to the snake fauna of Honduras. *Herpetological Review* **10:** 62.

WILSON, L. D., J. R. MCCRANIE, AND K. L. WILLIAMS. 1991. Additional departmental records for the herpetofauna of Honduras. *Herpetological Review* **22:** 69–71.

WILSON, L. D., AND J. R. MEYER. 1969. A review of the colubrid snake genus *Amastridium*. *Bulletin of the Southern California Academy of Sciences* **68:** 145–159.

WILSON, L. D., AND J. H. TOWNSEND. 2006. The herpetofauna of the rainforests of Honduras. *Caribbean Journal of Science* **42:** 88–113.

WILSON, L. D., AND J. H. TOWNSEND. 2007. Biogeography and conservation of the herpetofauna of the upland pine-oak forests of Honduras. *Biota Neotropica* **7:** 137–148.

WILSON, L. D., J. H. TOWNSEND, AND J. D. JOHNSON (EDS.). 2010. *Conservation of Mesoamerican Amphibians and Reptiles*. Eagle Mountain, Utah: Eagle Mountain Publishing LC.

## NOTE ADDED IN PROOF

While this manuscript was in press, we became aware of a report of *Norops wermuthi* Köhler and Obermeier in Honduras only some 3 m from the Nicaraguan border (Sunyer et al., 2013). The locality is Cerro Jesús, Cordillera de Dipilto, El Paraíso, 1802 m elev., 13°59′04.3″N, 86°11′24.1″W. That report brings the total known *Norops* species from Honduras to 39 and the total known species of anoles from the country to 40. Sunyer, J., R. García-Roa, and J. H. Townsend. 2013. First country record of Norops wermuthi Köhler & Obermeier, 1998, for Honduras. *Herpetozoa* **26:** 103–106.